FISH NUTRITION

Protein and Amino Acid Requirements

FISH NUTRITION

Protein and Amino Acid Requirements

Imtiaz Ahmed, PhD
Ishtiyaq Ahmad, PhD
Shabihul Fatma Sayed, PhD
Editors

APPLE ACADEMIC PRESS

First edition published 2026

Apple Academic Press Inc.
1265 Goldenrod Circle, NE,
Palm Bay, FL 32905 USA

760 Laurentian Drive, Unit 19,
Burlington, ON L7N 0A4, Canada

CRC Press
2385 NW Executive Center Drive,
Suite 320, Boca Raton FL 33431

4 Park Square, Milton Park,
Abingdon, Oxon, OX14 4RN, UK

Apple Academic Press exclusively co-publishes with CRC Press, an imprint of Taylor & Francis Group, LLC

Library and Archives Canada Cataloguing in Publication

CIP data on file with Canada Library and Archives

Library of Congress Cataloging-in-Publication Data

CIP data on file with US Library of Congress

ISBN: 978-1-998511-56-3 (hbk)
ISBN: 978-1-998511-57-0 (ebk)
DOI: 10.1201/9781998511570

About the Editors

Imtiaz Ahmed, PhD
Professor, Department of Zoology, Fish Nutrition Research Laboratory, University of Kashmir, Srinagar, Jammu and Kashmir, India

Imtiaz Ahmed, PhD, is a faculty member working as Professor in the Department of Zoology at the University of Kashmir, India. His pioneering contribution in the field of research is the determination of complete essential amino acid requirements of the Indian major carp, *Cirrhinus mrigala*, at the fingerling stage as part of his doctoral thesis, and the essential amino acid requirements of *Heteropneustes fossilis* at the fingerling stage as part of a research project. Key findings of his research projects have been submitted to the Ministry of Agriculture for proper implementation through DST–New Delhi. The feeding regime and appropriate ration size of some important cultured fish species have been determined and recommended for the proper utilization of feed. The protein requirement of two cultured fish species inhabiting Kashmir water bodies, i.e., Common Carp, *Cyprinus carpio communis*, and *Cyprinus carpio specularis*, was reported by him for the first time from Kashmir. Apart from protein and amino acid requirement studies, he has also established quantitative information on the vitamin requirements of different cultured fish species. He is the first to determine the feeding regime and protein and energy maintenance requirements of Rainbow Trout, *Oncorhynchus mykiss*, from Kashmiri waters under lab conditions. He has also established the nutritional value of indigenous Snow Trout, *Schizothorax*, of Kashmir and the seasonal variation in hemato-biochemical composition of fishes from the Kashmir Valley. Moreover, Dr. Ahmed has published more than 100 papers in well-reputed international journals and has several book chapters to his credit. Dr. Ahmed has supervised several M.Phil and doctoral students in the areas of fish nutrition, fish physiology, reproductive biology, and limnology.

Ishtiyaq Ahmad, PhD
Assistant Professor, Department of Zoology,
University of Kashmir, Srinagar, Jammu and Kashmir, India

Ishtiyaq Ahmad, PhD, is currently working as an Assistant Professor at the Department of Zoology, University of Kashmir, Srinagar, India. He earned his PhD in Zoology (Fish Nutrition) from the Fish Nutrition Research Laboratory, Department of Zoology, University of Kashmir, Srinagar, India, under the guidance of Prof. Imtiaz Ahmed. Previously, Dr. Ahmad worked as a Senior Research Fellow at the Division of Fish Genetics and Biotechnology, Faculty of Fisheries, Sher-e-Kashmir University of Agricultural Sciences and Technology of Kashmir (*SKUAST-K*), Rangil, Ganderbal, India, where his research was dedicated to developing genetically improved pure lines of rainbow trout to enhance aquaculture productivity. He specializes in fish nutrition, fish physiology and biochemistry, nutrigenomics, fish parasitology, and molecular biology. His major research contribution is the establishment of optimum branched-chain amino acid requirements for rainbow trout. He is the author of more than 45 research articles with international publishers, such as *Scientific Reports, Aquaculture Nutrition, Aquaculture Research, Aquaculture Reports, Journal of Applied Ichthyology, Frontiers in Marine Sciences,* etc. and several books and book chapters. Dr. Ahmad has participated in and presented his work at state, national, and international conferences, seminars, workshops, and symposia. He has received many awards, including a Young Scientist Award from a national organization and several Best Paper Presentation Awards. He is serving as an editorial member and review editor for several reputed international journals, such as *PLOS ONE, Frontiers in Nutrition,* etc. Dr. Ahmad has also worked as a Senior Research Fellow in a *DBT*-sponsored project on the nutrient requirements of rainbow trout fingerlings. In addition to this, Dr. Ahmad is the recipient of a *CSIR NET* award.

Shabihul Fatma Sayed, PhD

Associate Professor, Department of Nursing, Jazan University, Saudi Arabia

Shabihul Fatma Sayed, PhD, is currently working as an Associate Professor at Jazan University, Saudi Arabia. She is a passionate research scientist with over 13 years of international academic teaching and research experience in Biological Sciences (Zoology), mainly in protein biochemistry, dietary protein and amino acid requirements, vitamin and essential mineral requirements, biochemistry, nutritional physiology, aquatic toxicology, pesticides/insecticides and their impact on human health, as well as public health issues and challenges. She is an international researcher with extensive experience in working on challenging projects, building and developing collaborations, creating networks, and mentoring and training students. Her challenging postdoctoral work has fostered resource management skills in her and has honed her organizational and project management capabilities, enabling her to adapt to varied research environments. She is also a science communication enthusiast and a skilled scientific writer and reviewer for various academic presses, including Elsevier, Taylor & Francis, and Cambridge Press.

Contents

Contributors

Mustapha Aba
Independent Researcher and Aquaculture Expert, Fish Nutrition, Rabat, Morocco

Irfan Ahmad
Division of Fish Genetics and Biotechnology, Faculty of Fisheries, Sher-e-Kashmir University of Agricultural Sciences and Technology of Kashmir (SKUAST–Kashmir), Rangil, Ganderbal, Jammu and Kashmir, India

Ishtiyaq Ahmad
Department of Zoology, Fish Nutrition Research Laboratory, University of Kashmir, Hazratbal Srinagar, Srinagar, Jammu and Kashmir, India; Division of Fish Genetics and Biotechnology, Faculty of Fisheries, Sher-e-Kashmir University of Agricultural Sciences and Technology of Kashmir (SKUAST-Kashmir), Rangil, Ganderbal, Srinagar, Jammu and Kashmir, India

Imtiaz Ahmed
Department of Zoology, Fish Nutrition Research Laboratory, University of Kashmir, Hazratbal, Srinagar, Jammu and Kashmir, India

Saima Andleeb
Division of Fish Genetics and Biotechnology, Faculty of Fisheries, Sher-e-Kashmir University of Agricultural Sciences and Technology of Kashmir (SKUAST–Kashmir), Rangil, Ganderbal, Jammu and Kashmir, India

Rida Azam
Department of Zoology, Wildlife and Fisheries, University of Agriculture, Faisalabad, Punjab, Pakistan

Asim Iqbal Bazaz
Division of Fisheries Resource Management, Faculty of Fisheries, Sher-e-Kashmir University of Agricultural Sciences and Technology of Kashmir (SKUAST–Kashmir), Rangil, Ganderbal, Jammu and Kashmir, India

Bhavesh Choudhary
College of Fisheries, Central Agricultural University (Imphal), Lembucherra, Agartala, Tripura, India

Vinayaka Choudhary
Vidya Bhawan Rural Institute, Mohanlal Sukhadia University, Udaipur, Rajasthan, India; Regional Institute of Education (NCERT, New Delhi), Ajmer, Rajasthan, India

Suparna Deb
College of Fisheries, Central Agricultural University (Imphal), Lembucherra, Agartala, Tripura, India

João Batista K. Fernandes
Aquaculture Center, São Paulo State University (CAUNESP), Via de Acesso Professor Paulo Donato Castellane, Jaboticabal, São Paulo, Brazil

Mohammad Amin Hajam
Department of Biochemistry, University of Kashmir, Srinagar, Jammu and Kashmir, India

Aamina Hassan
Department of Zoology, Fish Nutrition Research Laboratory, University of Kashmir, Hazratbal Srinagar, Srinagar, Jammu and Kashmir, India

Neha Iqbal
Department of Zoology, Wildlife and Fisheries, University of Agriculture, Faisalabad, Punjab, Pakistan

Marium Jameel
Department of Zoology, Wildlife and Fisheries, University of Agriculture, Faisalabad, Punjab, Pakistan

Kousar Jan
Department of Zoology, Fish Nutrition Research Laboratory, University of Kashmir, Hazratbal, Srinagar, Jammu and Kashmir, India

Gautam Geeta Jiwatram
Department of Zoology, M.M.V, Banaras Hindu University, Varanasi, Uttar Pradesh, India

Vedika V. Kanoje
College of Fisheries, Dr. Balasaheb Sawant Konkan Krishi Vidyapeeth (DBSKKV), Shirgaon, Ratnagiri, Maharashtra, India

Kifayat Ullah Khan
Department of Zoology, Government College University, Lahore, Punjab, Pakistan; Aquaculture Center, São Paulo State University (CAUNESP), Via de Acesso Professor Paulo Donato Castellane, Jaboticabal, São Paulo, Brazil

Manas Kumar Maiti
Fish Nutrition, Biochemistry and Physiology Division,
ICAR–Central Institute of Fisheries Education, Versova, Mumbai, Maharashtra, India

Cleber Fernando M. Mansano
Graduate Program in Environmental Sciences, Brazil University, Fazenda Santa Rita, Fernandópolis, São Paulo, Brazil; and Graduate Program in Animal Production, Parque Universitário, Descalvado, São Paulo, Brazil

Sheza Shakeel Mir
Department of Biochemistry, University of Kashmir, Srinagar, Jammu and Kashmir, India

Mohd. Ashraf Rather
Division of Fish Genetics and Biotechnology, Faculty of Fisheries, Sher-e-Kashmir University of Agricultural Sciences and Technology of Kashmir (SKUAST–Kashmir), Rangil, Ganderbal, Jammu and Kashmir, India

Quseen Mushtaq Reshi
Department of Zoology, Fish Nutrition Research Laboratory, University of Kashmir, Hazratbal Srinagar, Srinagar, Jammu and Kashmir, India

Azra Shah
Division of Fish Genetics and Biotechnology, Faculty of Fisheries, Sher-e-Kashmir University of Agricultural Sciences and Technology of Kashmir (SKUAST–Kashmir), Rangil, Ganderbal, Jammu and Kashmir, India

Krishna Pada Singha
Department of Animal, Veterinary, and Food Sciences, Aquaculture Research Institute, University of Idaho, Moscow, Idaho, USA

Suman
Department of Animal Science, Volcani Agricultural Research Institute, Israel

Abbreviations

AA	amino acid
AAR	amino acid response
ADC	arginine decarboxylase
AGAT	arginine: glycine aminotransferase
AHAS	acetohydroxyacid synthase
AMP	adenosine 5'-monophosphate
AMPK	AMP-activated protein kinase
AVT	arginine-vasotocin
BCAA	branched-chain amino acids
BCAT	branched-chain amino acid transaminase
BCKDH	branched-chain ketoacid dehydrogenase
BH4	tetrahydrobiopterin
BPD	body protein deposition
BWG	body weight gain
CAT	catalase
CP	crude protein
CPS	carbamoyl phosphate synthase
CS	casein
DA	dietary protein
DDP	daily deposition
DL-HMTBa	DL-2-hydroxy-4-methylthiobutanoic acid
DST	Department of Science and Technology
EAA	essential amino acids
FAO	Food and Agriculture Organization
FBW	final body weight
FCR	feed conversion ratio

FER	feed ratio efficiency
FIGLU	formimino-glutamate
FM	feather meal
FM	fish meal
FRP	fiberglass reinforced plastics
FW	final weight
GABA	gamma-aminobutyric acid
GH	growth hormone
GHR1	growth hormone receptor 1
GK	glucokinase
GL	gelatin
GPT	glutamate pyruvate transaminase
GPx	glutathione peroxidase
GSH	glutathione
HRC	histidine-related compound
IDAA	integral quantifiable AA
IGF-1	insulin growth factor-1
IgM	immunoglobulin M
IL-1b	interleukin 1-beta
KIV	2-keto-isovalerate
LDH	lactate dehydrogenase
LDL-C	low-density lipoprotein cholesterol
Leu	leucine
LWG	live weight gain
MDA	malondialdehyde
Met-Met	DL-methionyl-DL-methionine
MHA	methionine hydroxy analogue
mTOR	mammalian target of rapamycin
NAH	N-(alpha)-acetylhistidine
NO	nitric oxide

NOS	nitric oxide synthase
NR	not recorded
NRC	National Research Council
PBM	poultry by product meal
PEPCK	phosphoenolpyruvate carboxykinase
PER	protein efficiency ratio
PH	plasma histidine
PK	pyruvate kinase
PPV	productive protein value
PRE	protein retention efficiency
RDM	red drum meal
RSM	response surface methodology
SAM	S-adenosylmethionine
SB	soybean meal
SGR	specific growth rate
SOD	superoxide dismutase
SR	survival
TCA	tricarboxylic acid
TOR	target of rapamycin
Val	valine
WB	wheat bran
WG	weight gain
WGM	wheat gluten meal

Preface

In the ever-evolving field of aquaculture, understanding the intricate nutritional needs of fish is paramount to ensuring their health, growth, and overall productivity. *Fish Nutrition: Protein and Amino Acid Requirements* delves into the essential aspects of fish nutrition, focusing specifically on the crucial roles of proteins and amino acids in their diet. This book is a culmination of extensive research and expertise from leading experts in the field of aquaculture nutrition. It provides a comprehensive overview of the protein and amino acid requirements of various fish species, shedding light on their specific dietary needs for optimal growth and performance.

Through detailed analyses and critical insights, this book explores the latest advancements in fish nutrition, highlighting key findings and recommendations for formulating balanced and nutritious diets. It serves as an invaluable resource for researchers, academics, and industry professionals seeking to enhance their understanding of fish nutrition and improve aquaculture practices.

We believe that *Fish Nutrition: Protein and Amino Acid Requirements* will serve as a cornerstone in the field of aquaculture nutrition, guiding future research and advancements in the sustainable production of high-quality fish protein for a growing global population.

—*Editors*

CHAPTER 1

Diving into the Depths: Exploring Proteins in Fish

SUMAN[1] and GAUTAM GEETA JIWATRAM[2,*]

[1]*Department of Animal Science, Volcani Agricultural Research Institute, Israel*

[2]*Department of Zoology, M.M.V, Banaras Hindu University, Varanasi, Uttar Pradesh, India*

*Corresponding author

ABSTRACT

Aquaculture stands out as one of the most rapidly expanding sectors in global food production, with various fish species serving as crucial components farmed across diverse regions worldwide. The primary component of fish flesh is water, comprising approximately 70–80% of the fillet's weight. The water in the muscle of fresh fish is closely bound to the proteins within its structure. Proteins are vital nutrients crucial for growth and as integral components of the body's cells. Amino acids (AAs) play a significant role as the fundamental building blocks of proteins. The conformational structure, as well as the chemical and biological properties of a protein, are influenced by the type and rank order of its AAs. Therefore, it is important to elucidate the protein content, its requirements, and its significant role in different species of fish. This book chapter elucidates a literature-based survey about the types of protein found, the approximate protein and AA content present in different fish species, and the requirement of AAs in different species of fish, which will be helpful for aquaculture.

Fish Nutrition: Protein and Amino Acid Requirements. Imtiaz Ahmed, Ishtiyaq Ahmad, and Shabihul Fatma Sayed (Eds.)

DOI: 10.1201/9781998511570-1

1.1 INTRODUCTION

Fish, being poikilothermic creatures, exhibit a metabolic rate influenced by water temperature, with nearly half of their dietary needs attributed to protein. Fish can fall into the categories of carnivores, omnivores, or herbivores, and while all hatchlings initially lean towards carnivorous tendencies, proteins represent the predominant organic component in fish tissue, constituting approximately 65–75% of the total on a dry-weight basis. Fish consume protein primarily for amino acid (AA) acquisition, utilizing excess AAs as an energy source. The digestion of protein releases free AAs, which are absorbed from the intestinal tract and employed by various tissues for synthesizing new proteins. Consistent AA intake is vital, as fish continually utilize them for protein synthesis. Inadequate protein in the diet leads to growth reduction and weight loss, while an excess of protein results in only a portion being used for synthesis, with the remainder converted into energy. Proteins are large biomolecules or biopolymers consisting of one or more chains of AAs. AA chains play an important role as building materials of proteins, and proteins are the building blocks of organisms, having essential bioactivities and functionalities in the growth and development of organisms. Fish protein contains all essential amino acids (EAAs) like tryptophan, cystine, lysine, methionine, and threonine. EAAs such as lysine, leucine, and methionine are much higher in aquatic protein than in terrestrial meat protein. Also, non-EAAs such as glutamic acid, alanine, and aspartic acid are abundantly present in aquatic sources, along with the AA-derived organic acid taurine. According to the World Health Organization, the digestibility of protein ranging from 77 to 98.7% was determined by comparing it with the standard protein. In fishes, protein content is seasonally dependent; like in most fish, protein content increases just before the spawning season and also during the spring season when food availability is abundant. The quality of protein and type of protein is different among all species of fish. It is reported that three kinds of muscles are present in fish, such as smooth, striated, and cardiac muscles, but the major one is striated muscles present in fish. So, on the basis of solubilization, fish muscle proteins are categorized into three types: structural proteins, sarcoplasmic proteins, and muscle proteins (Table 1.1).

TABLE 1.1 Types of Proteins in Fish

Type of Protein	Total Protein Content	Solubility
Structural protein (actomyosin, tropomyosin, Actin, myosin	70–80%	In natural salt solution in ionic ability (0.5 M)
Sarcoplasmic proteins (globulin, enzymes, Myoalbumin)	25–30%	In low ionic in salt solution (0.15 M)
Muscle protein (stroma, collagen, elastin)	3%	Insoluble

Source: Gokoglu and Yerlikaya (2015).

1.2 STRUCTURAL PROTEIN OR MYOFIBRILLAR PROTEIN

These types of proteins are abundant, comprising up to 70–80% of the fish muscle protein. They are soluble in natural salt solution with an ionic strength of 0.5 M. The proportion of these structural proteins in total muscle protein in fish is higher than in land animals. In the muscle contraction–relaxation cycle, myosin and actin are responsible. Thus, actin and myosin exist as an actin-myosin complex in post-mortem muscle. Myosin, as a thick filament, ranges from 50 to 60%, while actin filaments account for 15–20% and form thin filaments. The isoelectric points of myosin and the actin molecule are at pH 5.0–5.3 and pH 4.7, respectively. Tropomyosin, troponin, actinin, and C, I, and T proteins are other regulatory proteins that play an important role in muscle protein contraction and relaxation. Fish actomyosin proteins are easily changed and labile during processing and storage. Myosin protein in fish is unstable and very sensitive to denaturation, coagulation, and degrades easily due to chemical changes (Venugopal, 2009). These proteins play an important role in gelation, water holding, emulsifying, binding capacity, and gelation.

1.3 SARCOPLASMIC PROTEINS

Sarcoplasmic proteins form 25–30% of the total protein in fish muscle. They are readily soluble in low ionic salt solution (0.015 M). Several enzymes and pigmented proteins, such as hemoglobin, myoglobin, and other types of albumins, are categorized under sarcoplasmic proteins. Furthermore, this category encompasses antifreeze proteins and glycoproteins found in fish inhabiting cold water. Fish, in contrast to land animals, exhibit a higher presence of Ca^{2+}-binding proteins. The darker hue of the red muscle in fish results from a heightened concentration of myoglobin. Red muscle, essential for sustained aerobic metabolism of energy reserves, contains more mitochondria and less sarcoplasmic reticulum compared to white fibers. Pelagic fish muscles notably contain substantial amounts of dark muscle enriched with myoglobin, facilitating extended aerobic activity. In contrast, demersal fish, which drift with ocean currents, do not engage in active swimming for prolonged durations. Additionally, pelagic fish exhibit a higher content of sarcoplasmic protein than demersal fish. The myoglobin levels in muscle exhibit an increase with age and during migration seasons. The color characteristics of fish muscle are attributed to oxymyoglobin and oxyhemoglobin. When handled and stored, hemoglobin dissolves easily, while myoglobin is retained within the cell structure. Certain organisms like mollusks, crustaceans, and some colorless-blood Antarctic fish lack hemoglobin, relying instead on copper-containing proteins known as hemocyanins, especially in shellfish. The edible quality of fish is influenced by hydrolases, oxidoreductases, and transferase enzymes. Sarcoplasmic enzymes play a crucial role in the degradation of fish muscle. The presence of sarcoplasmic proteins adversely affects the strength and deformability of myofibrillar protein gels, and water-holding capacity. The relatively low gel strength observed in mackerel and sardine products can be attributed to their sarcoplasmic protein content. The content and composition of sarcoplasmic proteins can vary among species. Electrophoretic patterns of sarcoplasmic protein fractions serve as fingerprints for identifying fish species.

1.4 STROMA PROTEINS

The residual matter left after extracting sarcoplasmic and myofibrillar proteins from muscle is referred to as stroma or connective tissue proteins. Predominantly composed of collagen, with additional components of elastin and gelatin, stroma proteins are situated in the extracellular matrix, constituting 3% of the total muscle protein. However, elasmobranch fish such as sharks, rays, and skates may contain up to 10% stroma proteins, contributing to the soft texture of fish meat (Sivik, 2000). During chill storage, the myocommata of fish may fail to maintain the cohesion of muscle cells, leading to flesh gaping. Collagen, present not only in muscle tissue but also in fish skin, bones, and scales, is a major structural protein.

This triple-helix protein comprises repeated glycine-proline-hydroxyproline-glycine amino acid sequences. Fish muscle collagen is rich in essential amino acids, thermolabile, and features fewer but more labile cross-links than collagen from warm-blooded vertebrates. The thermal modification of collagen holds significance in processes such as hot smoking, canning, short-time sterilization, and the utilization of fish waste. In some squid species, the mantle muscle can exhibit toughness after cooking due to these thermal changes, and the quality alterations in fresh and frozen fish postmortem are attributed to collagen modifications. Fish proteins, including myofibrillar and sarcoplasmic proteins, serve as materials for film formation. Additionally, bioactive peptides isolated from different fish protein hydrolysates have demonstrated potential in numerous applications (Table 1.2).

TABLE 1.2 Protein Content in Different Parts of Fish

Major Organ	Protein Content %	References
Hairs	90	Pleer et al. (2011)
Fish skin	22–26	Mahboob et al. (2015)
Fish scale	21–24	Mahboob et al. (2015)
Meat	20	Nawaz et al. (2019)
Fish fins	16–18	Mahboob et al. (2015)
Fish waste	15–30	Ghaly et al. (2013)
Food waste	11.4	Pleeret et al. (2011)
Egg shells	3–6	Waheed et al. (2019)
Whey protein	2.5–4	Zhou et al. (2019)

Source: Khan et al. (2022).

1.5 FREE AMINO ACIDS (AAS)

Free AAs constitute the primary elements contributing to the flavor profile of fisheries, encompassing AAs, nucleotides, guanidine compounds, and quaternary ammonium compounds. Individual AAs like glycine, valine, alanine, and glutamic acid play a significant role in taste, along with the degradation components of nucleotides, such as inosine (Olafsdottir and Jonsdottir, 2010). The sweet taste discerned in fresh shrimp and crab can be attributed to their free glycine content. In comparison to finfish, shrimp, lobster, crab, squid, and other shellfish generally contain higher amounts of AAs like arginine, glutamic acid, glycine, and alanine. The elevated levels of these AAs during the winter season enhance the palatability of squids compared to those harvested in summer (Venugopal, 2009). Elasmobranchs seem to possess a higher quantity of free AA nitrogen content than teleosts (Sen, 2005). Distinct non-protein AAs, including taurine, β-alanine, methylhistidine, and proline, are prevalent in most fish. Taurine plays a crucial role in osmoregulation, acts as a food reserve, and participates in the Maillard browning reaction (Haard, 1995). It also holds importance in neural development. While adult humans can synthesize taurine in small amounts, mollusks like mussels and scallops are abundant in taurine, while crabs and certain fish species contain less (Spitze et al., 2003). The muscles of mollusks and crustaceans are rich in free AAs. Fish stands out among meat-producing animals for having free histidine in its muscle (Sen, 2005). Red muscles typically contain more histidine than white muscles. Scombroid fish tissues, such as tuna and mackerel, have high levels of free histidine, which may convert into histamine through the action of associated microorganisms. The levels of free AAs in fishery products generally increase during storage due to the activity of endogenous and exogenous proteases (Gökoğlu et al., 2004a).

1.6 PEPTIDES

In fish muscle, three fundamental dipeptides have been identified: carnosine (β-alanyl histidine), anserine (β-alanyl-1-methyl histidine), and balenine (β-alanyl-3-methyl histidine), the latter being a distinctive component found in whale muscle. The presence of these compounds is generally higher in dark muscles compared to white muscles. The ratio of carnosine to anserine is more pronounced in freshwater fish than in marine fish. Both anserine and carnosine have been reported to exhibit potent capabilities in eliminating hydroxyl radicals and singlet oxygens (Kikuchi et al., 2004) (Tables 1.3–1.5).

1.7 PROTEIN REQUIREMENTS IN FISH

Fish lack a specific, distinct protein requirement; instead, they necessitate a well-rounded combination of both indispensable and dispensable AAs. Numerous researchers have employed semi-purified and purified diets to

TABLE 1.3 Proximate Protein Content in Selected Bony Fish Species

Sl. No.	Fish Species	Protein Content %
1.	Sardine	18.10
2.	Mackerel	19.87
3.	Rainbow trout	19.80
4.	European eel	16.3
5.	Wild merge	22.10
6.	Cultured merge	20.40
7.	Wild seabass	17.64
8.	Cultured seabass	23.37
9.	*Thunnus albacares*	23.9
10.	*Harpadon nehereus*	8.2
11.	*Mugil cephalus*, *Euthynnus affinis*, *Etroplus suratensis*, *Stolephorus waitei*, *Tenualosa ilisha*, and *Lates calcarifer*	>20
12.	Carps, Catla, *Cirrhinus mrigala*, and *Labeo rohita*	15–16
13.	*Anabas testudineus*, *Amblypharyngodon mola* and *Puntius sophore*	<16
14.	*Sperata seenghala* and *Rita rita* (Catfishes)	> 19
15.	Indian prawn *Fennero penaeus indicus*	16.4
16.	Tiger prawn *Penaeus monodon*	19.4
17.	*Ailia coila*	12.9
18.	*Harpadon nehereus*	8.2
19.	*Heteropneustes fossilis*	16.12
20.	*Clarias batrachus*	16.26

Source: Luzzana et al. (2003); Gokog˘lu et al. (2004b); Periago et al. (2005); Cakli et al. (2006); Fernandes et al. (2014).

assess the protein needs of fish. The majority of these values have been derived from dose-response curves, determining the minimal dietary protein (DP) essential for optimal growth. It is worth noting that some of these requirement values might have been overestimated. The fish's DP requirement is intricately linked to factors such as the protein-to-energy balance in the diet, the AA composition, and the digestibility of the test substance. The quantity of non-protein energy sources within a test diet, along with the interplay of various proteins, plays a crucial role. Excessive energy in the test diet might restrict consumption, as fish, akin to other animals, consume to fulfill their energy needs (Lee and Putnam, 1973,

TABLE 1.4 Amino Acids Contents of Selected Bony Fish

Amino Acids (g/100 g)	Channel Catfish	Atlantic SALMON	Horse Mackerel	Rainbow Trout	Yellow Perch	Scabbard Fish	Mackerel	Herring	Heteropneustes fossilis
Threonine	4.41	4.95	4.28	4.76	3.19	4.46	4.24	4.55	9.4
Valine	5.15	5.09	4.53	5.09	4.12	5.73	5.41	4.21	9.2
Methionine	2.92	1.83	2.30	2.88	2.46	3.18	2.48	2.53	0.04
Isoleucine	4.29	4.41	4.57	4.37	3.43	5.10	4.77	4.22	0.079
Leucine	7.4	7.72	7.21	7.59	6.82	8.28	6.86	7.03	0.13
Phenylalanine	4.14	4.36	4.55	4.38	3.38	3.82	4.41	4.63	0.062
Lysine	8.51	9.28	6.82	8.49	6.06	10.19	8.02	7.05	0.069
Histidine	2.17	3.02	2.87	2.96	2.07	1.91	3.41	2.90	0.067
Arginine	6.67	6.61	5.31	6.41	5.03	6.37	7.07	5.13	0.072
Serine	4.89	4.61	4.52	4.66	3.31	3.82	4.83	4.20	0.087
Aspartic acid	9.74	9.92	9.61	9.94	8.13	10.19	9.05	9.60	19.9
Glutamic acid	14.39	14.31	12.02	14.22	12.23	15.29	13.91	13.36	–
Glycine	8.14	7.41	5.32	7.76	3.20	5.10	5.49	5.97	23.70
Alanine	6.31	6.52	5.51	6.57	4.37	6.37	5.22	5.19	10.30
Tyrosine	3.28	3.50	2.86	3.38	2.97	3.82	3.42	2.72	1.50
Proline	6.02	4.64	5.59	4.89	3.10	5.10	5.13	7.33	1.20
Cysteine	1.34	0.80	0.95	0.86	–	–	1.02	0.91	2.00
Tryptophan	0.78	0.93	–	0.93	1.06	–	–	–	1.00

Source: Mai et al. (1980); Bandarra et al. (2009); Sharma and Katz (2013); Oluwaniyi et al. (2010); Paul et al. (2016).

TABLE 1.5 Proximate Amino Acid Composition of Indian Fish Species

Common Name	Aspartic Acid	Threonine	Serine	Glutamic Acid	Glycine	Alanine	Valine	Methionine
				Marine Fish				
Seer	10.0	4.8	3.1	12.0	3.1	5.1	4.4	2.1
Hilsa	9.8	5.1	2.8	13.5	3.8	6.1	5.1	3.1
Anchovy	8.4	4.6	4.4	14.1	3.6	7.6	4.9	3.9
Pomfret (black)	9.4	3.9	4.9	15.4	4.1	6.1	5.6	3.7
Jew fish	8.9	4.9	3.9	16.0	3.7	4.9	5.8	4.0
Mullet	8.0	5.0	4.1	12.1	2.9	2.9	5.4	6.2
Congereel	8.0	4.2	4.2	14.0	2.9	7.0	5.7	2.9
Pomfret (White)	8.8	4.0	4.1	16.1	3.1	6.9	5.0	3.1
Trevally	8.8	3.9	3.9	13.1	2.9	5.8	6.1	3.6
Pink Perch	9.0	4.0	4.1	14.1	3.0	6.2	5.8	3.7
Lesser Sardine	10.0	5.0	3.9	16.0	3.6	7.2	5.9	2.8
Thread fins	9.4	4.9	4.0	13.4	3.7	6.7	5.0	2.9
Bombay duck	10.3	4.3	14.1	2.9	5.1	5.0	5.0	3.1
Giant Perch	9.0	4.1	2.9	16.0	3.0	6.4	5.1	3.1
Shark	8.4	3.9	2.4	13.9	3.9	5.0	5.0	4.1
				Freshwater Fish				
Murrel	8.4	3.9	3.8	13.2	2.8	8.5	4.4	1.9
Rohu	10.2	4.8	2.8	14.2	3.1	6.3	4.9	2.7
Catla	9.6	4.6	3.0	15.0	3.8	8.1	4.6	3.1

TABLE 1.5 *(Continued)*

Common Name	Isoleucine	Leucine	Tyrosine	Phenylalanine	Lysine	Histidine	Arginine	Proline
Marine Fish								
Seer	3.8	7.0	4.2	4.0	8.1	2.0	5.0	3.0
Hilsa	4.1	7.6	3.9	4.1	9.1	3.9	5.8	3.1
Anchovy	5.1	7.1	3.2	3.9	10.0	2.8	5.8	4.1
Pomfret (Black)	4.9	7.8	4.2	4.9	10.1	2.8	5.8	3.6
Jew fish	4.1	8.0	4.1	4.7	8.1	3.1	8.1	4.2
Mullet	2.9	4.4	8.0	3.8	4.1	4.0	6.1	3.8
Congereel	4.8	7.1	3.9	4.6	10.4	3.8	5.0	5.8
Pomfret (White)	4.7	7.6	4.1	5.0	10.8	3.1	5.9	5.4
Trevally	3.9	8.1	4.2	5.1	9.1	2.9	6.0	3.8
Pink Perch	4.0	8.0	3.9	4.8	8.9	4.0	5.9	4.1
Lesser Sardine	4.1	7.9	3.1	5.0	10.0	3.9	4.9	3.9
Thread fins	4.6	7.0	3.8	4.9	8.9	2.9	5.1	4.6
Bombay duck	5.1	7.0	4.1	5.0	8.9	2.9	5.9	4.8
Giant Perch	5.2	7.6	3.9	4.9	8.8	3.9	5.1	3.0
Shark	5.1	7.1	3.1	4.0	9.3	3.9	5.4	3.6
Freshwater Fish								
Murrel	3.8	8.8	4.0	4.4	8.0	3.1	5.2	4.0
Rohu	4.8	7.1	3.8	4.1	10.0	3.9	6.0	4.2
Catla	3.7	6.7	3.3	3.9	8.4	3.8	5.2	4.1

Source: Chandrashekhar and Deosthale (1993).

Page and Andrew, 1973). While many researchers assert the use of isoenergetic diets to establish protein requirements, it is important to note that the metabolizable energy of most ingredients remains undetermined for a majority of fish species. Various physiological fuel values have been employed, such as 4, 8, and 1.6 kcal/g for protein, fat, and carbohydrate in salmonids (Phillip, 1969). Divergent values are utilized for different fish species, like 3.5, 8.1, 2.5 (Post, 1977) and 4, 9, 4 (Galing and Wilson, 1976) for channel catfish, *Ictalurus punctatus*; 4, 8, 2 (Ogino and Takeuchi, 1976) and 4, 9, 4 (Zeitoun et al., 1973) for rainbow trout, *Salmo gairdneri*; 4.5, 9.0, 4.0 (Mazid et al 1979) and 4.5, 8.5, 3.5 (Jauncey, 1982) for *Tilapia zillii* and *T. mossambica*, respectively; 5, 9, 2 (Wee and Tacon, 1982) for snakehead, *Channa micropeltes*, and 5.7, 9.5, 4 (Cowey et al., 1972) for plaice, *Pleuronectes platessa*.

The impact of alterations in DP-to-energy ratios on growth and protein utilization has been evidenced in various fish species, including rainbow trout (Lee and Putnam, 1973; Takeuchi, 1978); yellowtail, *Seriola quinqueradiata* (Shimeno et al., 1980; Takeda, 1975); common carp, *Cyprinus carpio* (Takeuchi, 1979); channel catfish (Garling and Wilson, 1976; Page and Andrew, 1973); brook trout, *Salvelinus fontinalis* (Ringrose, 1971); *Tilapia aurea*, *T. mossambica* (Jauncey, 1982; Winfree and Stickney, 1981) and the estuary grouper, *Epinephelus salmoides* (Teng and Lim, 1978). Numerous studies underscore the significant sparing effect of non-protein energy sources on DP utilization. The efficacy of dietary carbohydrate utilization varies among species, with evidence of protein sparing in salmonids (Bergot, 1979; Buhler and Halver, 1961; Lee and Putnam, 1973; Piper and Pfeffer, 1979; Ringrose, 1971); plaice (Cowey et al., 1975); turbot, *Scophthalmus maximus* (Adron et al., 1976); channel catfish (Garling and Wilson, 1976, 1977); sea bass, *Dicentrarchus labrax* (Alliot et al., 1979); common carp (Ogino and Takeuchi, 1976; Sen et al., 1978; Takeuchi, 1979); and red sea bream, *Chrysophrys major* (Furuichi, 1971). Lipids have also been shown to spare protein and enhance protein utilization in salmonids (Lee and Putnam, 1973; Reinitz, 1978; Ringrose, 1971; Takeuchi et al., 1978; Viola and Rappaport, 1979; Yu and Sinnhuber, 1977, 1981); common carp (Sin, 1973; Viola and Rappaport, 1979); channel catfish (Garling and Wilson, 1976, 1977; Murray et al., 1977; Page and Andrew, 1973; Stickney and Andrew, 1972); turbot; striped bass, *Morone saxatilis* (Millikin, 1983); and *Tilapia aurea* (Winfree and Stickney, 1981). Furthermore, the quality of the DP utilized in test diets to establish protein requirements significantly influences the estimated requirement value.

1.8 FACTORS AFFECTING THE PROTEIN REQUIREMENTS

In general, the protein needs of fish tend to diminish as they grow larger and older. To illustrate, very young salmonids thrive on a diet with an optimal protein level of 45–50%, while juveniles find 40% sufficient, and yearlings require approximately 35% DP (Hilton and Slinger 1981). Likewise, channel catfish fry thrive on around 40% protein, transitioning to 30–35% protein for fingerlings, and larger fish demonstrate sufficient growth with a protein content ranging from 25–35% (Page and Andrew, 1973). Recommended DP levels have been suggested for common carp across different size categories: 43–47% for fry, 37–42% for fingerlings and subadults, and 28–32% for adult and brood fish. In a review by Balarin and Haller (1982) on tilapia, it was determined that fish weighing less than 1 g benefit from a protein content of 35–50%, while those in the 1–5 g range require 30–40% protein; for tilapia exceeding 25 g in weight, a recommended protein level falls within the range of 20–25%. The protein requirement in certain fish species has been observed to change with variations in water temperature. Chinook salmon, for instance, necessitate 40% protein at 8°C and 55% protein at 15°C (DeLong et al., 1958). Similarly, striped bass have been found to require 47% protein at 20°C and approximately 55% at 24°C (Millikin, 1982, 1983). Rainbow trout, when fed practical diets containing 35%, 40%, and 45% crude protein across temperatures ranging from 9 to 18°C, exhibited no apparent differences in protein requirements.

1.9 MAINTENANCE OF PROTEIN REQUIREMENTS

The protein requirement for maintenance can be determined by considering endogenous nitrogen excretion data and factoring in the digestibility and biological value of the test protein. Ogino and Chen (1973) derived a maintenance requirement of 0.95 g protein/kg body weight/day for carp when casein served as the exclusive protein source. Kaushik and Luquet (1981) approximated the maintenance requirement for rainbow trout at 1.6 g protein/kg body weight/day, utilizing data from fish fed fish meal as the sole protein source. In the case of channel catfish, a maintenance

TABLE 1.6 Important Amino acid Requirements[a,b]

Fish	Arginine	Histidine	Isoleucine	Leucine	Valine
Chinook salmon	6.0 (2.4/40) (Klein and Halver, 1970)	1.8 (0.7/40) (Klein and Halver, 1970)	2.2 (0.9/41) (Chance et al., 1964)	3.9 (1.6/41) (Chance et al., 1964)	3.2 (1.3140) (Chance et al., 1964)
Coho salmon	5.8 (2.3/40) (Klein and Halver, 1970)	1.8 (0.7/40) (Klein and Halver, 1970)	–	–	–
Common carp	4.3 (1.6138.5) (Nose, 1979)	2.1(0.8/38.5) (Nose 1979)	2.5(0.9/38.5) (Nose, 1979)	3.3(1.3/38.5) (Nose 1979)	3.6 (1.4138.5) (Nose 1979)
Japaneese eel	4.5 (I. 7/37.7) (Nose, 1979)	2.1(0.8/37.7) (Nose 1979)	4.0 (1.5/37.7) (Nose, 1979)	5.3(2.0/37.7) (Nose 1979)	4.0(1.5/37.7) (Nose 1979)
Channel fish	4.3 (1.03/24) (Robinson et al., 1981)	1.5(0.37/24) (Wilson et al., 1980)	2.6 (0.62124) (Wilson et al., 1980)	3.5(0.84124) (Wilson et al 1980)	3.0(0.71J24) (Wilson et al 1980)
				3.5(0.96127.6) (Hughes et al 1983)	3.3(0.78/23.7) (Hughes et al 1983)
Lake trout	–	–	2.6(0.72127.6) (Hughes et al 1983)		–
Rainbow trout	3.3(1.2136) (Kaushik 1979)	–	–	–	–
	4.0(1.4/35) (Kim et al 1983)				
	5.9(2.8/47) (Ketola 1983)				
Gilthead bream	5.0(1.7/34) (Luquet and Sabaut 1974)	–	–	–	–
Tilapia	4.01(1.59/40) (Jackson and Capper 1982)	–	–	–	–

TABLE 1.6 *(Continued)*

Fish	Lysine	Phenylalanine	Methionine	Threonine	Tryptophan
Chinook salmon	5.0 (2.0/40) (Halver, 1965)	6.5 (2.5/38.5) (Nose, 1979) Tyr = 0%	4.0 (1.6/40) (Halver, 1958) Cys = 1%	2.2 (0.9/40) (DeLong, 1958)	0.5 (0.2/40) (Halver, 1964)
Common carp	5.7 (2.2/38.5) (Nose, 1979)	6.5 (2.5/38.5) (Nose, 1979) Tyr = 0%	3.1 (1.2/38.5) (Nose, 1979) Cys = 0%	3.9 (1.5/38.5) (Nose, 1979)	0.8 (0.3/38.5) (Nose, 1979) 0.3 (0.13/42.5) (16)
Japaneese eel	5.3 (2.0137.7 (Nose, 1979)	5.8 (2.2/37.7) (Nose, 1979) Tyr = 0%	3. 2 (1. 2/37.7) (Nose, 1979) Cys = 0%	4.0 (1.5/37.7) (Nose, 1979)	1.1 (0.4137.7) (Nose, 1979)
Channel fish	5.1 (1.23/24) (Wilson et al., 1977)	5.0 (1.20/24) (Robinson et al 1980) Tyr = 0.3%	2.3 (0.56124) (Harding, 1977) Cys = 0%	2.0 (0.53/24) (Wilson et al., 1977)	0.5 (0.12124) (Wilson et al., 1977)
Rainbow trout	5.0 (1.5/30) (Robinson et al., 1980) 3.7(1.3/35) (Kim and Kayes, 1982) 4. 2 (1. 9/45) (Walton et al., 1984) 6.1 (2.9/47) (Ketola, 1983)	–	2.2 (1.0/46.4) (Walton et al., 1982) Cys = 0% 3.0 (1.1/35) (Rumsay et al., 1983) Cys = 0.3% 2.9 (1.0/35) (Kim et al., 1984) Cys = 0.5%	–	0.5 (0.25155) (Walton et al., 1984) 1.4 (0.58/42) (Poston and Rumsay, 1983)
Gilthead bream	5.0 (1.7/34) (Luquet and Sabaut, 1974)	–	4.0(1.4/34) (Luquet and Sabaut, 1974) Cys = not stated	–	0.6 (0.2/34) (Luquet and Sabaut,1974)

TABLE 1.6 *(Continued)*

Fish	Lysine	Phenylalanine	Methionine	Threonine	Tryptophan
Tilapia	4.1 (1.62/40) (Jackson and Capper, 1982)	–	3.2 (1.27/40) (Jackson and Capper, 1983) Cys = 0.7%	–	–
Sea bass	–	–	2.0 (1.0/50) (Thebault, 1983) Cys = not stated	–	–

[a]Requirements are expressed as percentage of protein. In parentheses, the numerators are requirements as percentage of diet and the denominators are percentage of total protein in diet. References, in parentheses, follow each notation.

[b]Requirements are expressed as percentage of protein. In parentheses, the numerators are requirements as percentage of diet and the denominators are percentage of total protein in diet. References, in parentheses, follow each notation. Cys and Tyr = percentage in the diet.

requirement of 1.3 g protein/kg body weight/day was established based on the growth rates of fish fed diets comprising 25% or 35% crude protein, generated from a casein-gelatin mixture. An alternative finding, relying on protein retention data from the aforementioned growth studies, determined the requirement to be approximately 1.0 g protein/kg body weight/day.

1.10 ARGININE

Table 1.6 summarizes the arginine requirement values for various fish species. Among them, salmon exhibit the highest requirement, accounting for approximately 6% of DP, while other species typically require around 4–5%. The arginine requirement in rainbow trout appears to be influenced by salinity, as reported by Kaushik (1979). In freshwater, the requirement is 3.3% of protein, decreasing to 2.8% at 20 ppt salinity and further to 2.2% of protein in full-strength seawater. These findings suggest a potential decrease in the protein requirement with increasing salinity. However, Zeitoun et al. (1973) contradicted this by reporting that rainbow trout fingerlings require 40% protein for optimal growth at 10 ppt salinity and 45% protein at 20 ppt salinity. It is essential to conduct additional studies to validate these observations before reaching conclusive results.

1.11 HISTIDINE

There is a remarkable consensus in the histidine requirements of various fish species, ranging from 1.5% to 2.1% of protein. Alliot et al. (1979) observed a positive response in serum free-histidine levels in channel catfish with increasing dietary histidine intake. The concentration of free histidine in serum showed a significant increase up to the dietary requirement, as determined by growth data, and remained constant at higher dietary intake. In the chinook salmon, dietary histidine was found to impact muscle carnosine concentration, as reported by previous studies (Lukton, 1958). When fed a histidine-deficient diet, muscle carnosine was depleted. Conversely, in the channel catfish, regardless of the dietary level of histidine, carnosine could not be detected in muscle tissue (Wilson, 1980). This observation might potentially serve as an additional indicator of histidine status in other fish species.

1.12 ISOLEUCINE

The isoleucine requirements for fish generally range from 2.0% to 2.6% of protein among the studied species, with the exception of the Japanese eel, which exhibits a significantly higher value. The initially reported value for lake trout, *Salvelinus namaycush*, was lower (1.54–2.06% of protein) compared to other fish species. Hughes et al. (1983) noted that the test diets were formulated to meet the required AA levels for a 35% protein diet. However, when the total protein level is recalculated based on the apparent nitrogen content of the test diets, a substantially lower DP level is obtained. Recalculating the requirement value based on the calculated nitrogen content aligns it within the range reported for other species. Wilson et al. (1980) investigated the impact of dietary isoleucine on serum free-isoleucine, leucine, and valine in channel catfish. Despite a slight increase in serum isoleucine with rising isoleucine intake, these data did not corroborate the requirement determined by growth data. The concentrations of serum free-leucine and valine seemed to follow a similar pattern to the serum free-isoleucine concentrations. Additionally, these researchers noted a mortality rate in their isoleucine-deficient fish that was significantly higher than anticipated.

1.13 LEUCINE

The leucine requirements for fish generally fall within the range of 3.3% to 4.0% of protein for the studied species, excluding the eel, which demonstrates a notably higher requirement. The reported value for lake trout (Hughes, 1983) initially appeared lower (2.74–3.66% of protein); however, upon recalculation, it aligns with values previously reported for other species. According to Wilson et al. (1980), the serum free-leucine level in channel catfish remained consistent regardless of dietary leucine intake. Nevertheless, there was a significant impact of dietary leucine on the serum free-isoleucine and valine levels. At the 0.7% dietary leucine level, both concentrations experienced a six-fold increase compared to the 0.6% leucine level. These heightened levels of isoleucine and valine persisted until a dietary level of 1.2% or above was administered. This observation suggests that leucine may play a role in facilitating the tissue uptake of branched-chain AAs and/or their intracellular metabolism.

1.14 VALINE

There is a generally consistent agreement in the reported valine requirements for the studied species, ranging from approximately 3% to 4% of protein. Initially, a notably lower value (1.77–2.23% of protein) was reported for lake trout (Hughes et al., 1983); however, upon recalculation, the revised value aligns with the range reported for other species. The impact of valine intake on serum valine levels in channel catfish was found to be comparable to that observed for isoleucine (Alliot et al., 1979).

1.15 ISOLEUCINE-LEUCINE-VALINE INTERACTIONS

Divergent isoleucine-leucine-valine interactions are evident among various fish species. Chance et al. (1969) noted a slight increase in the isoleucine requirement for Chinook salmon with rising dietary leucine levels. This phenomenon, however, was not observed in either common carp (Nose, 1979) or channel catfish (Robinson and Wilson, 1974). Nose (1979) did observe diminished growth rates in carp fed high levels of dietary isoleucine during his examination of leucine requirements. Interestingly, this reduction in growth was not evident when the study was replicated at lower isoleucine levels. Hughes et al. (1983) observed a distinct pattern in plasma free branched-chain AAs in lake trout when exposed to increasing valine levels, differing from observations in channel catfish (Wilson, 1980). Notably, no significant change in plasma free-valine concentrations occurred until after meeting the dietary requirement, after which it increased approximately 2.5-fold.

Plasma free-isoleucine and leucine exhibited elevation in valine-deficient fish, followed by a decrease as dietary valine levels increased. Robinson et al. (1978) concluded that there is a nutritional interrelationship among branched-chain AAs in the channel catfish. However, this interaction seems less severe compared to observations in certain other animals. Additionally, based on their findings, these researchers suggest that leucine may regulate either the tissue uptake or catabolism of valine and isoleucine in the channel catfish.

1.16 LYSINE

Table 1.6 provides the lysine requirement values, with similar values reported for Chinook salmon, Japanese eel, channel catfish, and gilthead bream.

Common carp exhibited a slightly higher value. The lower value for tilapia may be attributed to observed low growth rates and/or the specific test diet utilized. Rainbow trout's lysine requirement appears to be lower than that of other species. Walton et al. (1982) noted excellent agreement between requirement values determined through growth studies and AA oxidation studies in rainbow trout. However, the notably higher value reported by Ketola (1983) appears to deviate from the consensus. He observed significant mortality and caudal fin erosion in fish fed lysine-deficient diets. As caudal fin erosion was absent in fish fed diets deficient in arginine, the author specifically attributed the fin erosion to lysine deficiency. A recent report from the same laboratory also highlighted a high incidence of caudal fin erosion in tryptophan-deficient rainbow trout (Poston and Rumsey, 1983).

1.17 ARGININE-LYSINE INTERACTION

The documented dietary interplay between arginine and lysine, commonly referred to as lysine-arginine antagonism, is well-established in certain animals. However, Robinson et al. (1981) were unable to demonstrate this antagonism in channel catfish, whether excess lysine was provided in diets adequate or marginal in arginine, or vice versa. Similarly, the growth rates of rainbow trout fed low levels of arginine were unaffected by the feeding of excess lysine (Kim et al., 1983). Kaushik and Fauconneau (1984) presented biochemical evidence suggesting the potential existence of metabolic antagonism between lysine and arginine in rainbow trout. The increase in dietary lysine intake influenced plasma arginine and urea levels, as well as ammonia excretion. These changes were attributed to a decrease in the relative rate of arginine degradation with the increasing level of dietary lysine (Kaushik and Fauconneau, 1984).

1.18 PHENYLALANINE

Phenylalanine and tyrosine fall under the category of aromatic AAs, both crucial for proper protein synthesis and various physiological functions in fish. Fish possess the ability to convert phenylalanine to tyrosine or utilize dietary tyrosine to fulfill their metabolic requirements for this AA. Consequently, to establish the total aromatic AA requirement (comprising phenylalanine and tyrosine), the dietary requirement for phenylalanine is determined either in the absence of tyrosine or with test diets containing very low levels of tyrosine. The phenylalanine or total aromatic AA requirements are detailed in Table 1.6. Chinook salmon and channel catfish exhibit similar values, while the Japanese eel and common carp require slightly higher values.

1.19 METHIONINE

A parallel relationship, akin to that between phenylalanine and tyrosine, is observed between methionine and cystine. Cystine is considered dispensable as fish can synthesize it from the indispensable AA methionine. When methionine is provided without cystine, a portion is utilized for protein synthesis, while another portion is converted to cysteine for incorporation into proteins. The inclusion of cystine in the diet diminishes the dietary methionine requirement. Fish exhibit a total sulfur AA requirement rather than a distinct methionine requirement. The methionine or total sulfur AA requirement values (refer to Table 1.4) exhibit variations among species. Chinook salmon and gilthead bream necessitate the highest levels, approximately 4% of protein, while channel catfish have the lowest requirement at 2.3% of protein. Other species studied generally require around 3.0% of protein. Rainbow trout present a unique characteristic in which methionine deficiency leads to bilateral cataracts (Poston et al., 1977). They observed cataracts in rainbow trout fed diets containing isolated soybean protein, and this condition was prevented by supplementing the diet with methionine. Cataracts associated with methionine deficiency have also been reported by Walton et al. (1982) and Rumsey et al. (1983) in rainbow trout, with no such deficiency sign observed in other fish species.

Several studies indicate that the presence of dietary cystine reduces the required amount of dietary methionine for optimal growth. The cystine replacement value for methionine on a sulfur basis is approximately 60% for channel catfish (Harding et al., 1977) and 40% for rainbow trout (Kim et al., 1984). Robinson et al. (1978) investigated the utilization of various dietary sulfur compounds in channel catfish, finding that DL-methionine was as effective as L-methionine in promoting growth. Methionine hydroxy analogue was only about 25% as effective as L-methionine. Taurine or inorganic sulfate additions to the basal diet did not elicit significant growth responses. Similarly, Page et al. (1978) were unable to detect the utilization of taurine and inorganic sulfate as sulfur sources in rainbow trout. D-methionine has been demonstrated to replace L-methionine on an equal basis in rainbow trout (Kim et al., 1984).

1.20 THREONINE

The threonine requirement appears to be higher in the Japanese eel and common carp compared to the Chinook salmon and channel catfish. DeLong et al. (1962) discovered that the threonine requirement for young Chinook salmon remained constant when determined at rearing temperatures of 8° and 15°C. These findings were unexpected, as these researchers had previously reported an increase in the protein requirement from 40% at 8° to 55% at 15°C (DeLong et al., 1958). The data on threonine requirements suggest that the actual protein requirement for optimal protein synthesis does not change with increasing temperature.

1.21 TRYPTOPHAN

For most fishes, a tryptophan level of 0.5% of protein seems adequate, except for the Japanese eel and possibly the common carp. The value of 0.3% of protein reported by Dabrowski (1981) for common carp appears low and may have resulted from experimental conditions. The value obtained by Poston and Rumsey (1983) for rainbow trout appears too high when compared to other data reported for this species. Tryptophan deficiency results in anatomical deformities in certain salmonid species, but not in other fishes. Halver and Shanks (1960) observed scoliosis and lordosis in sockeye salmon, but not in chinook salmon fed tryptophan-deficient diets. Scoliosis and lordosis have also been observed in tryptophan-deficient rainbow trout (Shanks et al., 1962). These deformities were reversible when the fish were fed an adequate dietary tryptophan (Kloppel and Post, 1975; Shanks et al., 1962). Other signs of tryptophan deficiency in rainbow trout include renal calcinosis (Klioppel and Post, 1975), caudal fin erosion, cataracts, short gill opercula (Poston and Rumsey, 1983), and increased liver and kidney levels of calcium, magnesium, sodium, and potassium (Walton, 1984).

1.22 CONCLUSION

Understanding the chemical composition of fish is crucial for determining their nutritional properties, applications, and storage stability. The chemical composition of fish is subject to fluctuations based on factors such as age, season, feeding patterns, migratory swimming, sexual maturation, and environment (Rasmussen, 2001). In aquaculture, where conditions are more regulated, the composition of fish can be approximated by taking into account factors like feed composition, environment, fish size, and genetic characteristics. Numerous reports gathered in recent studies have effectively underscored the significance of fish proteins in terms of their functional attributes, including solubility, foaming properties, emulsifying properties, water-holding capacity, and fat-binding capacity. These appealing characteristics suggest that fish proteins could be employed as encapsulants or wall materials in the field of nanotechnology. The reciprocal impacts of one AA on another have been explored primarily in the context of cystine-methionine and tyrosine-phenylalanine. Arginine and methionine analogs have proven to be effective nitrogen sources for salmon. Experiments have measured isoleucine-leucine ratios, revealing some growth inhibition when the ratio exceeded 2/1 (Chance et al., 1964). Elevated levels of valine have also been observed to impede growth. Further investigation is required to fully understand the effects of subtle variations in AA ratios in the diet. Significant attention should be devoted to the crucial role of dispensable AAs in fish nutrition.

KEYWORDS

- **amino acid requirements**
- **aquaculture nutrition**
- **dietary amino acid balance**
- **fish muscle biochemistry**
- **fish protein composition**
- **functional properties of fish proteins**

REFERENCES

Adron, J. W., Blair, A., Cowey, C. B., & Shanks, A. M. (1976). Effects of dietary energy level and dietary energy source on growth, feed conversion and body composition of turbot (*Scophthalmus maximus* L.). *Aquaculture, 7*(2), 125–132. https://doi.org/10.1016/0044-8486(76)90046-6.

Alliot, E., Pastoureaud, A., & Nedelec, J. (1979). Étude de l'apport calorique et du rapport et la composition corporelle. In J. E. Halver & K. Tiews (Eds.), *Finfish Nutrition and Fish Feed Technology* (Vol. 1, pp. 241–252). Heenemann.

Balarin, J. A., & Haller, R. D. (1982). The intensive culture of tilapia in tanks, raceways and cages. In *Recent Advances in Aquaculture* (pp. 266–355).

Bandarra, N. M., Batista, I., & Nunes, M. L. (2009). Chemical composition and nutritional value of raw and cooked black scabbardfish (*Aphanopus carbo*). *Scientia Marina, 73*(1), 105–113. https://doi.org/10.3989/scimar.2009.73n1105.

Bergot, F. (1979). Carbohydrate in rainbow trout diets: Effects of the level and source of carbohydrate and the number of meals on growth and body composition. *Aquaculture, 18*(2), 157–167. https://doi.org/10.1016/0044-8486(79)90076-0.

Buhler, D. R., & Halver, J. E. (1961). Nutrition of salmonoid fishes: IX. Carbohydrate requirements of Chinook salmon. *The Journal of Nutrition, 74*(3), 307–318. https://doi.org/10.1093/jn/74.3.307.

Cakli, S., Dincer, T., Cadun, A., Saka, S., & Firat, K. (2006). Seasonal variation of proximate and fatty acid class composition of wild and cultured brown meagre (*Sciena umbra*), a new species for aquaculture. In J. B. Luten, C. Jacobsen, K. Bekaert, A. Saebo, & J. Oehlenschlager (Eds.), *Seafood Research from Fish to Dish* (pp. 469–476). Wageningen Academic Publishers.

Chance, R. E., Mertz, E. T., & Halver, J. E. (1964). Nutrition of salmonoid fishes: XII. Isoleucine, leucine, valine and phenylalanine requirements of Chinook salmon and interrelations between isoleucine and leucine for growth. *The Journal of Nutrition, 83*(3), 177–185. https://doi.org/10.1093/jn/83.3.177.

Chandrashekar, K., & Deosthale, Y. G. (1993). Proximate composition, amino acid, mineral, and trace element content of the edible muscle of 20 Indian fish species. *Journal of Food Composition and Analysis, 6*(2), 195–200. https://doi.org/10.1006/jfca.1993.1022.

Cowey, C. B., Adron, J. W., Brown, D. A., & Shanks, A. M. (1975). Studies on the nutrition of marine flatfish. The metabolism of glucose by plaice (*Pleuronectes platessa*) and the effect of dietary energy source on protein utilization in place. *British Journal of Nutrition, 33*(2), 219–231. https://doi.org/10.1079/BJN19750026.

Cowey, C. B., Pope, J. A., Adron, J. W., & Blair, A. (1972). Studies on the nutrition of marine flatfish. The protein requirement of plaice (*Pleuronectes platessa*). *British Journal of Nutrition, 28*(3), 447–456. https://doi.org/10.1079/BJN19720050.

Dabrowski, K. R. (1981). Tryptophan requirement of common carp (*Cyprinus carpio* L.) fry. *Zeitschrift für Tierphysiologie, Tierernährung und Futtermittelkunde, 46*(1–5), 64–71. https://doi.org/10.1111/j.1439-0396.1981.tb01277.x.

De Long, D. C., Halver, J. E., & Mertz, E. T. (1962). Nutrition of salmonoid fishes: X. Quantitative threonine requirements of Chinook salmon at two water temperatures. *The Journal of Nutrition, 76*(2), 174–178.

DeLong, D. C., Halver, J. E., & Mertz, E. T. (1958). Nutrition of salmonoid fishes: VI. Protein requirements of Chinook salmon at two water temperatures. *The Journal of Nutrition, 65*(4), 589–599.

Fernandes, C. E., Vasconcelos, M. A. S., Ribeiro, M. A., Sarubbo, L. A., Andrade, S. A. C., & Filho, A. B. M. (2014). Nutritional and lipid profiles in marine fish species from Brazil. *Food Chemistry, 160*, 67–71. https://doi.org/10.1016/j.foodchem.2014.03.055.

Furuichi, M. (1971). Studies on the nutrition of red sea bream. IV: Nutritive value of dietary carbohydrate. *Report of the Fisheries Research Laboratory Kyushu University, 1*, 75–81.

Garling, D. L., & Wilson, R. P. (1976). Optimum dietary protein to energy ratio for channel catfish fingerlings (*Ictalurus punctatus*). *The Journal of Nutrition, 106*(9), 1368–1375.

Garling, D. L., & Wilson, R. P. (1977). Effects of dietary carbohydrate-to-lipid ratios on growth and body composition of fingerling channel catfish. *The Progressive Fish-Culturist, 39*(1), 43–47. https://doi.org/10.1577/1548-8659(1977)39[43:EODCRT]2.0.CO;2.

Ghaly, A. E., Ramakrishnan, V. V., Brooks, M. S., Budge, S. M., & Dave, D. (2013). Fish processing wastes as a potential source of proteins, amino acids and oils: A critical review. *Journal of Microbial & Biochemical Technology, 5*(4), 107–129. https://doi.org/10.4172/1948-5948.1000110.

Gokoğlu, N., & Yerlikaya, P. (2015). *Seafood Chilling, Refrigeration and Freezing: Science and Technology*. John Wiley & Sons.

Gokoğlu, N., Yerlikaya, P., & Cengiz, E. (2004a). Changes in biogenic amine contents and sensory quality of sardine (*Sardina pilchardus*) stored at 4°C and 20°C. *Journal of Food Quality, 27*(3), 221–231. https://doi.org/10.1111/j.1745-4557.2004.tb00655.x.

Gokoğlu, N., Yerlikaya, P., & Cengiz, E. (2004b). Effects of cooking methods on the proximate composition and mineral contents of rainbow trout (*Oncorhynchus mykiss*). *Food Chemistry, 84*(1), 19–22. https://doi.org/10.1016/S0308-8146(03)00161-4.

Haard, N. F. (1995). Composition and nutritive value of fish proteins and other nitrogen compounds. In A. Ruiter (Ed.), *Fish and Fishery Products: Composition, Nutritive Properties and Stability* (pp. 20–45). CAB International.

Halver, J. E. (1958). Threonine and lysine requirements of Chinook salmon. *Federation Proceedings, 17*, 478.

Halver, J. E. (1959). Methionine and cystine requirements of Chinook salmon. *Federation Proceedings, 18*, 2076.

Halver, J. E. (1965). Tryptophan requirements of Chinook, sockeye and silver salmon. *Federation Proceedings, 24*, 169.

Halver, J. E., & Shanks, W. E. (1960). Nutrition of salmonoid fishes: VIII. Indispensable amino acids for sockeye salmon. *The Journal of Nutrition, 72*(3), 340–346. https://doi.org/10.1093/jn/72.3.340.

Harding, D. E., Allen, O. W., & Wilson, R. P. (1977). Sulfur amino acid requirement of channel catfish: L-methionine and L-cystine. *The Journal of Nutrition, 107*(11), 2031–2035. https://doi.org/10.1093/jn/107.11.2031.

Hilton, J. W., & Slinger, S. J. (1981). Nutrition and feeding of rainbow trout. *Canadian Special Publication of Fisheries and Aquatic Sciences, 55*, 1–15.

Hughes, S. G., Rumsey, G. L., & Nesheim, M. C. (1983). Dietary requirements for essential branched-chain amino acids by lake trout. *Transactions of the American Fisheries*

Society, 112(6), 812–817. https://doi.org/10.1577/1548-8659(1983)112<812:DRFEBA >2.0.CO;2.

Jackson, A. J., & Capper, B. S. (1982). Investigations into the requirements of the tilapia *Sarotherodon mossambicus* for dietary methionine, lysine and arginine in semi-synthetic diets. *Aquaculture, 29*(3, 4), 289–297. https://doi.org/10.1016/0044-8486(82)90141-0.

Jauncey, K. (1982). The effects of varying dietary protein level on the growth, food conversion, protein utilization and body composition of juvenile tilapias (*Sarotherodon mossambicus*). *Aquaculture, 27*(1), 43–54. https://doi.org/10.1016/0044-8486(82)90006-3.

Kaushik, S. (1979). Application of a biochemical method for the estimation of amino acid needs in fish: Quantitative arginine requirements of rainbow trout in different salinities. In J. E. Halver & K. Tiews (Eds.), *Finfish Nutrition and Fish Feed Technology* (Vol. 1, pp. 197–207). Heenemann GmbH.

Kaushik, S. J., Luquet, P., & Blanc, D. (1981). Usefulness of feeding protein and non-protein calories apart in studies on energy-protein interrelationships in rainbow trout. *Annales de Zootechnie, 30*(1), 3–11. https://doi.org/10.1051/animres:19810101.

Ketola, H. G. (1983). Requirement for dietary lysine and arginine by fry of rainbow trout. *Journal of Animal Science, 56*(1), 101–107. https://doi.org/10.2527/jas1983.561101x.

Khan, S., Rehman, A., Shah, H., Aadil, R. M., Ali, A., Shehzad, Q., & Xia, W. (2022). Fish protein and its derivatives: The novel applications, bioactivities, and their functional significance in food products. *Food Reviews International, 38*(8), 1607–1634. https://doi.org/10.1080/87559129.2020.1845283.

Kikuchi, K., Matahira, Y., & Sakai, K. (2004). Separation and physiological functions of anserine from fish extract. In J. L. M. Dekkers & S. De Roos (Eds.), *Developments in Food Science* (Vol. 42, pp. 97–105). Elsevier. https://doi.org/10.1016/S0167-4501(04)80038-2.

Kim, K. I., & Kayes, T. B. (1982). Test diet development and lysine requirement of rainbow trout. *Federation Proceedings, 41*(3), 392.

Kim, K. I., Kayes, T. B., & Amundson, C. H. (1983). Protein and arginine requirements of rainbow trout. *FAESEB Journal, 42*, 2198.

Kim, K. I., Kayes, T. B., & Amundson, C. H. (1984). Requirements for sulfur-containing amino acids and utilization of D-methionine by rainbow trout. *Federation Proceedings, 43*(4), 856.

Klein, R. G., & Halver, J. E. (1970). Nutrition of salmonid fishes: Arginine and histidine requirements of Chinook and Coho salmon. *The Journal of Nutrition, 100*(9), 1105–1109. https://doi.org/10.1093/jn/100.9.1105.

Kloppel, T. M., & Post, G. (1975). Histological alterations in tryptophan-deficient rainbow trout. *The Journal of Nutrition, 105*(7), 861–866. https://doi.org/10.1093/jn/105.7.861.

Lee, D. J., & Putnam, G. B. (1973). The response of rainbow trout to varying protein/energy ratios in a test diet. *The Journal of Nutrition, 103*(6), 916–922. https://doi.org/10.1093/jn/103.6.916.

Lukton, A. (1958). Effect of diet on imidazole compounds and creatine in Chinook salmon. *Nature, 182*(4641), 1019–1020. https://doi.org/10.1038/1821019a0.

Luquet, P. (1974). Nutrition azotée et croissance chez la daurade et la truite. In *Actes de Colloques sur l'Aquaculture, Brest* (Vol. 1, pp. 243–253).

Luzzana, U., Scolari, M., Campo Dall'Orto, B., et al. (2003). Growth and product quality of European eel (*Anguilla anguilla*) as affected by dietary protein and lipid sources. *Journal of Applied Ichthyology, 19*, 74–78. https://doi.org/10.1046/j.1439-0426.2003.00430.x.

Mahboob, S. (2015). Isolation and characterization of collagen from fish waste material—Skin, scales and fins of *Catla catla* and *Cirrhinus mrigala*. *Journal of Food Science and Technology, 52*, 4296–4305. https://doi.org/10.1007/s13197-014-1481-2.

Mai, J., Shetty, J. K., Kan, T., & Kinsella, J. E. (1980). Protein and amino acid composition of select freshwater fish. *Journal of Agricultural and Food Chemistry, 28*(5), 884–885. https://doi.org/10.1021/jf60230a048.

Mazid, M. A., Tanaka, Y., Katayama, T., Rahman, M. A., Simpson, K. L., & Chichester, C. O. (1979). Growth response of *Tilapia zillii* fingerlings fed isocaloric diets with variable protein levels. *Aquaculture, 18*(2), 115–122. https://doi.org/10.1016/0044-8486(79)90045-1.

Millikin, M. R. (1982). Effects of dietary protein concentration on growth, feed efficiency, and body composition of age-0 striped bass. *Transactions of the American Fisheries Society, 111*(3), 373–378. https://doi.org/10.1577/1548-8659(1982)111<373:EODPCO>2.0.CO;2.

Millikin, M. R. (1983). Interactive effects of dietary protein and lipid on growth and protein utilization of age-0 striped bass. *Transactions of the American Fisheries Society, 112*(2A), 185–193. https://doi.org/10.1577/1548-8659(1983)112<185:IEODPA>2.0.CO;2.

Murray, M. W., Andrews, J. W., & DeLoach, H. L. (1977). Effects of dietary lipids, dietary protein and environmental temperatures on growth, feed conversion and body composition of channel catfish. *The Journal of Nutrition, 107*(2), 272–280. https://doi.org/10.1093/jn/107.2.272.

Nawaz, A., Xiong, Z., Xiong, H., Chen, L., Wang, P. K., Ahmad, I., & Ali, S. W. (2019). The effects of fish meat and fish bone addition on nutritional value, texture and microstructure of optimized fried snacks. *International Journal of Food Science & Technology, 54*(4), 1045–1053. https://doi.org/10.1111/ijfs.14013.

Nose, T. (1978). Summary report on the requirements of essential amino acids for carp. *Bulletin of the Japanese Society of Scientific Fisheries, 44*, 1011–1017.

Nose, T. (1979). Summary report on the requirements of essential amino acids for carp. In J. E. Halver & K. Tiews (Eds.), *Finfish Nutrition and Fish Feed Technology* (pp. 145–156). Heenemann GmbH.

Ogino, C., & Chen, M. S. (1973). Protein nutrition in fish. 5. Relation between biological value of dietary proteins and their utilization in carp. *Bulletin of the Japanese Society of Scientific Fisheries, 39*(9), 955–959.

Ogino, C., & Takeuchi, T. (1976). Protein nutrition in fish. VI. Effects of dietary energy sources on the utilization of proteins by rainbow trout and carp. *Bulletin of the Japanese Society of Scientific Fisheries, 42*, 211–216.

Olafsdottir, G., & Jonsdottir, R. (2010). Volatile aroma compounds in fish. In L. M. L. Nollet & F. Toldrá (Eds.), *Handbook of Seafood and Seafood Products Analysis* (pp. 97–113). CRC Press.

Oluwaniyi, O. O., Dosumu, O. O., & Awolola, G. V. (2010). Effect of local processing methods (boiling, frying and roasting) on the amino acid composition of four marine fishes commonly consumed in Nigeria. *Food Chemistry, 123*(4), 1000–1006.

Page, J. W., & Andrews, J. W. (1973). Interactions of dietary levels of protein and energy on channel catfish (*Ictalurus punctatus*). *The Journal of Nutrition, 103*(9), 1339–1346.

Paul, B. N., Chanda, S., Sridhar, N., Saha, G. S., & Giri, S. S. (2016). Fatty acid, amino acid and vitamin composition of Indian catfish, magur (*Clarias batrachus*) and singhi (*Heteropneustes fossilis*). *SAARC Journal of Agriculture, 14*(2), 189–199.

Phillips, A. M. (1969). Nutrition, digestion, and energy utilization. In W. S. Hoar & D. J. Randall (Eds.), *Fish Physiology* (Vol. 1, pp. 391–432). Academic Press.

Pieper, A., & Pfeffer, E. (1978). Carbohydrates as possible sources of dietary energy for rainbow trout (*Salmo gairdnerii*, Richardson). In *Symposium on Finfish Nutrition and Feed Technology* (FAO, Hamburg, Germany, June 20, 1978).

Pleissner, D., Wimmer, R., & Eriksen, N. T. (2011). Quantification of amino acids in fermentation media by isocratic HPLC analysis of their α-hydroxy acid derivatives. *Analytical Chemistry, 83*(1), 175–181.

Post, G., Andrews Jr, J. W., Deyoe, C. W., Dupree, H. K., Lovell, R. T., Nose, T., & Wilson, R. P. (1977). Nutrient requirements of warmwater fishes. *FAO Technical Report Series*, FAO.

Poston, H. A., & Rumsey, G. L. (1983). Factors affecting dietary requirement and deficiency signs of L-tryptophan in rainbow trout. *The Journal of Nutrition, 113*(12), 2568–2577.

Reinitz, G. L., Orme, L. E., Lemm, C. A., & Hitzel, F. N. (1978). Influence of varying lipid concentrations with two protein concentrations in diets for rainbow trout (*Salmo gairdneri*). *Transactions of the American Fisheries Society, 107*(5), 751–754.

Ringrose, R. C. (1971). Calorie-to-protein ratio for brook trout (*Salvelinus fontinalis*). *Journal of the Fisheries Board of Canada, 28*(8), 1113–1117.

Robinson, E. H., Poe, W. E., & Wilson, R. P. (1984). Effects of feeding diets containing an imbalance of branched-chain amino acids on fingerling channel catfish. *Aquaculture, 37*(1), 51–62.

Robinson, E. H., Wilson, R. P., & Poe, W. E. (1980). Re-evaluation of the lysine requirement and lysine utilization by fingerling channel catfish. *The Journal of Nutrition, 110*(11), 2313–2316.

Robinson, E. H., Wilson, R. P., & Poe, W. E. (1980). Total aromatic amino acid requirement, phenylalanine requirement and tyrosine replacement value for fingerling channel catfish. *The Journal of Nutrition, 110*(9), 1805–1812.

Robinson, E. H., Wilson, R. P., & Poe, W. E. (1981). Arginine requirement and apparent absence of a lysine-arginine antagonist in fingerling channel catfish. *The Journal of Nutrition, 111*(1), 46–52.

Rumsey, G. L., Page, J. W., & Scott, M. L. (1983). Methionine and cystine requirements of rainbow trout. *The Progressive Fish-Culturist, 45*(3), 139–143.

Ryu, B., Shin, K. H., & Kim, S. K. (2021). Muscle protein hydrolysates and amino acid composition in fish. *Marine Drugs, 19*(7), 377.

Sen, D. P. (2005). *Advances in Fish Processing Technology* (p. 786). Allied Publishers.

Sen, P. R., Rao, N. G. S., Ghosh, S. R., & Rout, M. (1978). Observations on the protein and carbohydrate requirements of carps. *Aquaculture, 13*(3), 245–255.

Shanks, W. E., Gahimer, G. D., & Halver, J. E. (1962). The indispensable amino acids for rainbow trout. *The Progressive Fish-Culturist, 24*(2), 68–73.

Sharma, R., & Katz, J. (2013). Fish proteins in coronary artery disease prevention: Amino acid–fatty acid concept. In R. R. Watson & V. R. Preedy (Eds.), *Bioactive Food as Dietary Interventions for Cardiovascular Disease* (pp. 525–549). Elsevier.

Shimeno, S., Hosokawa, H., Takeda, M., & Kajiyama, H. (1980). Effects of calorie to protein ratios in formulated diet on the growth, feed conversion and body composition of young yellowtail. *Bulletin of the Japanese Society of Scientific Fisheries, 46*(9), 1083–1087.

Sin, A. W. (1973). The utilization of dietary protein for growth of young carp (*Cyprinus carpio*) in relation to variation in fat intake. *Hong Kong Fish Bulletin, 3*, 83–88.

Sivik, J. (2000). *Assessment of Protein Fingerprinting Method for Species Verification of Meats* (Master's thesis, University of Guelph, 171 pp.).

Spitze, A. R., Wong, D. L., Rogers, Q. R., & Fascetti, A. J. (2003). Taurine concentrations in animal feed ingredients; cooking influences taurine content. *Journal of Animal Physiology and Animal Nutrition, 87*(7, 8), 251–262.

Stickney, R. R., & Andrews, J. W. (1972). Effects of dietary lipids on growth, food conversion, lipid and fatty acid composition of channel catfish. *The Journal of Nutrition, 102*(2), 249–257.

Takeda, M. (1975). The effect of dietary calorie-to-protein ratio on the growth, feed conversion and body composition of young yellowtail. *Nippon Suisan Gakkaishi, 41*, 443–447.

Takeuchi, T. (1978). Optimum ratio of dietary energy to protein for rainbow trout. *Bulletin of the Japanese Society of Scientific Fisheries, 44*, 729–732.

Takeuchi, T. (1979). Optimum ratio of dietary energy to protein for carp. *Bulletin of the Japanese Society of Scientific Fisheries, 45*, 983–987.

Takeuchi, T., Watanabe, T., & Ogino, C. (1978). Supplementary effect of lipids in a high protein diet of rainbow trout. *Bulletin of the Japanese Society of Scientific Fisheries.*

Teng, S. K., Chua, T. E., & Lim, P. E. (1978). Preliminary observation on the dietary protein requirement of estuary grouper, *Epinephelus salmoides* Maxwell, cultured in floating net-cages. *Aquaculture, 15*(3), 257–271.

Thebault, H. (1983). *Étude du besoin en methionine chez le loop, Dicentrarchus labrax en milieu contrôlé* (Doctoral dissertation, Université d'Aix-Marseille II).

Venugopal, V. (2009). Polyunsaturated fatty acids and their therapeutic functions. In *Marine Products for Healthcare* (pp. 51–101). CRC Press.

Viola, S., & Rappaport, U. (1979). The "extra calorie effect" of oil in nutrition of carp. *Bamidgeh, 31*(3), 51–69.

Waheed, M., Butt, M. S., Shehzad, A., Adzahan, N. M., Shabbir, M. A., Suleria, H. A. R., & Aadil, R. M. (2019). Eggshell calcium: A cheap alternative to expensive supplements. *Trends in Food Science & Technology, 91*, 219–230. https://doi.org/10.1016/j.tifs.2019.07.021.

Walton, M. J., Coloso, R. M., Cowey, C. B., Adron, J. W., & Knox, D. (1984). The effects of dietary tryptophan levels on growth and metabolism of rainbow trout (*Salmo gairdneri*). *British Journal of Nutrition, 51*(2), 279–287.

Walton, M. J., Cowey, C. B., & Adron, J. (1984). The effect of dietary lysine levels on growth and metabolism of rainbow trout (*Salmo gairdneri*). *British Journal of Nutrition, 52*(1), 115–122.

Walton, M. J., Cowey, C. B., & Adron, J. W. (1982). Methionine metabolism in rainbow trout fed diets of differing methionine and cystine content. *The Journal of Nutrition, 112*(8), 1525–1535.

Wee, K. L., & Tacon, A. G. (1982). A preliminary study on the dietary protein requirement of juvenile snakehead. *Fishery Science*, *48*(10), 1463–1468.

Wilson, R. P., & Halver, J. E. (1986). Protein and amino acid requirements of fishes. *Annual Review of Nutrition*, *6*(1), 225–244.

Wilson, R. P., Allen Jr, O. W., Robinson, E. H., & Poe, W. E. (1978). Tryptophan and threonine requirements of fingerling channel catfish. *The Journal of Nutrition*, *108*(10), 1595–1599.

Wilson, R. P., Harding, D. E., & Garling, D. L. (1977). Effect of dietary pH on amino acid utilization and the lysine requirement of fingerling channel catfish. *The Journal of Nutrition*, *107*(1), 166–170.

Wilson, R. P., Poe, W. E., & Robinson, E. H. (1980). Leucine, isoleucine, valine and histidine requirements of fingerling channel catfish. *The Journal of Nutrition*, *110*(4), 627–633.

Winfree, R. A., & Stickney, R. R. (1981). Effects of dietary protein and energy on growth, feed conversion efficiency and body composition of *Tilapia aurea*. *The Journal of Nutrition*, *111*(6), 1001–1012.

Yu, T. C., & Sinnhuber, R. O. (1981). Use of beef tallow as an energy source in coho salmon (*Oncorhynchus kisutch*) rations. *Canadian Journal of Fisheries and Aquatic Sciences*, *38*(3), 367–370.

Yu, T. C., Sinnhuber, R. O., & Putnam, G. B. (1977). Use of swine fat as an energy source in trout rations. *The Progressive Fish-Culturist*, *39*(2), 95–97.

Zeitoun, I. H., Halver, J. E., Ullrey, D. E., & Tack, P. I. (1973). Influence of salinity on protein requirements of rainbow trout (*Salmo gairdneri*) fingerlings. *Journal of the Fisheries Board of Canada*, *30*(12), 1867–1873.

Zhou, X., Hua, X., Huang, L., & Xu, Y. (2019). Bio-utilization of cheese manufacturing wastes (cheese whey powder) for bioethanol and specific product (galactonic acid) production via a two-step bioprocess. *Bioresource Technology*, *272*, 70–76.

CHAPTER 2

Fish Digestion and Protein Utilization

MOHAMMAD AMIN HAJAM[1,*] and ISHTIYAQ AHMAD[2]

[1]*Department of Biochemistry, University of Kashmir, Srinagar, Jammu and Kashmir, India*

[2]*Department of Zoology, Fish Nutrition Research Laboratory, University of Kashmir, Srinagar, Jammu and Kashmir, India; Division of Fish Genetics and Biotechnology, Faculty of Fisheries, Sher-e-Kashmir University of Agricultural Sciences and Technology of Kashmir (SKUAST-Kashmir), Rangil, Ganderbal, Srinagar, Jammu and Kashmir, India*

*Corresponding author

ABSTRACT

Aquaculture production is becoming more demanding with each passing day to meet the increasing demands for fish protein sources. Around 17% of the animal protein supply worldwide and 7% of all the protein consumed by humans comes from fish, and it is expected that seafood will be considered the future trustworthy food source for the population living below the poverty line. Maintaining the sustainability of aquaculture production is a challenging task because the different challenges faced in aquaculture include feeding efficiency, overfeeding, food wastage, nutrient loss, and disease-free conditions to maintain health, which can be achieved by providing an adequate amount of feed in a completely balanced diet. The variations in dietary requirements arise due to different species, sex, developmental stages of fish, and the environmental conditions in which the fish are living. The important aspect of feed is to meet the nutritional requirements and be easily digestible to promote fish growth. Consumption of feed also depends on taste to prompt fish to consume feed and

Fish Nutrition: Protein and Amino Acid Requirements. Imtiaz Ahmed, Ishtiyaq Ahmad, and Shabihul Fatma Sayed (Eds.)

DOI: 10.1201/9781998511570-2

maintain digestibility, because feed ingredients influence the sensory and qualitative characteristics of fish, which in turn meet consumer expectations. Fish protein diets containing essential amino acids (EAAs), amino acid composition, and protein structural configuration are the best choice for feed preparations for optimal growth and reproduction, as well as other physiological processes of fish. Dietary nutrients that are eliminated through feces are included in the undigested components but not in the fish's nutrition; thus, an important condition of the diet is to provide a high level of digestibility. The energy digestibility coefficient is considered when designing complete feeds.

2.1 INTRODUCTION

Proteins are biomolecules comprising long chains of amino acids (AAs). The unique characteristics and functions of proteins depend upon the AA sequences in the proteins. Proteins play a vital role in the physiological functions of an organism, from providing a structural framework of the cell to all other enzymatic reactions within the cell. AAs that an organism is capable of synthesizing within the body are called non-EAAs, and those which are needed to be taken in the diet are called EAAs (Table 2.1).

TABLE 2.1 Dietary Classification of Amino Acids

Essential Amino Acids	Non-Essential Amino Acids
Arginine	Alanine
Histidine	Asparagine
Isoleucine	Aspartic acid
Leucine	Cystine
Lysine	Glutamic acid
Methionine	Glutamine
Phenylalanine	Glycine
Threonine	Proline
Tryptophan	Serine
Valine	Tyrosine

The nutritional value of proteins depends upon composition as well as the proportion of the AAs in the proteins. Proteins possessing a higher proportion of EAAs do have higher nutritional values. Thus, to maintain the healthy condition of an organism, EAAs should be present in adequate amounts, and deficiency of any of these EAAs limits protein synthesis and therefore causes weight loss. There is strong evidence that aquatic animals perform well in processes of growth, development, and reproduction when provided an adequate supply of nutritionally EAAs and non-EAAs in their diets. The effect is observed in the developmental stages as well, for example, during larval to metamorphosis stages. Animals perform the processes of digestion, absorption, and metabolism to absorb and use protein and other dietary AAs. The manufacturing procedures and AA degradation in the stomach affect the quantity of dietary AAs that enter the bloodstream; the digestibility and bioavailability of AAs should be carefully assessed. The absorbed AAs are used to synthesize proteins, peptides, AAs, and metabolites of AAs including nucleotides, to produce biological oxidation and ATP. The intermediate metabolites of AAs are used to carry out processes like gluconeogenesis and lipogenesis and control immunological responses, and antioxidative activity, and help in acid-base balance. Without taking into account the availability of AAs in the fish diet, fish farmers often concentrate on the amount or digestibility of dietary crude protein.

In the present scenario, little understanding is known regarding the impact of feed processing techniques on the digestion and utilization of AAs in various fish species or the precise cell and tissue metabolism of AAs. To support the global community and fish producers, much understanding is needed to create eco-friendly aquafeeds for aquaculture at an affordable price.

2.2 DIETARY PROTEIN REQUIREMENT OF FISHES

Around 17% of the animal protein supply worldwide and 7% of all the protein consumed by humans comes from fish (Food and Agriculture Organization, 2020). For optimal growth and reproduction, and other physiological processes, fishes require a diet rich in protein content containing EAAs, as they cannot synthesize and require the incorporation of these EAAs via an external diet (Weatherley et al., 1987). It is important to consider that a balanced mixture of AAs is provided in the diet because dietary protein (DP) is economically expensive, and it makes no sense to use excessive protein content. If a diet contains higher protein content, then subsequently higher nitrogenous wastes are eliminated from the body. The protein requirement in the diet is determined by various factors like the species of the fish, the age of the fish, the living environmental condition of the fish including the dissolved oxygen content, pH of water, and ammonia levels, availability of natural food organisms, feeding rate, stocking density, and protein/energy (P/E) ratio. Presence of natural food organisms and the other dietary contents in the food, like vitamins and minerals, if adequate, also determine the DP requirements. The AA composition and protein structural configuration also play a role in the dietary requirements.

2.3 AGE

The age of fish determines the requirement of DPs. In general, protein requirement decreases with the arrival of sexual maturity, as has been registered in different fish species. In the juvenile phase, fish requires 35% of DP, and in the adult phase, it is 25% (Wilson, 1986). In fingerlings, DP requirement is more than in juveniles, which decreases as the fish grows older. This is because, in early growth stages, a rapid increase of most tissues requires more protein, and relatively higher metabolic conversion takes place (Watanabe et al., 1997). It has been estimated in many species that DP intake is higher in the juvenile stage of fish. Proteins are digested through two metabolic phases: catabolism and anabolism. Catabolism involves the production of energy for maintenance, and anabolism involves the synthesis of other proteins and AA-related metabolic products. For proper growth and development, AA intake in the diet is crucial for tissue formation (protein deposition). This balance is thus maintained by the breakdown of AAs and rates of protein synthesis. However, any dysregulation in protein digestion and metabolism directly affects the physiological processes of fish (Millward, 1989). The excessive intake of protein diet is indicated by the nitrogen balance, which may be positive or negative. The incorporation of DP is necessary because its synthesis in fish has a major metabolic impact, as the use of 42% of the energy produced within the body by food intake is utilized for protein synthesis (Carter and Houlihan, 2001). This is because protein synthesis is a high-energy-demanding process. The maximum amount that a species may utilize proteins for growth is another factor to take into account when looking at proteins as macronutrients in fish diets. Once the proteins are deaminated during the process of metabolism, excessive AAs taken in the diet are converted to other forms as stored lipids for energy sources (Hepher, 1989).

2.4 LIVING ENVIRONMENTAL CONDITION

The final product of AA metabolism is ammonia, which occurs through the urea cycle. Ammonia production has a direct relationship with protein intake, and excessive intake of proteins thus leads to excessive ammonia. Fishes feeding on an excessive protein diet as an energy source cause a phenomenon known as eutrophication of the surrounding water wherein the fishes are residing. This excessive production of ammonia deteriorates the water quality and hence increases production costs. How can this phenomenon be lessened? By replacing the protein diet with nonprotein energy sources (Weatherley and Gill, 1987). This balance is necessary to be maintained to meet the energy requirements for growth and can be maintained by energy produced through lipid and carbohydrate diets. Dissolved oxygen content also affects protein consumption since oxygen has a lower solubility in water and varies with temperature and the salinity of water. Higher temperature reduces the content of dissolved oxygen, while cold freshwater has higher oxygen content. The higher oxygen content in water supports the metabolic rate in the fish's blood; this is because oxygen moves by diffusion through the fish's gills. Various factors are responsible for the diffusion of oxygen, such as the gill area, the distance between the gill epithelia, the diffusion constant, as well as the oxygen partial pressure difference across the gills. In an experiment carried out to check the effect of dissolved oxygen on growth performance, the observations concluded that hypoxia adversely affected feed intake and feed conversion ratio (FCR), and the fish growth improved with an increase in DP levels (Ngoepe et al., n.d.).

2.5 AVAILABILITY OF NATURAL FOOD ORGANISMS

Natural food organisms include phytoplanktons and zooplanktons. The presence of these organisms in the culture decreases the protein requirements of fishes in the farm ponds because these natural sources act as easily available protein materials for fishes, hence decreasing the DP requirements (Kvale et al., 2007).

2.6 FEEDING RATE

Fishes are raised both in intensive and ultra-intense farms, and such farming requires pellets and other feeds in higher amounts because of their higher DP needs. Those cultured in large systems make use of the naturally occurring living food organisms in the pond; as a result, the feeding rate in such naturally occurring farming ponds indicates lesser protein requirements (Dato-Cajegas and Yakupitiyage, 1996; Luo et al., 2014).

2.7 STOCKING DENSITY

It refers to the number of animals per unit area, which directly affects the DP requirements of the shellfish or the finfish. The greater the stocking density of the pond, the greater the protein requirements. This increase in protein requirements is observed due to multiple behavioral characteristics of the farm fish species. The higher stocking density makes it difficult for the feed to be uniformly distributed in the pond (Fenderson and Carpenter, 1971; Keenleyside and Yamamoto, 1962; Refstie and Kittelsen, 1976; Symons, 1970).

2.8 PROTEIN/ENERGY (P/E) RATIO

The amount of energy in the meal influences the amount of fish's protein needs. To enhance muscle growth, fish use protein to meet metabolic energy needs in case of diet's non-protein energy levels, that is, lipids and carbohydrates. It leads to increased ammonia excretion in the water, which directly stresses fish. Fish will not consume enough protein to meet their demand, which is stated as a percentage of dry matter and is influenced by the temperature of water, as already discussed in the above headings in detail. A higher amount of such non-protein energy-yielding compounds is also problematic because it inhibits food intake, although the growth rate may enhance more quickly because of a higher feed conversion efficiency.

2.9 GENERAL CHARACTERISTICS IN FISH DIGESTION

The general feature of digestion in all living organisms starts from the ingestion of food, which undergoes digestion and absorption. The absorbed nutrients from the diet are channeled into metabolic routes. Digestion involves a series of coordinated processes, both physical, chemical, and enzymatic. The physical process involves searching for food, capturing, and breaking down food by chewing, crushing, pulsating, and then chyme churning. The secretion of enzymes in the mouth starts the chemical processes, and the release of hydrochloric acid in the stomach or bicarbonate in the intestines. The chemical action of these secretions causes the hydrolysis of ingested diet components. The hydrolysis of proteins forms free AAs and small peptides, carbohydrates into simple sugars, and fats into free fatty acids and glycerol. The mouth, oral cavity, esophagus, stomach, and intestine with the pyloric cecum and accompanying organs (liver, gallbladder, and pancreas), rectum, and anus are the primary digestive system portions in fish. The structure of the mouth and tooth varies in fish, depending upon the type of food eaten by fish. Fishes which are predacious and feed on small invertebrates have simple conical teeth on the jaws, some bones on the roof of the mouth, and some on special gill arch structures near the esophagus. The majority of predacious fishes swallow whole prey, and teeth are used for holding and grasping the prey. The division shows many adaptations or modifications in different fish species depending upon the evolutionary position and the feeding habits of fish species. Fish species like sharks and piranhas possess biting teeth that pierce chunks out of their victim. Some fishes have modified teeth for crushing the food, like parrot fishes for coral breaking. Similarly, some fishes have brush-like teeth that scrape plant and animal growth from rocks. The digestion of proteins begins in the stomach, which contains a low pH due to the secretion of hydrochloric and proteolytic enzyme pepsin. The food leaving the stomach (chyme) is acidic and is neutralized by the fluids in the intestine; the further secretion of enzymes by the pancreas and intestines reduces the acidic nature of chyme. These enzymes break the complex food components carbohydrates, proteins, and lipids, which are further absorbed into the blood.

2.10 ENZYMES FOR PROTEIN DIGESTION

Protein digestion is carried out by various proteases, which act on the polypeptide chain. Three important and major enzymes involved are pepsin, trypsin, and chymotrypsin, which have great relevance in the gastrointestinal tract of fish (Carter and Houlihan, 2001). These enzymes act as endo- and exo-peptidases on polypeptides and oligopeptides and produce component AAs, which are ultimately absorbed (Carter and Houlihan, 2001). Enzyme pepsin usually begins the digestion of proteins in the stomach, which are supplemented by intestinal alkaline proteases such as trypsin and chymotrypsin (Jónás et al., 1983), breaking down the polypeptides and oligopeptides. Pancreatic proteases like collagenases and elastases are additionally secreted and are included in the endopeptidases of the serine family. These proteases show catalytic activity in the serine groups. These proteases show complementary action and act on different peptide bonds. Since different AA sequences are recognized as active sites in the polypeptide chains, this results in different peptides of many different sizes and are subsequently hydrolyzed by different oligopeptidases (Guillaume and Choubert, 2001). Oligopeptidases, which cleave peptide bonds on the amino-terminal, are called aminopeptidases, and those which cleave at the carboxy-terminal are called carboxypeptidases; for example, pancreatic lysosomal enzymes secreted in inactive forms as zymogens are amino-terminal peptidases and carboxy-terminal peptidases. The internal bonds of other dipeptides are cleaved by dipeptidases. Almost all the enzymes have an optimum activity temperature that varies between 30 and 40°C. But the exception lies in the cold-water fishes, which show an enzyme activity at low temperatures.

2.10.1 PEPSIN

The amino-terminal side of the aromatic AA residues, which include tyrosine, tryptophan, and phenylalanine, are the sites of cleavage where the enzyme pepsin acts. The breaking of peptide bonds in these sites causes the polypeptide chain to undergo cleavage, and smaller peptides are formed (Fänge and Grove, 1979). Pepsinogen is an inactive form of pepsin and is activated in the presence of acidic conditions. Pepsin is active in acidic pH in the gastric portion of the stomach and hydrolyzes many proteins except a few like mucins, spongins, conchiolin, keratin, and peptides with low molecular weight (Fänge and Grove, 1979). Pepsin causes initial and partial hydrolysis of proteins in the stomach in acidic conditions facilitated by the secretion of hydrochloric acid; the process is completed in the intestines by the combined action of trypsin and chymotrypsin (Sabapathy and Teo, 1993). The DP levels influence the activity of pepsin in different ways. In some fish species, the DP intake induces pepsin production, whereas in some species, pepsin production is not under the influence of DP levels (Moraes and de Almeida, 2020). Upon reaching the small intestines, trypsin and chymotrypsin act as major digestive enzymes, which are secreted in the anterior portions of the small intestines.

2.10.2 TRYPSIN

Trypsin is stored as an inactive zymogen in the pancreatic cells as proenzyme granules. Upon entry into the intestinal lumen, these zymogens are transformed into active enzymes, and trypsinogen is transformed into trypsin by the action of proteases produced by mucosal cells of the intestines. Once trypsin is activated, other pancreatic zymogens are in turn activated by trypsin. Trypsin causes the breakdown of proteins further into much smaller peptides by severing the peptide bonds upon hydrolysis. The peptide bond that is present between the carboxylic acid group of lysine or arginine and the amino group of the next AA is cleaved by the action of trypsin. Aromatic AAs—tyrosine, tryptophan, and phenylalanine—are the residues that are substituted for the carbamyl activity in peptides by chymotrypsin. Amides and esters of these aromatic AAs are also hydrolyzed by chymotrypsin. In comparison to trypsin, chymotrypsin shows much higher proteolytic activity than that of trypsin (Falcón-Hidalgo et al., 2011).

2.10.3 CHYMOTRYPSIN

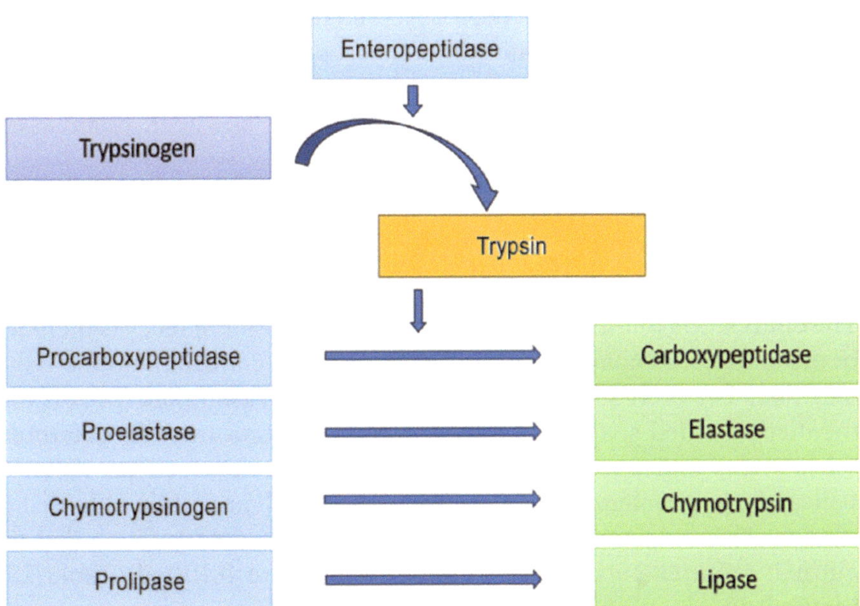

Chymotrypsinogen is another zymogen, and chymotrypsin in its active form is an endopeptidase secreted by the pancreatic cells for protein digestion. Chymotrypsin has three variants: chymotrypsin A, B, and C, which vary in their solubility, electrophoretic mobility, isoelectric point, and different cleavage sites. In the case of fishes, only chymotrypsin A and B are found. The specificity of chymotrypsin has been shown to have a higher specific activity in the case of cold-water fishes. Chymotrypsin also shows variable specificity at low pH and temperature tolerance. 7.5–11 is the optimal pH range of fish chymotrypsin, whereas mammalian chymotrypsin has a lower optimal pH; that is, fish chymotrypsin shows more stability than in acidic conditions. Factors like the age of fish, fish weight, temperature of water, age of fishes, and the extent of starvation regulate the concentration and the activity of chymotrypsin in fishes (Zhou et al., 2011). Chymotrypsin is comprised of 245 AAs with 5 pairs of disulfide linkages and a molecular mass of 24 kDa (Haard and Simpson, 2000). Chymotrypsin attacks the peptide bonds with carbonyl form side chains of aromatic AAs phenylalanine, tryptophan, and tyrosine. At the early life stages of fishes, both trypsin and chymotrypsin are normally detected (Jimenez-Martinez et al., 2012). It has been observed that animal and fish-isolated chymotrypsin show four major differences, and the differences lie in pH and temperature stability, enzyme activity, and molecular weight of the enzyme. The AA composition of chymotrypsin, however, remains the same (Heu et al., 1995).

2.11 PROTEIN UTILIZATION BY FISHES

The elimination of dietary nutrients and energy through feces is included in the undigested components and thus does not contribute to the fish's nutrition. The prerequisite condition desirable for complete feeds is that the diet provided shall have a high level of digestibility. To determine the nutrient and energy digestibility coefficient of feed provided to fishes, the percentage of ingested nutrients which are retained by fishes is used. This helps in the formulation of specific feedstuffs so that the nutrient requirements of the cultured species are fulfilled (Table 2.2; Figure 2.1).

TABLE 2.2 Commercial Fish Species with their Protein Retention Efficiency

Fish Species	Protein Retention Efficiency	References
Chanos chanos	21ssss	Lim et al. (1979)
Cyprinus carpio	30	Takeuchi et al. (1979)
Cyprinus carpio	40	Ogino et al. (1976)
Cyprinus carpio	48	Murai et al. (1986)
Ictalurus punctatus	40	Garling and Wilson (1976)
Micropterous dolumieu	27	Anderson et al. (1981)
Micropterous salmoides	31	Anderson et al. (1981)
Morone saxatilis	25	Mallikin (1983)
Pleuronectes platessa	23	Cowey et al. (1972)
Salvenius alpinus	33	Jobling and Wandsvik (1983)
Salmo gairdneri	40	Ogino et al. (1976)
Sarstherdon mosssambicus	25	Jauncey (1982)
Sciaenops ocellatus	32	Daniels and Robinson (1986)
Scopthalonus maximus	29	Boorman (1980)
Tiapia nilotica	40	De Silva and Perera (1985)
Tiapia nilotica	26	Santiago et al. (1982)
Tilapia nilotica × *T. aureus*	31	Viola and Zohar (1984)

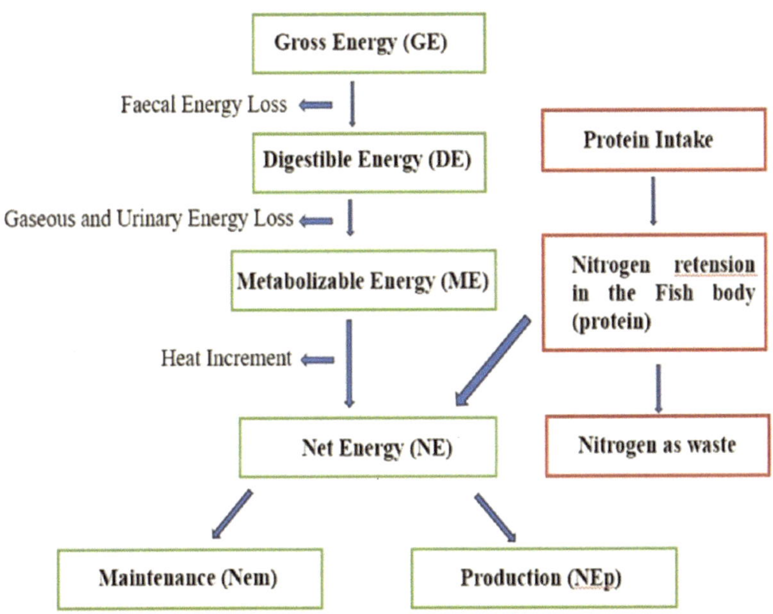

FIGURE 2.1 Energy partition and protein utility in fish body.

2.12 METHODS USED TO DETERMINE PROTEIN UTILIZATION

There is limited research on how the dietary AA balance affects feed intake. It has not been thoroughly examined if a dietary AA surplus or shortage causes an increase in voluntary feed consumption over time. According to preliminary investigation, the AA deficits of methionine, leucine, arginine, and lysine reduce feed intake (Lim et al., 1979). The principle for determining the DP utilization by fish is the same as that of terrestrial farm animals. Utilizing balance techniques for aquatic animals involves significant technical challenges. The two most common indicators of feed use efficiency are the FCR, calculated by dividing feed applied by net fish production, and the feed ratio efficiency (FER), which is the inverse of the FCR. Reliable results can be achieved by employing simple growth rate tests, which include the determination of protein efficiency ratio (PER) and productive protein value (PPV) with the help of nitrogen increase. PER can be calculated as:

PER = Increase of body weight in grams/protein intake in grams

Productive protein value (PPV) not only considers the increase in body weight but also considers the protein increment of fish. PPV is calculated as:

PPV = increase in the body protein/protein intake × 100

Crude protein concentrations vary greatly in aquaculture feeds because protein diets for different species vary; omnivorous and carnivorous species, herbivorous species require less protein. Aquaculture feeds vary greatly in crude protein concentration, because herbivorous species require less protein in their diets than do omnivorous and carnivorous species. The majority of feeds will include between 25 and 40% crude protein. As per the Alltech Global Feed Survey conducted in the year 2018, a total of 40 million metric tons of aquaculture feed was produced in the year 2017. Out of these 40 million metric tons likely contained at least 12 million metric tons of protein.

Fishes require a higher amount of DP to fulfill the need for EAAs than mammals and birds. Ogino and Saito in 1970 demonstrated the significance of DP content for juvenile carp's consumption of protein, and casein was used as the sole source of protein. Based on their observations, it was concluded that growth rates rose when DP intake was increased by 55% and that a DP level of 38% resulted in maximum protein increment (Table 2.3).

TABLE 2.3 Influence of Dietary Protein Level (Casein) on PPV of Juvenile Carp

Dietary Protein Level %	PPV	PPV x Dietary Protein Level
4.6	19.6	90
13.0	29.8	387
21.4	34.4	736
29.8	29.7	885
38.2	28.4	1,084
46.6	23.5	1,095
55.0	19.6	1,078

Source: Ogino and Saito (1970).

KEYWORDS

- **aquaculture feed efficiency**
- **dietary protein requirements**
- **essential amino acids**
- **feed conversion ratio**
- **fish nutrition and growth**
- **protein digestibility in fish**

REFERENCES

Anderson, R. J., Kienholz, E. W., & Flickinger, S. A. (1981). Protein requirements of smallmouth bass and largemouth bass. *The Journal of Nutrition, 111*(1), 1885–1897.

Boorman, K. N. (1980). Dietary constraints on nitrogen retention. In P. J. Buttery & D. B. Lindsay (Eds.), *Protein deposition in animals* (pp. 147–166). Butterworths.

Carter, C. G., & Houlihan, D. F. (2001). Protein synthesis. In P. A. Wright & P. M. Anderson (Eds.), *Nitrogen Excretion* (Vol. 20, pp. 31–75).

Cowey, C. B., Pope, J. A., Adron, J. W., & Blair, A. (1972). Studies on the nutrition of marine flatfish: The protein requirement of plaice (*Pleuronectes platessa*). *British Journal of Nutrition, 28*, 447–456.

Daniels, W. H., & Robinson, E. H. (1986). Protein and energy requirements of juvenile red drum (*Sciaenops ocellatus*). *Aquaculture, 53*, 243–252.

Dato-Cajegas, C. R. S., & Yakupitiyage, A. (1996). The need for dietary mineral supplementation for Nile tilapia, *Oreochromis niloticus*, cultured in a semi-intensive system. *Aquaculture, 144*(1–3), 227–237.

De Silva, S. S., & Perera, M. K. (1985). Effects of dietary protein level on growth, food conversion, and protein use in young *Tilapia nilotica* at four salinities. *Transactions of the American Fisheries Society, 114*, 584–589.

Falcón-Hidalgo, B., Forrellat-Barrios, A., Carrillo Farnés, O., & Ubieta Hernández, K. (2011). Digestive enzymes of two freshwater fishes (*Limia vittata* and *Gambusia punctata*) with different dietary preferences at three developmental stages. *Comparative Biochemistry and Physiology Part B: Biochemistry and Molecular Biology, 158*(2), 136–141. https://doi.org/10.1016/j.cbpb.2010.11.004.

Fänge, R., & Grove, D. (1979). Digestion. In W. S. Hoar, D. J. Randall, & J. R. Brett (Eds.), *Fish Physiology: Bioenergetics and Growth* (Vol. 8, pp. 161–260). Academic Press.

Fenderson, O. C., & Carpenter, M. R. (1971). Effects of crowding on the behavior of juvenile hatchery and wild landlocked Atlantic salmon (*Salmo salar* L.). *Animal Behavior, 19*(3), 439–447.

Food and Agriculture Organization. (2020). *The State of World Fisheries and Aquaculture 2020: Sustainability in Action*. FAO. https://doi.org/10.4060/ca9229en.

Garling, D. L., Jr., & Wilson, R. P. (1976). Optimum dietary protein to energy ratio for channel catfish fingerlings (*Ictalurus punctatus*). *The Journal of Nutrition, 106*(9), 1368–1375.

Guillaume, J., & Choubert, G. (2001). Digestive physiology and nutrient digestibility in fishes. In J. Guillaume, S. Kaushik, P. Bergot, & R. Metailler (Eds.), *Nutrition and Feeding of Fish and Crustaceans* (pp. 27–58). Springer.

Haard, N. F., & Simpson, B. K. (2000). *Seafood Enzymes: Utilization and Influence on Postharvest Seafood Quality*. Marcel Dekker.

Hepher, B. (1989). Principles of fish nutrition. In M. Shilo & S. Sarig (Eds.), *Fish Culture in Warm Water Systems: Problems and Trends* (pp. 121–141). CRC Press.

Heu, M. S., Kim, H. R., & Pyeun, J. H. (1995). Comparison of trypsin and chymotrypsin from the viscera of anchovy, *Engraulis japonicus*. *Comparative Biochemistry and Physiology Part B: Biochemistry and Molecular Biology, 112*, 557–567. https://doi.org/10.1016/0305-0491(95)00072-0.

Jauncey, K. (1982). The effects of varying dietary protein level on the growth, food conversion, protein utilization and body composition of juvenile tilapias (*Sarotherodon mossambicus*). *Aquaculture, 27*, 43–54.

Jimenez-Martinez, L. D., Alvarez-González, C. A., Tovar-Ramírez, D., et al. (2012). Digestive enzyme activities during early ontogeny in common snook (*Centropomus undecimalis*). *Fish Physiology and Biochemistry, 38*(2), 441–454. https://doi.org/10.1007/s10695-011-9525-9.

Jobling, M., & Wandsvik, A. (1983). Quantitative protein requirements of Arctic charr, *Salvelinus alpinus* (L.). *Journal of Fish Biology, 22*, 705–712.

Jónás, E., Rágyanszki, M., Oláh, J., & Boross, L. (1983). Proteolytic digestive enzymes of carnivorous (*Silurus glanis* L.), herbivorous (*Hypophthalmichthys molitrix* Val.), and omnivorous (*Cyprinus carpio* L.) fishes. *Aquaculture, 30*(1–4), 145–154.

Keenleyside, M. H., & Yamamoto, F. T. (1962). Territorial behavior of juvenile Atlantic salmon (*Salmo salar* L.). *Behavior, 19*(1, 2), 139–168.

Kvale, A., Nordgreen, A., Tonheim, S. K., & Hamre, K. (2007). The problem of meeting dietary protein requirements in intensive aquaculture of marine fish larvae, with emphasis on Atlantic halibut (*Hippoglossus hippoglossus* L.). *Aquaculture Nutrition, 13*(3), 170–185.

Lim, C., Sukhawongs, S., & Pascual, F. P. (1979). A preliminary study on the protein requirements of *Chanos chanos* (Forskal) fry in a controlled environment. *Aquaculture, 17*, 195–201.

Luo, G., Gao, Q., Wang, C., Liu, W., Sun, D., Li, L., & Tan, H. (2014). Growth, digestive activity, welfare, and partial cost-effectiveness of genetically improved farmed tilapia (*Oreochromis niloticus*) cultured in a recirculating aquaculture system and an indoor biofloc system. *Aquaculture, 422*, 1–7.

Mallikin, M. R. (1983). Interactive effects of dietary protein and lipid on growth and protein utilization of age-0 striped bass. *Transactions of the American Fisheries Society, 112*, 185–193.

Millward, D. J. (1989). The nutritional regulation of muscle growth and protein turnover. *Aquaculture, 79*(1–4), 1–28.

Moraes, G., & de Almeida, L. C. (2020). Nutritional and functional aspects of digestion in fish. In V. R. de Almeida-Val, G. Moraes, & A. Chippari-Gomes (Eds.), *Biology and Physiology of Freshwater Neotropical Fish* (pp. 251–271). Academic Press.

Murai, T., Ogata, H., Kosutarak, P., & Arai, S. (1986). Effects of amino acid supplementation and methanol treatment on utilization of soy flour by fingerling carp. *Aquaculture, 56*, 197–204.

Ngoepe, T. K., Qiang, J., Chen, D., Tao, Y., Bao, J., Mamuru, G. L. S., Zhu, H., Tadese, D. A., & Xu, P. (n.d.). The effects of dissolved oxygen and dietary protein levels on growth performance, physiological parameters and the immune response of the genetically improved farmed tilapia juveniles (*Oreochromis niloticus*).

Ogino, C., Chiow, J. Y., & Takeuchi, T. (1976). Protein nutrition in fish—VI. Effects of dietary energy on the utilization of proteins by rainbow trout and carp. *Bulletin of the Japanese Society of Scientific Fisheries, 42*, 213–218.

Refstie, T., & Kittelsen, A. (1976). Effect of density on growth and survival of artificially reared Atlantic salmon. *Aquaculture, 8*(4), 319–326.

Sabapathy, U., & Teo, L. H. (1993). A quantitative study of some digestive enzymes in the rabbitfish, *Siganus canaliculatus* and the sea bass, *Lates calcarifer*. *Journal of Fish Biology, 42*(5), 595–602.

Santiago, C. B., Aldaba, M. B., & Laron, M. A. (1982). Dietary crude protein requirement of *Tilapia nilotica* fry. *Kalikasan: The Philippine Journal of Biology, 11*, 255–265.

Symons, P. E. (1970). The possible role of social and territorial behavior of Atlantic salmon parr in the production of smolts. *Fisheries Research Board of Canada, Biological Station.*

Takeuchi, T., Watanabe, T., & Ogino, C. (1979). Optimum ratio of dietary energy to protein for carps. *Bulletin of the Japanese Society of Scientific Fisheries, 45*, 983–987.

Viola, S., & Zohar, G. (1984). Nutrition studies with market size hybrids of tilapia (*Oreochromis*) in intensive culture: 3. Protein levels and sources. *Bamidgeh, 36*, 3–15.

Watanabe, T., Kiron, V., & Satoh, S. (1997). Trace minerals in fish nutrition. *Aquaculture, 151*(1–4), 185–207. https://doi.org/10.1016/S0044-8486(96)01503-7.

Weatherley, A. H., & Gill, H. S. (1987). *The Biology of Fish Growth.* Academic Press.

Weatherley, A. H., Gill, H. S., & Casselman, J. M. (1987). *The Biology of Fish Growth.* Academic Press.

Wilson, R. P. (1986). Protein and amino acid requirements of fishes. *Annual Review of Nutrition, 6*, 225–244. https://doi.org/10.1146/annurev.nu.06.070186.001301.

Zhou, L., Budge, S. M., Ghaly, A. E., Brooks, M. S., & Dave, D. (2011). Extraction, purification and characterization of fish chymotrypsin: A review. *American Journal of Biochemistry and Biotechnology*, *7*(3), 104–123. https://doi.org/10.3844/ajbbsp.2011.104.123.

CHAPTER 3

Introduction to Proteins in Fish Feed and Protein Requirements of Different Species of Cultured Fish

KIFAYAT ULLAH KHAN,[1,2,*] RIDA AZAM,[3] NEHA IQBAL,[3] MARIUM JAMEEL,[3] CLEBER FERNANDO M. MANSANO,[4] and JOÃO BATISTA K. FERNANDES[2]

[1]*Department of Zoology, Government College University, Lahore, Punjab, Pakistan*

[2]*Aquaculture Center, São Paulo State University (CAUNESP), Via de Acesso Professor Paulo Donato Castellane, Jaboticabal, São Paulo, Brazil*

[3]*Department of Zoology, Wildlife and Fisheries, University of Agriculture, Faisalabad, Punjab, Pakistan*

[4]*Graduate Program in Environmental Sciences, Brazil University, Fazenda Santa Rita, Fernandópolis, São Paulo, Brazil; and Graduate Program in Animal Production, Parque Universitário, Descalvado, São Paulo, Brazil*

*Corresponding author

ABSTRACT

In fish farming, nutrition is usually reflected as one of the most critical factors because it is directly related to fish performance and health, as well as total production costs. To avoid reduced or unhealthy growth and high feeding costs, and ultimately lower gross margins, nutritional requirements should be accurately determined for commercially important fish species in different growth stages, considering the end product of farming.

Fish Nutrition: Protein and Amino Acid Requirements. Imtiaz Ahmed, Ishtiyaq Ahmad, and Shabihul Fatma Sayed (Eds.)

DOI: 10.1201/9781998511570-3

Proteins are among the essential nutrients and are considered an integral part of fish feed due to their significant role in body maintenance, growth, and health status of fish. According to most of the available literature, total protein requirements have been extensively studied in fish nutrition. First, optimal levels of protein content are usually established in fish feeds to achieve maximum performance of different species of fish. Second, most of the available literature has reported crude protein content, and little attention has been paid to improving the quality of dietary protein (DP). Third, protein requirements are established for feeding fish that are usually marketed as whole fish, while other marketing methods are completely neglected. All these three concerns about protein requirements need to be properly addressed because this important nutrient is composed of amino acids (AAs) that are needed in appropriate proportions, particularly EAAs, for the synthesis of optimal body protein. Its extra amounts and/or unbalanced forms in feed are utilized as an energy source, which may result in high fat deposition or high nitrogen excretion in water. Consequently, the growth and health of fish, as well as the total costs of production, and hence net profits, are adversely affected. In addition, feeding fish with DPs without considering the way the harvested product would be sold could also result in significantly lower net profits. Taking into account all these scenarios, this chapter stresses the establishment of DP requirements through appropriate models by considering multiple factors such as DP quality, energy content, sustainable growth, species with high market demand, growth stage, cost-benefit analysis, and the way the end product is marketed. Comprehending sustainable and ecosystem-supporting fisheries and aquaculture practices is suggested in accordance with the recommendations of the "Food and Agriculture Organization Code of Conduct for Responsible Fisheries" to promote sustainable and healthy fish farming to ensure long-term food security, food safety, and livelihoods of fishermen and consumers.

3.1 FISH FARMING AND SOCIO-ECONOMIC GROWTH

The fisheries and aquaculture industry produces more than 90 million metric tons of fish per year as a source of nutritious food for the growing population worldwide (Food and Agriculture Organization, 2018). Fish provide biologically valuable nutrients, particularly proteins, which could help to reduce malnutrition (Albert et al., 2015; Bennett et al., 2018). In developing countries, most people get approximately 30% of their total animal proteins from aquatic products (Food and Agriculture Organization, 2018). In the current scenario of various health disorders, particularly cardiovascular diseases, acquiring only food security is not sufficient, so producing nutritionally healthy products for human consumption through sustainable aquaculture is a practical and worthwhile approach. Additionally, the production of healthy aquatic products using inland fisheries could help reduce the burden on marine stocks, which are greatly utilized for human consumption as well as medicinal purposes and preparation of fish meal and fish oil. Strengthening inland fisheries by adopting sustainable and healthy approaches may help ensure food security and livelihood of the community by providing inexpensive, protein-rich, and healthy food and employment (Khan et al., 2022). According to Yang et al. (2002), Eroldogan et al. (2008), and NRC (2011), improving feed utilization efficiency through balanced protein diets could greatly increase sustainable production of healthy fish (Khan et al., 2019; Pelletier et al., 2018). Balanced nutrition helps avoid the additional protein supplementation, economize additional feeding costs, and reduces high nitrogen excretion in water (Chen et al., 2012; Deng et al., 2011; El-Abed et al., 2013; El-Saidy and Gaber, 2005; Kaushik and Seiliez, 2010; Khan et al., 2019; Marammazi and Kahkesh, 2011). Given that, recent research should focus on promoting sustainable aquaculture to support the promotion of socio-economic growth (Food and Agriculture Organization, 2018; Pelletier et al., 2018; Sargent and Tacon, 1999).

3.2 FISH FEEDING AND NUTRITION

Feed, feeding, and nutrition are critical factors in aquaculture which have a great impact on fish production and gross margins. Proteins, amino acids (AAs), and energy are essential and important components of fish feed due to their significant role in the formation of body structure and the regulation of metabolism (Lee et al., 2001, 2003; Ma et al., 2019; National Research Council, 2011). Fishes require an adequate amount of protein (amino acids) content in their diets for maintenance purposes and to gain optimal growth. On the other hand, the excess dietary supply of this important nutrient can result in its conversion into energy in the liver and, therefore, causes the production of fatty fish in combination with increased nitrogen excretion in water (McGoogan and Gatlin, 1999; Wood, 1993). Increasing or decreasing the amount of energy in feeds has a significant influence on fish growth and health. In addition to being essential macronutrients, proteins and energy are expensive components of fish feed, leading to higher feed costs and ultimately lower gross margins (National Research Council, 2011).

The success of aquaculture production systems and the profitability of the final product are closely related to adequate feeding and nutrition, including food quality and effective feeding programs (Besson et al., 2014). The amount of food that a fish actually needs should be provided at appropriate times and intensity. Before feeding, exploring the eating behavior of fish is extremely important to know about feeding regimes (composition of feed, amount, and nutritional intake) to achieve sustainable performance without any loss in net profit or reduction in growth (El-Saidy and Gaber, 2005). This technique is helpful in avoiding extra protein supplementation in feed, and its application in fish farming could greatly reduce feed costs and nitrogen losses in water (Hardy, 2002). Feeding fish with an adequate amount of nutritionally balanced diets at the right time is essential to healthier fish production at sustainable feeding costs (Cho and Bureau, 2001; Eroldoğan et al., 2008; Halver and Hardy, 2002; Houlihan et al., 2001; National Research Council, 2011). According to the available literature, feed containing balanced protein content provided with optimum energy when offered to fish (usually conventional feeding) according to the apparent satiation strategy could greatly enhance the feed intake and hence the performance of fish. This way, fish try to eat food according to their genetic potential to show the maximum possible nutritional responses (Khan et al., 2020a, b; 2022).

3.3 FEED FORMULATIONS AND FEEDING STRATEGIES

Working on appropriate feed formulations and feeding strategies (regimes) could help prevent dietary proteins (DP) from being used as an energy source and make it convenient to accurately determine protein requirements. Formulating diets with an AA balance and providing optimal energy is necessary for proper use of the available DP only for muscle growth (Gao et al., 2011; National Research Council, 2011).

Diets with a balance of AAs are usually called balanced protein diets, and such diets could greatly improve the body composition (meat quality) of the animal and can avoid supplementation with a higher protein content of the diet (Gous and Morris, 1985). On the other hand, diets formulated without providing optimal AA balance and energy could negatively affect the body composition of fish and generally produce fatty fish, which are less acceptable in the consumer market due to their poor taste and lower nutritional quality. Also, such diets may lead to the determination of inaccurate protein requirements because, in the absence of AA balance and sufficient energy, fish utilize available DP as an energy source rather than for muscle growth. Increased feed costs and high nitrogen losses have prompted the feed industries to pay attention to developing nutritionally balanced and economically sustainable diets. These diets may play an important role in increasing production rates and net profits at sustainable feeding costs (Food and Agriculture Organization, 2018; Melo et al., 2006; Toan et al., 2011; Tacon and Metian, 2015).

Nutritionists and feed industries are trying to develop size (growth stage) specific diets for commercially important fish species through efficient and dynamic formulation approaches (McGoogan and Gatlin, 2000; Webb Jr. and Gatlin, 2003; Wu and Gatlin, 2014). In fish nutrition, since a long time ago the graded supplementation method is widely used to formulate experimental and practical diets for several fish species (Alam et al., 2008; Alvarez-González et al., 2001; Árnason et al., 2009; Bicudo et al., 2010; Elangovan and Shim, 1997; Fernandes et al., 2000; Fernandes et al., 2001; Hung and Quy, 2013; Kim and Lee, 2005; Luo et al., 2004; Tacon et al., 2020; Zhang et al., 2010; Ye et al., 2017). However, this method has a serious problem as the relationship between EAA does not remain constant throughout consecutive diets, which may negatively influence the nutritional responses and thus ultimately lead to the determination of inaccurate protein requirements. On the other hand, the diet-dilution method developed by Fisher and Morris (1970) allows maintaining an EAA balance across experimental diets such that only the protein level would increase but the relationships between EAAs in relation to lysine may remain constant (Abboudi et al., 2007; Gous and Morris, 1985; Liebert and Benkendorff, 2007). There are two approaches to providing AA balance to fish feeds. In one method, the entire AA profile could be achieved through synthetic AAs, while in the second approach the AA profile is mostly attained through intact protein sources. In case of the later method, some of the most EAAs, such as lysine, methionine, and threonine, could be added in synthetic form at a rationale proportion just to fix the desired profile and to avoid the limitations of these important AAs in the case of using plant-based feedstuffs. This approach provides a practical and efficient basis for the development of a nutritionally balanced and cost-effective aquafeed. The absolute addition of crystalline AAs has several disadvantages, such as increased feed costs, restricted feed intake, reduced growth, and net profitability together with higher nitrogen excretions (Eits et al., 2005; Food and Agriculture Organization, 2018; Gous and Morris, 1985; Gous et al., 2018; Kaushik, 2013; Robb and Crampton, 2013; Ye et al., 2017; Shipton and Hecht, 2013; Tidwell and Allan, 2001; Vandergeest, 2007; White, 2013).

3.4 CLASSICAL AND ADVANCED APPROACHES TO PROTEIN RESEARCH IN FISH

Proteins are the basic and necessary structural and functional components of the fish body. They are found, in combination with other compounds, in muscles, bones, skin, and almost every other part or tissue of the body. They make up the enzymes that conduct several chemical reactions and the hemoglobin that transports oxygen in blood. They are composed of building blocks called AAs that are classified as non-essential and essential. The 10 EAAs play an important role in promoting body growth and maintaining health conditions. They must come from food because most teleost fishes either could not synthesize absolutely or synthesize insignificant amounts of EAAs (Food and Agriculture Organization, 2018). The available literature NRC (2011) specifies that actually it is the protein source or package and the protein composition (protein quality) rather than amount of protein that conceivably marks the difference for fish growth and health. The evidence-based takeaway is that consuming a healthy protein diet is considerably important to produce healthy fish products for human consumption and use in the aquafeed industry or medicinal purposes. Eating healthier fish products regularly could greatly reduce the risk of several diseases, especially cardiovascular disease and early death (Food and Agriculture Organization, 2018; National Research Council, 2011).

Protein is one of the most studied nutrients, and its requirements have been extensively studied for different fish species due to its significant role in the growth and health of fish (Halver and Hardy, 2002; Kim and Lee, 2009; Meyer and Fracalossi, 2004; Sá et al., 2008; Yang et al., 2003). Being a rich source of all essential and non-EAAs, supplementation with DP improves the performance, feeding aspects, and body composition of fish. Supplementation of a sufficient amount and quality of protein in fish feeds enhances the efficiency of fish feed utilization (El-Sayed and Teshima, 1992; Satpathy et al., 2003; Shiau and Lan, 1996; Yang et al., 2002). On the other hand, its inadequate or surplus amounts, and/or poor quality may result in unfitting feed utilization efficiencies, increased feeding costs, and high nitrogenous waste production coupled with reduced growth (Abdel-Tawwab et al., 2010; Abimorad and Carneiro, 2004; Bicudo et al., 2010; Engin and Carter, 2001; Monentcham et al., 2010).

The rearing environment, the genetic pattern of the species, the growth phase, feeding rates, the available source of food, and feeding habits are the main factors that influence protein requirements (Craig and Helfrich, 2017; Kaushik, 2013). Most previous studies have focused on the determination of protein requirements only for juvenile fish. From the market point of view, large fish should be properly focused on to reduce extra nitrogen supplementation that may cause additional feeding costs and exert negative effects on net profit, fish health, and water quality of the rearing system. Estimation of fish DP requirements in different growth phases can help feed industries establish species and growth stage-specific balanced protein diets for fish species with high market demand to facilitate farmers to provide an accurate amount of DP to fish to get maximum profits with sustainable growth (Deng et al., 2011; De Silva and Anderson, 1995; Nelson and Cox, 2005).

For sustainable and profitable aquaculture, nutrition and physiology-oriented research in fish needs to be improved according to newly introduced practicable approaches, keeping in view growth response and total costs of production. Protein requirements should be accurately optimized using balanced diets provided with an EAA balance and optimum energy (Tacon and Metian, 2015). Fish farmers generally prefer to feed fish with diets that result in maximum net profits with sustainable growth at lower feeding costs (Aliu et al., 2017). Diets with high protein content, lack of AA balance, and insufficient energy, along with irregular feeding frequencies and mismanaged feeding practices, can lead to increased feeding costs and higher nitrogen excretions. Consequently, the net profit of the final product declines, and both the aquatic environment and fish health are negatively affected (Abdel-Tawwab et al., 2010; National Research Council, 2011; Zeitoun et al., 1976). Feeding costs generally account for about 50 to 65% of the total production costs in intensive aquaculture systems, and if such costs are properly managed, maximum profits could be obtained without any adverse environmental effects, even from small investments (Abdel-Tawwab et al., 2010; Ali et al., 2008; El-Sayed, 2019; El-Saidy and Gaber, 2005; Glencross et al., 2007; Nguyen et al., 2009).

3.4.1 QUALITY OF DIET PROTEINS AND THE CONCEPT OF BALANCED PROTEIN

Low-quality DP usually has low digestibility due to a lack of AA balance, which negatively influences the determination of optimum DP (Gous et al., 2018; Hamre et al., 2003; Sadasivam et al., 2010). Diets that contain low-quality food in terms of AA profile can result in the supplementation of higher amounts of DP (Fortes-Silva et al., 2012; Wethli et al., 1975). Such diets generally adversely affect food intake and nutritional responses and lead to the estimation of inadequate protein requirements (Thoman et al., 1999). This problem can be solved by using good-quality protein ingredients and maintaining the balance between EAAs in diets (El-Saidy and Gaber, 2005; Halver and Hardy, 2002; Jauncey and Ross, 1982). When formulating diets according to the concept of balanced protein, lysine is used as the reference AA to estimate the ratios of the other EAAs. These ratios are kept constant between EAAs in all test diets, and the DP content is only gradually increased according to the desired level (Eits et al., 2005; Mack et al., 1999).

3.4.2 BALANCE OF ESSENTIAL AMINO ACIDS (EAA)

Diets provided with the balance of EAA show better quality and palatability (Furuya et al., 2015; Gous and Morris, 1985; Houlihan et al., 2001). EAA patterns determined in dose-response trials should be preferably maintained in test diets, but if such data are unavailable for a species, then the whole-body AA composition-based estimated EAA pattern can be considered to develop test diets (Khan et al., 2020c; Wilson, 2002). In previous studies, a close correlation has been determined between AA requirements estimated in dose-response trials and those estimated by using the whole-body AA composition (Abimorad et al., 2010; Khan et al., 2020c; Wilson and Poe, 1985). Since dose-response trials are expensive and time-consuming, whole-body AA-based estimated EAA patterns have been determined for several culturable fish species. The benefits of the latter approach are that it is comparatively less expensive and more practical, as well as the data could be achieved in a short time. Although dose response-based data are available for some of the most limiting EAAs such as lysine, methionine, and threonine, the requirements of all EAA are still unavailable for most species of fish (Abimorad et al., 2010; Bicudo and Cyrino, 2009; Gurure et al., 2007; Portz and Cyrino, 2004; Wilson, 2002; Wilson and Poe, 1985). So, until dose-response data are available, estimated requirements based on whole-body AA patterns are recommended to be used in the formulation of practical and experimental diets for relevant fish species (Abimorad et al., 2010; Kaushik, 1998; Khan et al., 2019, 2020c; National Research Council, 2011; Tacon, 1992; Wilson and Poe, 1985).

The balance of EAAs can be achieved through supplementation with crystalline AAs or protein-bound AAs, that is, by using protein-rich ingredients. The addition of large amounts of synthetic AAs in fish diets has several drawbacks, such as some synthetic AAs being very costly and increasing feeding costs. Moreover, leaching losses can reduce AA efficiency, and the use of any coating substance, such as gelatin or agar-agar, could probably lead to poor texture and palatability of the food (Ambardekar and Reigh, 2007; Lall and Anderson, 2005). Fishes do not accept such diets favorably, and the feed intake could be restricted, causing a great decline in growth and net profit of the product. Another serious concern among aqua nutritionists regarding synthetic AAs is that they are generally absorbed more rapidly than protein-bound EAAs, thus limiting the utilization efficiency of fish due to higher catabolism rates (Cowey, 1995; Zarate and Lovell, 1997). Thus, maintaining the balance of EAA by using protein-rich feedstuffs with the slight addition of some most limiting EAAs (such as lysine, methionine, and threonine in the case of working with plant feedstuffs) is a great alternative to the graded supplementation method. The latter approach (diet-dilution) seems more practical and efficient and facilitates preparing aquaculture diets to get sustainable and healthy growth at feasible feeding costs. If necessary, that is, in the case of using plant-based protein, some of the most limiting EAAs, such as lysine, methionine, and threonine, could be slightly added in synthetic form, just to fix the desired EAA profile, to diets (Ambardekar et al., 2009; Gous and Morris, 1985; Lemme, 2010; Schuhmacher et al., 1997).

3.4.3 ENERGY IN RESPONSE TO PROTEIN CONTENT IN FISH FEEDS

Effective protein-to-energy ratios must be maintained in diets to achieve better performance and health. Providing optimum energy in the diet facilitates the estimation of accurate protein requirements (Bicudo et al., 2010; Gous et al., 2018; Khan et al., 2019). When high protein content is added to feed with an insufficient energy level, fish utilize a part of the available DP as an energy source to meet their energy needs rather than fueling muscle growth. On the other hand, if high energy is provided with lower protein contents, energy will probably show the protein sparing effect. Such diets generally lead to the determination of inaccurate protein requirements and low-quality product (Ai et al., 2004; Gous et al., 2018; Meer et al., 1995; Sweilum et al., 2005). Using the optimal amounts of energy boosts the proper metabolism of the available DP and allows one to determine the actual protein requirements of fish (Catacutan and Coloso, 1995; Wang et al., 2013). In this context, maintaining optimum protein-to-energy ratios in fish diets is a key factor in effective protein research (Cho and Bureau, 2001; El-Abed et al., 2013; Lemos et al., 2014; Wang et al., 2013).

3.4.4 ESTABLISHING DIETARY PROTEIN REQUIREMENTS OF FISH SPECIES EITHER FOR MAXIMUM PERFORMANCE OR MAXIMUM ECONOMIC RETURN

To achieve sustainable performance with maximum net profit at harvest, the formulation of economically sustainable and nutritionally balanced diets is gaining popularity in fish farming (Kim and Lee, 2005; Sá et al., 2008; Zhang et al., 2010). Balanced diets generally have a balance of EAAs coupled with optimum energy (Catacutan and Coloso, 1995; El-Saidy and

Gaber, 2005; Eroldoğan et al., 2008; Khan et al., 2019; National Research Council, 2011; Okorie et al., 2007). In poultry and pigs, proper biological and economic models are available to determine protein requirements, taking into account diverse biological and economic responses (Eits et al., 2005; Gous and Morris, 1985; Gonzalez et al., 1992; Gous et al., 2018; Hurwitz et al., 1985). In fish, although biological models have been widely used to estimate protein requirements for performance (Khan et al., 2019), no suitable economic models are available to model protein requirements for maximum economic return with sustainable performance. Diets that contain protein content established for maximum performance usually cost more and could lead to a great decrease in net profit at harvest. Thus, in fish nutrition, we need to accurately differentiate the DP contents required for performance and economic returns.

For the formulation of experimental diets in dose-response studies, so far two methods have been practiced in animal and fish nutrition. One is the graded supplementation method, commonly practiced in animal nutrition, including fish (Abimorad et al., 2007; Bicudo et al., 2010; Fernandes et al., 2000, 2001; Kpogue et al., 2013; Klein et al., 2014; Sankian et al., 2017; Wang et al., 2013), and the other is the diet-dilution method. The latter method has been well studied in animals such as pigs and poultry (Fisher and Morris, 1970; Gous and Morris, 1985; Gous et al., 2018), but in fish nutrition, it has been reported in few studies, such as by Abboudi et al. (2007) and Liebert and Benkendorff (2007).

The graded supplementation method has been widely used for the formulation of protein diets to determine the protein requirements of different species of fish (Abimorad et al., 2007; Bicudo et al., 2010; Fernandes et al., 2000, 2001; Mohanta et al., 2007; Ozório et al., 2006; Signor et al., 2010; Sweilum et al., 2005; Yang et al., 2002; Ye et al., 2017). This method is based on the principle of graded supplementation of a test nutrient to a base diet. The main disadvantage of this method is that a systematic change occurs in the balance of EAAs between different protein diets of the same test series, which can negatively affect the growth response in animals. If growth were compromised, net profits would also remain low (Fisher and Morris, 1970; Gous and Morris, 1985; Gous et al., 2018).

A good alternative to the graded supplementation method is the diet dilution method, which was established in principle to evaluate growth responses to methionine intake in laying hens by Fisher and Morris (1970). Pilbrow and Morris (1974) have used the same method with some modifications to determine the response of laying hens to dietary lysine intakes. Freeman (1979) used it to determine the requirement for chick tryptophan. Similarly, Gous and Morris (1985) used the same technique to assess broiler chicken responses to increasing levels of lysine. Later, it was also tested in fish species such as Atlantic salmon (*Salmo salar* L.) to determine its maintenance threonine requirements for maintenance by Abboudi et al. (2007) and in *Oreochromis niloticus* to determine its lysine requirements by Liebert and Benkendorff (2007). This concept is based on the hypothesis that AA utilization is independent of the DP level and is based on the principle of diluting a high protein diet with a protein-free or low-protein diet to achieve intermediate protein levels (Fisher and Morris, 1970). According to this method, the high protein diet usually contains all EAAs in excess of their optimal requirement, and the low protein diet provides an amount of EAA lower than their requirement. Both basal diets are provided with the same amount of energy, vitamins, minerals, and other necessary supplements. This method has an advantage over the previously mentioned method (graded supplementation) because it avoids the problem of a systematic change in EAA balance among different dietary treatments of the same test series. By exercising this method, serial increases in DP levels could be achieved by maintaining the same EAA balance in all test diets (Gous and Morris, 1985). It focuses on supplementing protein-bound AAs rather than synthetic AAs, and thus avoids the problems of high feeding costs and restricted feed intake and low feed utilization efficiency which may occur if large amounts of synthetic AAs are added to diets (Gous and Morris, 1985; Hauler and Carter, 2001). Therefore, the diet dilution method allows the development of well-balanced protein diets at sustainable cost (Abboudi et al., 2007; Fisher and Morris, 1970; Gous and Morris, 1985; Hardy, 2002; Halver and Hardy, 2002; Liebert and Benkendorff, 2007).

Diets formulated according to the diet dilution approach considerably improve fish feed intake and growth performance and can reduce nitrogen leaching (Abboudi et al., 2007; Fisher and Morris, 1970; Gous and Morris, 1985). The energy level should be maintained at optimal so that fish use the available DP only for muscle growth and maintenance purposes rather than as an energy source (Gous et al., 2018; Khan et al., 2019). Based on existing data, in the case of future studies that focus on determining protein requirements for maximum biological performance with sustainable growth, the diet dilution method is an important and efficient option for the formulation of economically sustainable and nutritionally balanced protein diets for both experimental and practical purposes.

3.4.5 DETERMINING DP REQUIREMENTS TAKING INTO ACCOUNT MULTIPLE FACTORS SUCH AS REARING SYSTEM, GROWTH STAGE, COST-BENEFIT ANALYSIS AND HARVEST MARKETING STRATEGIES

Several factors affect protein requirements, such as species type, diet composition (quality), rearing conditions, and size, that is, growth stage (National Research Council, 2011). It has been reported in recent literature that the target of fish production, that is, producing fish for food (consumer market) or to supply quality seed to farmers or experimental purposes, as well as marketing strategies used to sell the harvest, could greatly influence the economic returns and ultimately the protein requirements (Khan et al., 2020a, b; 2022). Feeding fish with nutritionally balanced feeds according to their growth stage, good management practices and effective feeding programs, and sustainable marketing strategies are important factors to promote sustainable fish farming in the long run. Juvenile fish require relatively high protein content in their diet compared to larger fish (Ye et al., 2017). Therefore, protein requirements should be determined in different growth phases. Moreover, using appropriate statistical analysis and mathematical models are the need of modern animal production science to calculate a more realistic and sustainable amount of for economically important species of fish. On the other hand, the improper addition of protein contents to fish feeds not only increases feeding costs but also leads to increased nitrogen excretions, low nutritional quality, and hence unsustainable net profitability (Jiang et al., 2016; Ma et al., 2019; Ye et al., 2017).

Most of the previously reported nutritional data for different species of fish are usually determined for fishes during their early stages of life, while little information is available for fishes in the adult and/or mature stages (National Research Council, 2011). Still, farmers are feeding different species of fishes and/or all sizes of a fish species with the same type of diet. The concept of "protein sparing effect" is still functional among farmers. Moreover, raising fish without taking into account the marketing strategy exercised at harvest is also a common practice among farmers. All these factors, if not properly addressed, could lead to a great decline in product nutritional quality and net profit. In this context, determining protein requirements of fish by considering the mentioned concerns is very important to avoid unhealthy growth and undesirable loss in economic return (Abdel-Tawwab et al., 2010; Ali et al., 2008; Dumas et al., 2010; El-Saidy and Gaber, 2005; Jauncey and Ross, 1982; National Research Council, 2011; Nguyen et al., 2009; Ye et al., 2017).

3.4.6 DIETARY PROTEIN REQUIREMENTS DETERMINED FOR DIFFERENT SPECIES OF CULTURED FISH (1994–2022)

In fish farming, fish nutrition is generally reflected as the most critical factor because it is directly related to fish performance and total production costs (Gatlin, 2002). To avoid reduced and/or unhealthy growth and high feeding costs, nutritional requirements should be accurately determined for fish in different growth stages by considering all important factors, which are discussed earlier in this chapter. Protein is one of the most essential and important nutrients and is considered an integral part of fish feed due to its significant role in body maintenance, growth, and health (Hatlen et al., 2005; Kim et al., 2003; National Research Council, 2011). Supplementation of this important nutrient in balanced form and at optimal levels significantly improves fish performance. Otherwise, fish performance and net profitability greatly decline (Cho and Bureau, 2001; Gatlin et al., 2007). There is a need to educate fish farmers to adjust their farming practices to the available scientific information in true spirit and letter in order to produce healthy fish and get maximum economic benefits in response to their important investments. This way, food security, food safety, and the livelihood of the farmers and consumers could be secured in the long run. In this chapter, we have tried to summarize and present the available data on protein requirements for important culturable fish species from 1994 to 2022 (Table 3.1).

The literature mentioned (Tables 3.1–3.5) clearly shows that DP requirements of several important species of cultured fish are mainly estimated in their early stages of life. These data provide important insight into the deficiencies and flaws that still exist in protein research. Another important point is that several studies have been conducted with protein levels lower than four, which is below the standard experimental designs required for nutritional studies. In nutritional requirements studies to accurately determine nutritional responses of a species, at least four test

TABLE 3.1 Dietary Protein Requirements of Different Species of Cultured Fish (1994–1995)

Common Name	Scientific Name	Protein Levels Tested	Growth Stage	Optimum Protein Levels Estimated	Response Parameters Assessed	Culture System Used	Reported Temperature (°C)	Trial Duration	Reference
Atlantic salmon	*Salmo salar*	42%, 39%, 37%, 35% (crude protein)	Smolt	39%	Growth performance, feed utilization, dressing-out percentage	Seawater cages	4–13	07 months	Hillestad and Johnsen (1994)
Asian sea bass	*Lates calcarifer*	35%, 42.5% or 50% (crude protein)	Juvenile	42.5%	Growth performance	Flow-through tanks	29	54 days	Catacutan and Coloso (1995)
Catla	*Catla catla* (Ham.)	25%, 30%, 35% (crude protein)	Fingerlings	30 and 35% with fat and carbohydrate levels of 4 and 35%, respectively. Carbohydrates shown protein sparing effect.	Growth performance	Glass aquarium tank	27.2 ± 0.96	40 days	Seenappa and Devaraj (1995)

TABLE 3.2 Dietary Protein Requirements of Different Species of Cultured Fish (2000–2005)

Common Name	Scientific Name	Protein Levels Tested	Growth Stage	Optimum Protein Levels Estimated	Response Parameters Assessed	Culture System Used	Reported Temperature (°C)	Trial Duration	References
Nile tilapia	*Oreochromis niloticus* L.	15% to 50%, with 5% increments (crude protein)	Fry	35%	Weight gain, protein and energy deposition	Flow-through system	26.8	10 weeks	Abdelghany (2000)
American eel	*Anguilla rostrata*	35%, 39%, 43%, 47%, 51% (crude protein)	Juvenile	47%	Growth, feed utilization, nutrient digestibility	cylindrical fiber glass tanks	22 ± 0.2	84 days	Tibbetts et al. (2000)
Common carp	*Cyprinus carpio* L.	35%, 40%, 45% (crude protein)	Juvenile	40%	Growth, feed efficiency	water-recirculating system glass rectangular tanks	18.1–21.0	11 weeks	Cho et al. (2001)
Atlantic salmon	*Salmo salar* L.	34%, 39%, 44% (crude protein)	Larva (0.93 6 ± 0.10 g)	34%	Experiment-1: Growth, feed conversion ratio	Sea cages	7.5	Experiment-1: 14 months	Hillestad et al. (2001)
					Experiment-2: Slaughter quality		11.2	Experiment-II: A & B: 06 months	
Mahseer	*Tor putitora* (Hamilton)	Within a range of 20–50% with 5% increments (crude protein)	Fingerlings	40%	Growth performance	Static rearing system, glass aquaria	24.4–32.10 Authors supposed that temperature above 30°C is not suitable for its growth.	50 days	Hossain et al. (2002)
Nile tilapia	*Oreochromis niloticus*	25%, 30%, 35%, 40% (crude protein)	Brood stock	40%	Spawning performance	Glass aquaria	30	195 days	El-Sayed et al. (2003)
Pacu	*Piaractus mesopotamicus*	10%, 20%, 30%, 40% (crude protein)	Juvenile	30%	Dig. Enzyme response, acid protease (AcP), alkaline protease (AkP)	Indoor tanks	25 ± 2	15 days	Moraes and Bidinotto (2004)

TABLE 3.2 *(Continued)*

Common Name	Scientific Name	Protein Levels Tested	Growth Stage	Optimum Protein Levels Estimated	Response Parameters Assessed	Culture System Used	Reported Temperature (°C)	Trial Duration	References
Mahseer	*Tor putitora* Hamilton	25% to 50%, with 5% increments (crude protein)	Fingerlings	45%, 50% (45.3%)	Feed utilization and growth performance	Earthen ponds	29.5	120 days	Islam and Tanaka (2004)
Nile tilapia	*Oreochromis niloticus* (L)	25% or 30% (crude protein)	Juvenile	25%	Growth performance, production traits (economic evaluation)	Concrete tank recirculation system	27.5 ± 0.7	28 weeks	El-Saidy and Gaber (2005)
Rohu	*Labeo rohita*	20%, 25%, 30%, 35%, 40% (crude protein)	Yearlings of female	25% crude protein	Reproductive performance, egg quality, and composition	Outdoor concrete tank	Not reported	360 days	Khan et al. (2005)
				> 250% crude protein	Somatic growth				
Humpback grouper	*Cromileptes altivelis*	40%, 46% or 52% (digestible protein)	Fingerling	51% digestible protein	Weight gain, Specific growth rate (%), feed conversion ratio	Floating net sea cage	Not reported	180 days	Oesman et al. (2005)
Bagrid catfish	*Pseudobagrus fulvidraco*	22%, 32%, 42%, 52% (crude protein)	Fingerlings	42% crude protein	Growth and protein utilization,	Plastic tanks	25 ± 1.0	06 weeks	Kim and Lee (2005)

TABLE 3.3 Dietary Protein Requirements of Different Species of Cultured Fish (2006–2010)

Common Name	Scientific Name	Protein Levels Tested	Growth Stage	Optimum Protein Levels Estimated	Response Parameters Assessed	Culture System Used	Reported Temperature (°C)	Trial Duration	References
	Cirrhinus mrigala (Hamilton, 1822)	40%, 35%, 30% (crude protein)	Fry	30% crude protein	Growth performance	Glass aquaria	26.2–27	60 days	Singh et al. (2006)
Rohu	*Labeo rohita*	25%, 30%, 35%, 40% (crude protein)	Fingerlings	30% crude protein	Liver, plasma free amino acids, plasma protein fractions (total protein, albumin globulin)	Plastic tanks, recirculation salt–water system.	28–30	45 days	Debnath et al. (2007)
Malabar grouper	*Epinephelus malabaricus*	44%, 49%, 55%, 60% (crude protein)	Fingerlings	55% crude protein	Feed utilization, Energy Retention, growth	Rectangular polyethylene tank	29 ± 0.7	08 weeks	Tuan and Williams (2007)
Tiger puffer	*Takifugu rubripes*	35%, 40%, 45%, 50%, 55% (crude protein)	Juvenile	41% crude protein	Growth and physiological performances	Tank, Flow-through system	22 &17.5	08 weeks	Kim and Lee (2008)
Rohu	*Labeo rohita* (Hamilton)	25%, 35%, 40%, 45% (crude protein)	Fingerling	40% crude protein	Growth and nutrient utilization	Flow-through fiber-glass tanks	29–32	60 days	Satpathy and Ray (2009)
Nile tilapia	*Oreochromis niloticus* L.	25%, 35%, 45% dry matter basis (crude protein)	Fry Fingerlings Juvenile	45% crude protein 35% crude protein 35% crude protein	Growth performance and feed utilization	Glass aquaria	27 ± 1	10 weeks	Abdel-Tawwab (2010)
Blue streak hap	*Labidochromis caeruleus*	30%, 35%, 40%, 45% (crude protein)	Juvenile	35% crude protein	Growth performance	Aquarium	27.5	08 weeks	Ergün et al. (2010)
Blunt snout bream	*Megalobrama amblycephala*	27%, 31%, 35% (crude protein)	Fingerlings	31% crude protein	Growth and effective protein utilization	Floating cages	24–31	08 weeks	Li et al. (2010)
Silver pomfret,	*Pampus argenteus*	35%, 40%, 45%, 50%, 55% (crude protein)	Juvenile	49% crude protein	Growth, %WG	Flow-through system fiberglass tanks	26.0–29.5	06 weeks	Hossain et al. (2010)
Snakehead	*Channa striatus*	35%, 40%, 45%, 50% (crude protein)	Fry	45% crude protein	Growth performance, nutrient utilization, survival	Re-circulatory water system	28–30	08 weeks	Aliyu-Paiko et al. (2010)

TABLE 3.4 Dietary Protein Requirements of Different Species of Cultured Fish (2011–2015)

Common Name	Scientific Name	Protein Levels Tested	Growth Stage	Optimum Protein Levels Estimated	Response Parameters Assessed	Culture System Used	Reported Temperature (°C)	Trial Duration	References
Pacific threadfin	*Polydactylus sexfilis*	25%, 30%, 35%, 40%, 45% (crude protein)	Juvenile	41% crude protein 35% crude protein	Growth performance Protein utilization	Indoor flow-through system	26	08 weeks	Deng et al. (2011)
African catfish	*Clarias gariepinus* (Burchell)	20%, 25%, 30%, 35%, 40%, 45%, 50% (crude protein)	Fingerlings	Range of 34.4% to 39.6% crude protein	Growth, feed conversion and nutrient retention	Circular PVC troughs water flow-through system	27.5–28	08 weeks	Farhat and Khan (2011)
Atlantic salmon	*Salmo salar* L.	29%, 33%, 35% (crude protein)	Adult (2053 g)	29% protein (low protein) with 38% lipid (high lipid) at high water temperature	Growth, Feed conversion ratio and chemical composition	Sea cages	11.6 ± 1.1	10 weeks	Karalazos et al. (2011)
Totoaba	*Totoaba macdonaldi*	43%, 48%, 52% (crude protein)	Juveniles	52% crude protein	Growth performance	Fiberglass tanks recirculation system	25.0 ± 1	10 weeks	López et al. (2011)
Grass carp	*Ctenopharyngodon idella*	33% and 37% (digestible protein)	Juvenile	33% digestible protein	Growth and nutrient utilization	Glass aquaria	26 ± 20	90 days	Köprücü (2012)
Genetically improved farmed tilapia	*Oreochromis niloticus*	Range of 25–50% (crude protein)	Juvenile	40.3% crude protein at 29.9 °C 37.9% crude protein at 29.5 °C	Growth and feed efficiency Somatotropic gene expression levels	Plastic buckets	20–34	56 days	Qiang et al. (2012)
Hybrid sturgeon, ♂	*Acipenser baerii* ♀ × *A. gueldenstaedtii*	25%, 30%, 35%, 40%, 45%, 50%, 55% (crude protein)	Juvenile	34% and 37% crude protein	Growth performance	Concrete tanks	21.4 ± 0.5	08 weeks	Guo et al. (2012)

TABLE 3.4 *(Continued)*

Common Name	Scientific Name	Protein Levels Tested	Growth Stage	Optimum Protein Levels Estimated	Response Parameters Assessed	Culture System Used	Reported Temperature (°C)	Trial Duration	References
Sharpsnout sea bream	*Diplodus puntazzo*, Cetti 1777	15 to 55%	Juveniles	42.9% crude protein 43.8% crude protein	Maximal gain in body weight Maximum nitrogen retention	Semi-recirculation sys cylindrical fiberglass tanks	22 ± 0.5	11 weeks	Coutinho et al. (2012)
Gilthead sea bream	*Sparus aurata*	35%, 38%, 41%, 44%, 47%, 50%, 53% (crude protein)	Juveniles	41% crude protein	Total protease activity, l-lysine absorption capacity	Fiberglass tanks	18–28	12 weeks	García-Meilán et al. (2013)
Brown-marbled grouper	*Epinephelus fuscoguttatus*	45%, 50%, 55% (crude protein)	Juvenile (8.8 ± 1.0 g)	50% crude protein	Growth performance, feed utilization	Flow-through seawater system Fiber glass tanks	28.5–29.5	08 weeks	Shapawi et al. (2014)
Red drum	*Sciaenops ocellatus*	30%, 35% 40% dry matter basis (crude protein)	Juvenile	35 or 40% daily or weekly alternate basis	Weight gain, ammonia production	Aquaria recirculating system.	26–28	08 weeks	Wu and Gatlin III. (2014)
Nile tilapia	*Oreochromis niloticus*	25%, 30%, 35%, 40% (crude protein)	Juveniles	35% crude protein	Growth performance, Feed utilization	Indoor tanks	28 ± 0.43	42 days	Daudpota et al. (2014)
Rice field eel	*Monopterus albus*	35%, 40%, 45% (crude protein)	Juvenile	45% crude protein	Growth performance	Net cage	28 ± 4	10 weeks	Ma et al. (2014)
Patagonian blennie	*Eleginops maclovinus*	Range of 9%–44% dry weight basis (crude protein)	Juvenile	34.7% crude protein 35.7% crude protein	Maximum growth Maximum nitrogen retention	Flow-through marine water system, tanks	14 ± 1	98 days	Sa et al. (2014)
Grouper hybrid	(*Epinephelus fuscoguttatus* (*Epinephelus lanceolatus*)	40%, 45%, 50%, 55% (crude protein)	Juvenile	50% crude protein	Growth performance	Cages	30.5 ± 0.7	8 weeks	Rahimnejad et al. (2015)

TABLE 3.4 *(Continued)*

Common Name	Scientific Name	Protein Levels Tested	Growth Stage	Optimum Protein Levels Estimated	Response Parameters Assessed	Culture System Used	Reported Temperature (°C)	Trial Duration	References
Grass carp	*Ctenopharyngodon idella*	20%, 25%, 30%, 35%, 40%, 45% dry diet basis (crude protein	Juvenile	40% crude protein	Growth performance, Feed utilization, gene expression, metabolic products	Recirculating system, glass-fiber tanks	26.9 ± 0.5	8 weeks	Jin et al. (2015)

TABLE 3.5 Dietary Protein Requirements of Different Species of Cultured Fish (2016–2022)

Common Name	Scientific Name	Protein Levels Tested	Growth Stage	Optimum Protein Levels Estimated	Response Parameters Assessed	Culture System Used	Reported Temperature (°C)	Trial Duration	Reference
Hybrid grouper	*Epinephelus fuscoguttatus* ♀ × *Epinephelus lanceolatus* ♂	38%, 42%, 46%, 50%, 54%, 58%, 62% dry matter basis (crude protein)	Juveniles	53.5% crude protein	Weight gain	Floating cages	27–28	7 weeks	Jiang et al. (2016)
Grass carp	*Ctenopharyngodon idella*	14.31%, 17.67%, 21.72%, 25.75%, 29.22%, 32.28% (digestible protein)	Young	25.07% g digestible protein 25.54% digestible protein	Growth Performance Immunity against enteritis	Cages	28 ± 2	8 weeks	Xu et al. (2016)
Red spotted grouper	*Epinephelus akaara*	32%, 38%, 44%, 50%, 56%, 62% (crude protein)	Juvenile	50.83% crude protein	Weight gain	Net cages	27.5 ± 2.3	8 weeks	Wang et al. (2016)
Northern snakehead fish	*Channa argus*	45%, 48%, 51% (crude protein)	Juvenile	48% crude protein	Growth, Feed utilization	In-door concrete flow-through tanks	23–26.5	8 weeks	Sagada et al. (2017)

TABLE 3.5 *(Continued)*

Common Name	Scientific Name	Protein Levels Tested	Growth Stage	Optimum Protein Levels Estimated	Response Parameters Assessed	Culture System Used	Reported Temperature (°C)	Trial Duration	Reference
Hybrid snakehead	*Channa maculata* ♀ × *Channa argus* ♂	34%, 40%, 46%, 52% or 57% (crude protein)	Fingerlings	47.9% crude protein at lipid level 12% and 50.5% crude protein at lipid level 6.5%	Growth Performance, target of rapamycin signaling	Fiberglass tanks	27.1 ± 1.6	8 weeks	Zhang et al. (2017)
Common carp	*Cyprinus carpio*	20.88% 25.12% (digestible protein)	Juvenile (56.97 ± 7.13 g)	20.88% digestible protein	Growth and feed utilization efficiency	Tanks (semi-static conditions)	23.0 ± 0.2	56 days	Heinitz et al. (2018)
Striped catfish	*Pangasianodon hypophthalmus*	20% 25% 30%, 35%, 40%, 45%, (crude protein)	Juvenile	37.1% crude protein	Growth	Circular tanks (semi-static conditions)	27.2–28.1	60 days	Jayant et al. (2018)
Minor / Reba carp	*Cirrhinus reba*	20%, 25%, 30%, 35%, 40% (crude protein)	Fingerlings	30% crude protein	Overall performance	Indoor circular Fiberglass Reinforced Plastics (FRP) tanks	16.0–23.0	105 days	Yadav et al. (2019)
				28.148 to 28.253% crude protein	Growth performance				
Rainbow trout	*Oncorhynchus mykiss*	48%, 45%, 42%, 39% (crude protein)	Juvenile (35 g)	42% crude protein	Growth performance, body composition, metabolic amino acid sufficiency, tolerance to chronic stress	Tanks	15	15 weeks	Lee et al. (2019)
Rainbow trout	*Oncorhynchus mykiss*	Range of 30% to 55% (crude protein)	Fingerlings	45%–47.1% crude protein	Weight gain, protein deposition	Circular trough, flow-through system	14.5–16.5	10 weeks	Ahmed and Ahmad (2020)

TABLE 3.5 *(Continued)*

Common Name	Scientific Name	Protein Levels Tested	Growth Stage	Optimum Protein Levels Estimated	Response Parameters Assessed	Culture System Used	Reported Temperature (°C)	Trial Duration	Reference
Grass carp	*Ctenopharyngodon idellus*	25%, 28%, 31%, 34% (crude protein)	133.05 ± 2.88 g	30.32% to 30.41% crude protein with 8% lipid	Growth performance, feed utilization, protein and lipid deposition and promote digestive system repair	Outdoor cemented pool system	15–25	8 weeks	Wu et al. (2020)
Red swamp crayfish	*Procambarus clarkia*	26%, 28%, 30%, 32%, 34% (crude protein)	Juvenile (5.17±0.32 g)	31% crude protein	Weight gain, specific growth rate (%)	Re-circulating water system	22.5–25.5	8 weeks	Miao et al. (2020)
Mrigal	*Cirrhinus mrigala*	25.52%, 28.18%, 31.36%, 33.91%, 37.28% (crude protein)	Fingerlings	31.36% crude protein	Growth performance and feed utilization	Fiberglass tanks, flow-through water system	22.5 to 30	8 weeks	Alam et al. (2020)
Pacu	*Piaractus mesopotamicus* Holmberg	16.3%, 20.1%, 23.8%, 27.2%, 31.5% and 34.8% dry matter basis (balanced digestible protein)	Juvenile	32.6% digestible protein for maximum performance	Body weight gain	Concrete tanks (recirculating water system)	27.1 ± 0.24	45 days	Khan et al. (2020a)
				311 balanced digestible protein	Whole-body fish				
				317 balanced digestible protein	Eviscerated fish				
				319 balanced digestible protein	Sliced fish				

TABLE 3.5 *(Continued)*

Common Name	Scientific Name	Protein Levels Tested	Growth Stage	Optimum Protein Levels Estimated	Response Parameters Assessed	Culture System Used	Reported Temperature (°C)	Trial Duration	Reference
Pacus	*Piaractus mesopotamicus*	16.3%, 20.1%, 23.8%, 27.2%, 31.5% and 34.8% dry matter basis (balanced digestible protein)	Fattening (490.60 ± 14.39 g)	292 g kg^{-1} diet (dry matter)	Maximum body weight gain	Concrete tanks (recirculating aquaculture system)	28.2 ± 0.35	7 weeks	Khan et al. (2022)
				249	Profit margin of harvest whole-body				
				256	Eviscerated				
				260	Sliced				
Pacu	*Piaractus mesopotamicus*	16.3%, 20.1%, 23.8%, 27.2%, 31.5% and 34.8% dry matter basis (balanced digestible protein)	Adult	23.2% and 24.5% dry matter basis (balanced digestible protein)	Final body weight (FBW) and body weight gain (BWG)	Concrete tanks (recirculating water system)	28.6 ± 0.5	44 days	Khan et al. (2020b)
				215	Whole-body				
				217	Eviscerated				
				222 g kg^{-1}	Sliced				
Pacu	*Piaractus mesopotamicus*	6.45%, 11.18%, 16.46%, 21.71%, 26.44%, 31.62% 36.95% (digestible protein)	Juveniles	31.62% and 36.95% digestible protein	Weight gain, Feed intake	Tanks (Flow through system)	28.70 ± 0.86	120 days	Mansano et al. (2021)
				18.7%, 21.7%, 25.6%	Maximum nitrogen retention				
Grass carp	*Ctenopharyngodon idella*	21, 25, 29, 33, 37, 41% (crude protein)	Juveniles	38.63% crude protein	Growth performance	Tanks (semi-static conditions)	29 ± 01	56 days	Wang et al. (2022)

levels are required in addition to control, which is the minimum criterion (National Research Council, 2011). The third important point is that most of the mentioned data (Tables 3.1–3.5) being recruited are based on experimental diets prepared according to the graded supplementation method, which is a classical approach. In this method, the diets lack an EAA balance. Protein diets without an EAA balance could adversely affect the feed intake, and hence performance, which could ultimately lead to the determination of inaccurate protein levels. On the other hand, diets prepared through the diet dilution method contain balanced protein that is EAA balance (advanced approach), could greatly improve the feed intake of the animals as they try to consume food according to their maximum genetic potential. This strategy could help farmers to measure the nutritional responses and ultimately the protein requirements in an accurate way. Otherwise, the determined protein requirements would be compromised due to the restricted feed intake, which usually happens in the case of diets being prepared using the graded supplementation method. The fourth critical point is that most of the above-mentioned protein requirements (Tables 3.1–3.5) are determined for raising fish that are only sold as a whole-body, while other marketing strategies are almost ignored, except the three recent studies (Khan et al., 2020a, b; 2022). In general, optimizing DP requirements by taking into account performance and cost-benefit analyses data, through appropriate models, can result in the establishment of nutritionally balanced and economically sustainable diets to promote sustainable and healthy fish farming (Khan et al., 2020a, b; 2022; Sá et al., 2008) as highlighted by Gous (2007) in poultry.

The summarized data (Tables 3.1–3.5) could greatly facilitate farmers in feeding their fish with appropriate protein diets to produce good quality product and to sell them by adopting an efficient marketing strategy to avoid any unsustainable decline in their net profit. These data present precise reference-line information which could greatly help researchers and nutritionists in designing future experiments to explore further understandings of proteins in fish feeds.

3.5 CONCLUSIONS

During the past several decades, global marine fisheries catches have increased tremendously in response to the growing need for healthy animal protein. In addition to human consumption, marine fish stocks are under intense pressure from several industries such as pharmaceuticals, surgery, oil, and aquafeed. The inland fisheries and aquaculture industry (particularly intensive agriculture) needs rapid expansion to reduce the burden on marine fish stocks. However, the increasing intensive inland agriculture is also placing intense pressure on terrestrial resources, which could threaten other industries, especially poultry farming. The changing climatic conditions could also cause severe damage to intensive farming. Given that, the implementation of ecosystem-based inland fisheries management, which is an important component of the "Food and Agriculture Organization Code of Conduct for Responsible Fisheries" in spirit and letter, could help to adequately address the mentioned issues in the long run. In the context of increasing feed costs due to the unavailability of several feedstuffs coupled with high nitrogen excretions in water due to the use of feedstuffs of poor nutritional quality, the optimization of fish nutrition (nutrient utilization, particularly protein and energy) should be properly focused on in order to promote healthier production while gaining maximum economic return at harvest. Such efforts are essential and important in the development of sustainable and healthy intensive fish farming. Compared to many terrestrial farmed animals, fish require higher levels of protein supply and lower energy in feed due to their aquatic mode of life. Based on the advanced literature, the performance and nutritional quality of fish could be improved through the supplementation of balanced protein feed. In this context, the establishment of environment-supporting inland intensive rearing systems with the execution of responsible fisheries management and balanced and sustainable nutrition could be an efficient step towards producing healthier aquatic food through sustainable fisheries and aquaculture.

KEYWORDS

- **dietary protein optimization**
- **ecosystem-based fisheries management**
- **essential amino acids in aquaculture**
- **fish feed formulation**
- **nutrient utilization in fish farming**
- **sustainable aquaculture**

REFERENCES

Abboudi, T., Ooghe, W., Larondelle, Y., & Rollin, X. (2007). Determination of the threonine requirement for maintenance in Atlantic salmon (*Salmo salar* L.) fry with the diet dilution procedure. *Aquaculture Nutrition, 13*(4), 281–290. https://doi.org/10.1111/j.1365-2095.2007.00475.x.

Abdelghany, A. E. (2000). Optimum dietary protein requirements for *Oreochromis niloticus* L. fry using formulated semi-purified diets. In K. Fitzsimmons & J. C. Filho (Eds.), *Tilapia Aquaculture in the 21st Century: Proceedings from the 5th International Symposium on Tilapia Aquaculture* (Vol. 1, pp. 101–108). Rio de Janeiro, Brazil.

Abdel-Tawwab, M., Ahmad, M. H., Khattab, Y. A. E., & Shalaby, A. M. E. (2010). Effect of dietary protein level, initial body weight, and their interaction on the growth, feed utilization, and physiological alterations of Nile tilapia, *Oreochromis niloticus*. *Aquaculture, 298*(3, 4), 267–274. https://doi.org/10.1016/j.aquaculture.2009.10.027.

Abimorad, E. G., & Carneiro, D. J. (2004). Métodos de coleta de fezes e determinação dos coeficientes de digestibilidade da fração protéica e da energia de alimentos para o pacu, *Piaractus mesopotamicus* (Holmberg, 1887) [Methods for feces collection and determination of digestibility coefficients of protein fraction and energy in feeds for pacu]. *Revita Brasileira de Zootecnia, 33*(5), 1101–1109. https://doi.org/10.1590/S1516-35982004000500001.

Abimorad, E. G., Carneiro, D. J., & Urbinati, E. C. (2007). Growth and metabolism of pacu (*Piaractus mesopotamicus* Holmberg 1887) juveniles fed diets containing different protein, lipid and carbohydrate levels. *Aquaculture Research, 38*(1), 36–44. https://doi.org/10.1111/j.1365-2109.2006.01621.x.

Abimorad, E. G., Favero, G. C., Squassoni, G. H., & Carneiro, D. (2010). Dietary digestible lysine requirement and essential amino acid to lysine ratio for pacu *Piaractus mesopotamicus*. *Aquaculture Nutrition, 16*(4), 370–377. https://doi.org/10.1111/j.1365-2095.2009.00674.x.

Ahmed, I., & Ahmad, I. (2020). Effect of dietary protein levels on growth performance, hematological profile and biochemical composition of fingerlings rainbow trout, *Oncorhynchus mykiss* reared in Indian Himalayan region. *Aquaculture Reports, 16*, 100268. https://doi.org/10.1016/j.aqrep.2019.100268.

Ai, Q., Mai, K., Li, H., Zhang, C., Zhang, L., Duan, Q., Tan, B., Xu, W., Ma, H., Zhang, W., & Liufu, Z. (2004). Effects of dietary protein to energy ratios on growth and body composition of juvenile Japanese seabass, *Lateolabrax japonicus*. *Aquaculture, 230*(1–4), 507–516. https://doi.org/10.1016/j.aquaculture.2003.09.040.

Alam, M. S., Liang, X. F., Liu, L., He, S., Liu, Y., & Li, J. (2020). Effect of dietary protein-to-energy ratios on growth performance, body composition, feed utilization and nitrogen metabolism enzyme of *Cirrhinus mrigala*. *Aquaculture Research, 51*(12), 5056–5064. https://doi.org/10.1111/are.14844.

Alam, M. S., Watanabe, W. O., & Carroll, P. M. (2008). Dietary protein requirements of juvenile black sea bass Centropristis striata. *Journal of the World Aquaculture Society, 39*(5), 656–663. https://doi.org/10.1111/j.1749-7345.2008.00204.x.

Albert, S., Aswani, S., Fisher, P. L., & Albert, J. (2015). Keeping food on the table: Human responses and changing coastal fisheries in Solomon Islands. *PLoS ONE, 10*(7), e0130800. https://doi.org/10.1371/journal.pone.0130800.

Ali, A., Al-Ogaily, S. M., Al-Asgah, N. A., Goddard, J. S., & Ahmed, S. I. (2008). Effect of feeding different protein to energy (P/E) ratios on the growth performance and body composition of *Oreochromis niloticus* fingerlings. *Journal of Applied Ichthyology, 24*(1), 31–37. https://doi.org/10.1111/j.1439-0426.2007.00897.x.

Aliu, B. S., Afolabi, J. O., Egwenomhe, M., & Esume, A. C. (2017). Survey on nutrition and feeding practices of fish by farmers in Ondo State, Nigeria. *Asian Journal of Biology, 2*(1), 1–9. https://doi.org/10.9734/AJOB/2017/31605.

Aliyu-Paiko, M., Hashim, R., & Shu-Chien, A. C. (2010). Influence of dietary lipid/protein ratio on survival, growth, body indices and digestive lipase activity in snakehead (*Channa striatus*, Bloch 1793) fry reared in recirculating water system. *Aquaculture Nutrition, 16*(5), 466–474. https://doi.org/10.1111/j.1365-2095.2009.00683.x.

Alvarez-González, C. A., Civera-Cerecedo, C. R., Ortiz-Galindo, J. L., Dumas, S., Moreno-Legorreta, M., & Alamo, T. G. (2001). Effect of dietary protein level on growth and body composition of juvenile spotted sand bass, *Paralabrax maculatofasciatus*, fed practical diets. *Aquaculture, 194*(1, 2), 151–159. https://doi.org/10.1016/S0044-8486(00)00512-3.

Ambardekar, A. A., & Reigh, R. C. (2007). Sources and utilization of amino acids in channel catfish diets: A review. *North American Journal of Aquaculture, 69*(2), 174–179. https://doi.org/10.1577/A06-007.1.

Ambardekar, A. A., Reigh, R. C., & Williams, M. B. (2009). Absorption of amino acids from intact dietary proteins and purified amino acid supplements follows different time-courses in channel catfish (*Ictalurus punctatus*). *Aquaculture, 291*(3, 4), 179–187. https://doi.org/10.1016/j.aquaculture.2009.02.044.

Árnason, J., Imsland, A. K., Gústavsson, A., Gunnarsson, S., Arnarson, I., Reynisson, H., Jónsson, A. F., Smáradóttir, H., & Thorarensen, H. (2009). Optimum feed formulation for Atlantic halibut (*Hippoglossus hippoglossus* L.): minimum protein content in diet for maximum growth. *Aquaculture, 291*(3, 4), 188–191. https://doi.org/10.1016/j.aquaculture.2009.03.025.

Bennett, A., Patil, P., Kleisner, K., Rader, D., Virdin, J., & Basurto, X. (2018). Contribution of fisheries to food and nutrition security: Current knowledge, policy, and research (NI Report 18–02). Duke University. https://nicholasinstitute.duke.edu/sites/default/files/publications/contribution_of_fisheries_to_food_and_nutrition_security_0.pdf (accessed on 3 June 2025).

Besson, M., Komen, H., Aubin, J., de Boer, I. J. M., Poelman, M., Quillet, E., Vancoillie, C., Vandeputte, M., & Arendonk, J. A. M. V. (2014). Economic values of growth and feed efficiency for fish farming in recirculating aquaculture system with density and nitrogen output limitations: A case study with African catfish (*Clarias gariepinus*). *Journal of Animal Science, 92*(12), 5394–5405. https://doi.org/10.2527/jas.2014-8266.

Bicudo, A. J. A., Sado, R. Y., & Cyrino, A. J. P. (2010). Growth performance and body composition of pacu *Piaractus mesopotamicus* (Holmberg 1887) in response to dietary protein and energy levels. *Aquaculture Nutrition, 16*(2), 213–222. https://doi.org/10.1111/j.1365-2095.2009.00653.x.

Bicudo, J. A. A., & Cyrino, J. E. P. (2009). Estimating amino acid requirement of Brazilian freshwater fish from muscle amino acid profile. *Journal of the World Aquaculture Society, 40*(6), 818–823. https://doi.org/10.1111/j.1749-7345.2009.00303.x.

Catacutan, M. R., & Coloso, R. M. (1995). Effect of dietary protein to energy ratios on growth, survival, and body composition of juvenile Asian sea bass, *Lates calcarifer*. *Aquaculture, 131*(1, 2), 125–133. https://doi.org/10.1016/0044-8486(94)00358-U.

Chen, G., Feng, L., Kuang, S., Liu, Y., Jiang, J., Hu, K., Jiang, W., Li, S., Tang, L., & Zhou, X. (2012). Effect of dietary arginine on growth, intestinal enzyme activities and gene expression in muscle, hepatopancreas and intestine of juvenile Jian carp (*Cyprinus carpio* var. Jian). *British Journal of Nutrition, 108*(2), 195–207. https://doi.org/10.1017/S0007114511005459.

Cho, C. Y., & Bureau, D. P. (2001). A review of diet formulation strategies and feeding systems to reduce excretory and feed wastes in aquaculture. *Aquaculture Research, 32*(S1), 349–360. https://doi.org/10.1046/j.1355-557x.2001.00027.x.

Cho, S. H., Jo, J. Y., & Kim, D. S. (2001). Effects of variable feed allowance with constant energy and ratio of energy to protein in a diet for constant protein input on the growth of common carp *Cyprinus carpio* L. *Aquaculture Research, 32*(5), 349–356. https://doi.org/10.1046/j.1365-2109.2001.00564.x.

Coutinho, F., Peres, H., Guerreiro, I., Pousão-Ferreira, P., & Oliva-Teles, A. (2012). Dietary protein requirement of sharpsnout sea bream (*Diplodus puntazzo*, Cetti 1777) juveniles. *Aquaculture, 356, 357*, 391–397. https://doi.org/10.1016/j.aquaculture.2012.04.037.

Cowey, C. B. (1995). Protein and amino acid requirements: A critique of methods. *Journal of Applied Ichthyology, 11*(3, 4), 199–204. https://doi.org/10.1111/j.1439-0426.1995.tb00019.x.

Craig, S., & Helfrich, L. (2017). *Understanding Fish Nutrition, Feeds, And Feeding* (Revised by D. Kuhn & M. H. Schwarz). Virginia Cooperative Extension. https://www.pubs.ext.vt.edu/FST/FST-269/FST-269.html (accessed on 3 June 2025).

Daudpota, A. M., Siddiqui, P. J., Abbas, G., Narejo, N. T., Shah, S. S. A., Khan, N., & Dastagir, G. (2014). Effect of dietary protein level on growth performance, protein utilization and body composition of Nile tilapia cultured in low salinity water. *International Journal of Interdisciplinary and Multidisciplinary Studies, 2*(2), 135–147. http://www.ijims.com/uploads/7b410e7906e614a47110zppd_pak_fi.pdf (accessed on 3 June 2025).

De Silva, S. S. D., & Anderson, T. A. (1995). *Fish Nutrition in Aquaculture*. Chapman and Hall.

Debnath, D., Pal, A. K., Sahu, N. P., Yengkokpam, S., Baruah, K., Choudhury, D., & Venkateshwarlu, G. (2007). Digestive enzymes and metabolic profile of *Labeo rohita* fingerlings fed diets with different crude protein levels. *Comparative Biochemistry and Physiology Part B: Biochemistry and Molecular Biology, 146*(1), 107–114. https://doi.org/10.1016/j.cbpb.2006.09.008.

Deng, D., Yong, Z., Dominy, W., Murashige, R., & Wilson, R. P. (2011). Optimal dietary protein levels for juvenile Pacific threadfin (*Polydactylus sexfilis*) fed diets with two levels of lipid. *Aquaculture, 316*(1–4), 25–30. https://doi.org/10.1017/S0007114511005459.

Dumas, A., France, J., & Bureau, D. (2010). Modelling growth and body composition in fish nutrition: Where have we been and where are we going? *Aquaculture Research, 41*(2), 161–181. https://doi.org/10.1111/j.1365-2109.2009.02323.x.

Eits, R. M., Giesen, G. W. J., Kwakkel, R. P., Verstegen, M. W. A., & Hartog, L. A. D. (2005). Dietary balanced protein in broiler chickens. 2. An economic analysis. *British Poultry Science, 46*(3), 310–317. https://doi.org/10.1080/00071660500126482.

Eits, R. M., Kwakkel, R. P., Verstegen, M. W. A., & Hartog, L. A. D. (2005). Dietary balanced protein in broiler chickens. 1. A flexible and practical tool to predict dose-response curves. *British Poultry Science, 46*(3), 300–309. https://doi.org/10.1080/00071660500126656.

El-Abed, F. H., Salam, M., & El-Dahhar, A. A. (2013). Effect of dietary protein and metabolizable energy levels on growth and feed utilization of sea bass (*Dicentrarchus labrax*) larvae. *Journal of the Arabian Aquaculture Society, 8*(1), 1–18. https://core.ac.uk/reader/357318510 (accessed on 3 June 2025).

Elangovan, A., & Shim, K. F. (1997). Growth response of juvenile *Barbodes altus* fed isocaloric diets with variable protein levels. *Aquaculture, 158*(3, 4), 321–329. https://doi.org/10.1016/S0044-8486(97)00199-3.

El-Saidy, D. M. S. D., & Gaber, M. M. A. (2005). Effect of dietary protein levels and feeding rates on growth performance, production traits and body composition of Nile tilapia, *Oreochromis niloticus* (L.) cultured in concrete tanks. *Aquaculture Research, 36*(2), 163–171. https://doi.org/10.1111/j.1365-2109.2004.01201.x.

El-Sayed, A. F. M. (2019). *Tilapia Culture* (2nd ed.). CABI Publishing.

El-Sayed, A. F. M., & Teshima, S. I. (1992). Protein and energy requirements of Nile tilapia (*Oreochromis niloticus*) fry. *Aquaculture, 103*(1), 55–63. https://doi.org/10.1016/0044-8486(92)90278-S.

El-Sayed, A. F. M., Mansour, C. R., & Ezzat, A. A. (2003). Effects of dietary protein level on spawning performance of Nile tilapia (*Oreochromis niloticus*) broodstock reared at different water salinities. *Aquaculture, 220*(1–4), 619–632. https://doi.org/10.1016/S0044-8486(02)00221-1.

Engin, K., & Carter, C. G. (2001). Ammonia and urea excretion rates of juvenile Australian short-finned eel (*Anguilla australis australis*) as influenced by dietary protein level. *Aquaculture, 194*(1, 2), 123–136. https://doi.org/10.1016/S0044-8486(00)00506-8.

Ergün, S., Güroy, D., Tekeşoğlu, H., Güroy, B., Çelik, İ., Tekinay, A. A., & Bulut, M. (2010). Optimum dietary protein level for blue streak hap, *Labidochromis caeruleus*. *Turkish Journal of Fisheries and Aquatic Sciences, 10*(1), 27–31. https://doi.org/10.4194/trjfas.2010.0104.

Eroldoğan, O. T., Taşbozan, O., & Tabakoğlu, S. (2008). Effects of restricted feeding regimes on growth and feed utilization of juvenile gilthead sea bream, *Sparus aurata*. *Journal of the World Aquaculture Society, 39*(2), 267–274. https://doi.org/10.1111/j.1749-7345.2008.00157.x.

Farhat, & Khan, M. A. (2011). Growth, feed conversion, and nutrient retention efficiency of African catfish, *Clarias gariepinus* (Burchell) fingerling fed diets with varying levels of protein. *Journal of Applied Aquaculture, 23*(4), 304–316. https://doi.org/10.1080/10454438.2011.626370.

Fernandes, J. B. K., Carneiro, D. J., & Sakomura, N. K. (2000). Sources and levels of crude protein in diets for pacu (*Piaractus mesopotamicus*) fingerlings [In Portuguese]. *Revita Brasileira de Zootecnia, 29*(3), 646–653. https://doi.org/10.1590/S1516-35982000000300002.

Fernandes, J. B. K., Carneiro, D. J., & Sakomura, N. K. (2001). Fontes e níveis de proteína bruta em dietas para juvenis de pacu (*Piaractus mesopotamicus*) (In Portuguese). *Sources and Levels of Crude Protein in Diets for Pacu (Piaractus mesopotamicus) juveniles* (English translation). *Revita Brasileira de Zootecnia, 30*(3), 617–626. https://doi.org/10.1590/S1516-35982001000300003.

Fisher, C., & Morris, T. R. (1970). The determination of the methionine requirement of laying pullets by a diet dilution technique. *British Poultry Science, 11*(1), 67–82. https://doi.org/10.1080/00071667008415793.

Food and Agriculture Organization. (2018). *Achieving Food and Nutrition Security Through Fisheries and Aquaculture.* Committee on Fisheries, 33rd session, 9–13 July, Austria, Room. http://www.fao.org/webcast/home/en/item/4787/icode/ (accessed on 3 June 2025).

Fortes-Silva, R., Rosa, P. V., Zamora, S., & Sánchez-Vázquez, F. J. (2012). Dietary self-selection of protein-unbalanced diets supplemented with three essential amino acids in Nile tilapia. *Physiology & Behavior, 105*(3), 639–644. https://doi.org/10.1016/j.physbeh.2011.09.023.

Freeman, C. P. (1979). The tryptophan requirement of broiler chicks. *British Poultry Science, 20*(1), 27–37. https://doi.org/10.1080/00071667908416546.

Furuya, W. M., Michelato, M., Salaro, A. L., Cruz, T. P. D., & Barriviera-Furuya, V. R. (2015). Estimation of the dietary essential amino acid requirements of colliroja *Astyanax fasciatus* by using the ideal protein concept. *Latin American Journal of Aquatic Research, 43*(5), 888–894. https://doi.org/10.3856/vol43-issue5-fulltext-8.

Gao, W., Liu, Y. J., Tian, L. X., Mai, K. S., Liang, G. Y., Yang, H. J., Huai, M. Y., & Luo, W. J. (2011). Protein-sparing capability of dietary lipid in herbivorous and omnivorous freshwater finfish: A comparative case study on grass carp (*Ctenopharyngodon idella*) and tilapia (*Oreochromis niloticus* × *O. aureus*). *Aquaculture Nutrition, 17*(1), 2–12. https://doi.org/10.1111/j.1365-2095.2009.00698.x.

García-Meilán, I., Valentín, J. M., Fontanillas, R., & Gallardo, M. A. (2013). Different protein to energy ratio diets for gilthead sea bream (*Sparus aurata*): Effects on digestive and absorptive processes. *Aquaculture, 412, 413*, 1–7. https://doi.org/10.1016/j.aquaculture.2013.06.031.

Gatlin, D. M. III, Barrows, F. T., Brown, P., Dabrowski, K., Gaylord, T. G., Hardy, R. W., Herman, E., Gongshe, H., Krogdahl, A., Nelson, R., Overturf, K., Rust, M., Sealy, W., Skonberg, D., Souza, E. J., Stone, D., Wilson, R., & Wurtele, E. (2007). Expanding the utilization of sustainable plant products in aquafeeds: A review. *Aquaculture Research, 38*(6), 551–579. https://doi.org/10.1111/j.1365-2109.2007.01704.x.

Gatlin, D. M. III. (2002). Nutrition and fish health. In J. E. Halver & R. W. Hardy (Eds.), *Fish Nutrition* (3rd ed., pp. 671–702). Academic Press.

Glencross, B. D., Booth, M., & Allan, G. L. (2007). A feed is only as good as its ingredients–A review of ingredient evaluation strategies for aquaculture feeds. *Aquaculture Nutrition, 13*(1), 17–34. https://doi.org/10.1111/j.1365-2095.2007.00450.x.

Gonzalez, M. J., Pesti, G. M., & Dorfman, J. H. (1992). Formulating feed for minimum cost versus maximum profits. In *Proceedings of the Georgia Nutrition Conference for the Feed Industry* (pp. 87–99). Atlanta.

Gous, R. M. (2007). Predicting nutrient responses in poultry: Future challenges. *Animal, 1*(1), 57–65. https://doi.org/10.1017/S1751731107657784.

Gous, R. M., & Morris, T. R. (1985). Evaluation of a diet dilution technique for measuring the response of broiler chickens to increasing concentrations of lysine. *British Poultry Science, 26*(2), 147–161. https://doi.org/10.1080/00071668508416799.

Gous, R. M., Faulkner, A. S., & Swatson, H. K. (2018). The effect of dietary energy:protein ratio, protein quality and food allocation on the efficiency of utilization of protein by broiler chickens. *British Poultry Science, 59*(1), 100–109. https://doi.org/10.1080/00071668.2017.1390211

Guo, Z., Zhu, X., Liu, J., Han, D., Yang, Y., Lan, Z., & Xie, S. (2012). Effects of dietary protein level on growth performance, nitrogen and energy budget of juvenile hybrid sturgeon, *Acipenser baerii* ♀ × *A. gueldenstaedtii* ♂. *Aquaculture, 338–341*, 89–95. https://doi.org/10.1016/j.aquaculture.2012.01.008.

Gurure, R., Atkinson, J., & Moccia, R. D. (2007). Amino acid composition of Arctic charr, *Salvelinus alpinus* (L.) and the prediction of dietary requirements for essential amino acids. *Aquaculture Nutrition, 13*(4), 266–272. https://doi.org/10.1111/j.1365-2095.2007.00473.x.

Halver, J. E., & Hardy, R. W. (2002). *Fish Nutrition* (3rd ed.). Academic Press.

Hamre, K., Ofsti, A., Naess, T., Nortvedt, R., & Holm, J. C. (2003). Macronutrient composition of formulated diets for Atlantic halibut (*Hippoglossus hippoglossus* L.) juveniles. *Aquaculture, 227*(1–4), 233–244. https://doi.org/10.1016/S0044-8486(03)00506-4.

Hardy, R. W. (2002). Rainbow trout, *Oncorhynchus mykiss*. In C. D. Webster & C. Lim (Eds.), *Nutrient Requirements and Feeding of Finfish for Aquaculture* (pp. 184–202). CABI Publishing.

Hatlen, B., Grisdale-Helland, B., & Helland, S. J. (2005). Growth, feed utilization and body composition in two size groups of Atlantic halibut (*Hippoglossus hippoglossus*) fed diets differing in protein and carbohydrate content. *Aquaculture, 249*(1–4), 401–408. https://doi.org/10.1016/j.aquaculture.2005.03.040.

Hauler, R. C., & Carter, C. G. (2001). Re-evaluation of the quantitative dietary lysine requirements of fish. *Reviews in Fisheries Science, 9*(3), 133–163. https://doi.org/10.1080/20016491101735.

Heinitz, M. C., Figueiredo Silva, C., Schulz, C., & Lemme, A. (2018). The effect of varying dietary digestible protein and digestible non-protein energy sources on growth, nutrient utilization efficiencies and body composition of carp (*Cyprinus carpio*) evaluated with a two-factorial central composite study design. *Aquaculture Nutrition, 24*(2), 723–740. https://doi.org/10.1111/anu.12601.

Hillestad, M., & Johnsen, F. (1994). High-energy/low-protein diets for Atlantic salmon: Effects on growth, nutrient retention and slaughter quality. *Aquaculture, 124*(1–4), 109–116. https://doi.org/10.1016/0044-8486(94)90366-2.

Hillestad, M., Johnsen, F., & Åsgård, T. (2001). Protein to carbohydrate ratio in high-energy diets for Atlantic salmon (*Salmo salar* L.). *Aquaculture Research, 32*(7), 517–529. https://doi.org/10.1046/j.1365-2109.2001.00569.x.

Hossain, M. A., Almatar, S. M., & James, C. M. (2010). Optimum dietary protein level for juvenile silver pomfret, *Pampus argenteus* (Euphrasen). *Journal of the World Aquaculture Society, 41*(5), 710–720. https://doi.org/10.1111/j.1749-7345.2010.00413.x.

Hossain, M., Hasan, N., Shah, A., & Hussain, M. (2002). Optimum dietary protein requirement of mahseer, *Tor putitora* (Hamilton) fingerlings. *Asian Fisheries Science, 15*(3), 203–214. https://doi.org/10.33997/j.afs.2002.15.3.002.

Houlihan, D., Bouiard, T., & Jobling, M. (2001). *Food Intake in Fish*. Blackwell Science Ltd. https://doi.org/10.1002/9780470999516.

Hung, L. T., & Quy, O. M. (2013). On-farm feeding and feed management in whiteleg shrimp (*Litopenaeus vannamei*) farming in Viet Nam. In M. R. Hasan & M. B. New (Eds.), *On-Farm Feeding and Feed Management in Aquaculture* (FAO Fisheries and Aquaculture Technical Paper No. 583, pp. 337–357). Rome: FAO. https://www.fao.org/

publications/card/zh/c/ec9080d6-8ceb-5952-afdf-a6605a9d6ebb (accessed on 3 June 2025).

Hurwitz, S., Talpaz, H., & Waibel, P. E. (1985). The use of simulation in the evaluation of economics and management of turkey production: Dietary nutrient density, marketing age and environmental temperature. *Poultry Science, 64*(8), 1415–1423. https://doi.org/10.3382/ps.0641415.

Islam, M. S., & Tanaka, M. (2004). Optimization of dietary protein requirement for pond-reared mahseer *Tor putitora* Hamilton (Cypriniformes: Cyprinidae). *Aquaculture Research, 35*(13), 1270–1276. https://doi.org/10.1111/j.1365-2109.2004.01149.x.

Jauncey, K., & Ross, B. (1982). *A Guide to Tilapia Feeds and Feeding*. Institute of Aquaculture, University of Stirling. https://books.google.com.pk/books/about/A_Guide_to_Tilapia_Feeds_and_Feeding.html?id=2x93AAAACAAJ (accessed on 3 June 2025).

Jayant, M., Muralidhar, A. P., Sahu, N. P., Jain, K. K., Pal, A. K., & Srivastava, P. P. (2018). Protein requirement of juvenile striped catfish, *Pangasianodon hypophthalmus*. *Aquaculture International, 26*(1), 375–389. https://doi.org/10.1007/s10499-017-0216-0.

Jiang, S., Wu, X., Luo, Y., Wu, M., Lu, S., Jin, Z., & Yao, W. (2016). Optimal dietary protein level and protein to energy ratio for hybrid grouper (*Epinephelus fuscoguttatus* ♀ × *Epinephelus lanceolatus* ♂) juveniles. *Aquaculture, 465*, 28–36. https://doi.org/10.1016/j.aquaculture.2016.08.030.

Jin, Y., Tian, L. X., Xie, S. W., Guo, D. Q., Yang, H. J., Liang, G. Y., & Liu, Y. J. (2015). Interactions between dietary protein levels, growth performance, feed utilization, gene expression and metabolic products in juvenile grass carp (*Ctenopharyngodon idella*). *Aquaculture, 437*, 75–83. https://doi.org/10.1016/j.aquaculture.2014.11.031.

Karalazos, V., Bendiksen, E. Å., & Bell, J. G. (2011). Interactive effects of dietary protein/lipid level and oil source on growth, feed utilization and nutrient and fatty acid digestibility of Atlantic salmon. *Aquaculture, 311*(1–4), 193–200. https://doi.org/10.1016/j.aquaculture.2010.11.022.

Kaushik, S. J. (1998). Nutritional bioenergetics and estimation of waste production in non-salmonids. *Aquatic Living Resources, 11*(4), 211–217. https://doi.org/10.1016/S0990-7440(98)89003-7.

Kaushik, S. J. (2013). Feed management and on-farm feeding practices of temperate fish with special reference to salmonid. In M. R. Hasan & M. B. New (Eds.), *On-Farm Feeding and Feed Management in Aquaculture* (FAO Fisheries and Aquaculture Technical Paper No. 583, pp. 519–551). FAO. https://www.fao.org/fishery/docs/CDrom/T583/root/20.pdf (accessed on 3 June 2025).

Kaushik, S. J., & Seiliez, I. (2010). Protein and amino acid nutrition and metabolism in fish: Current knowledge and future needs. *Aquaculture Research, 41*(3), 322–332. https://doi.org/10.1111/j.1365-2109.2009.02174.x.

Khan, K. U., Gous, R. M., Mansano, C. F. M., Nascimento, T. M. T., Romaneli, R. S., Rodrigues, A. T., Sakomura, N. K., & Fernandes, J. B. K. (2020a). Response of juvenile pacu (*Piaractus mesopotamicus* Holmberg, 1887) to balanced digestible protein. *Aquaculture Research, 51*(9), 3729–3740. https://doi.org/10.1111/are.14722.

Khan, K. U., Gous, R. M., Nascimento, T. M. T., Sakomura, N. K., Mansano, C. F. M., Queiroz, D. M. A., Rodrigues, A. T., Boaratti, A. Z., Romaneli, R. S., & Fernandes, J. B. K. (2022). Modeling of nutritional behavior and harvest profit margin in fattening Pacu

(*Piaractus mesopotamicus*) fed on balanced digestible protein contents and raised in an intensive recirculating aquaculture system. *Animal Feed Science and Technology, 292*, 115430. https://doi.org/10.1016/j.anifeedsci.2022.115430.

Khan, K. U., Gous, R. M., Sakomura, N. K., Azevedo, J. M., Nascimento, T. M. T., Mansano, C. F. M., Romaneli, R. S., Oliveira, T. S., Boaratti, A. Z., & Fernandes, J. B. K. (2020b). Modeling optimum levels of balanced digestible protein for adult Pacu (*Piaractus mesopotamicus*) and the development of an economic optimization procedure for different marketing strategies of the final yield. *Aquaculture, 526*, 735368. https://doi.org/10.1016/j.aquaculture.2020.735368.

Khan, K. U., Mansano, C. F. M., Nascimento, T. M. T., Boaratti, A. Z., Rodrigues, A. T., Queiroz, D. M. A., Romaneli, R. S., Sakomura, N. K., & Fernandes, J. B. K. (2020c). Whole-body amino acid pattern of juvenile, preadult, and adult pacu, *Piaractus mesopotamicus*, with an estimation of its dietary essential amino acid requirements. *Journal of the World Aquaculture Society, 51*(1), 224–234. https://doi.org/10.1111/jwas.12600.

Khan, K. U., Rodrigues, A. T., Mansano, C. F. M., Queiroz, D. M. A., Sakomura, N. K., Romaneli, R. S., do Nascimento, T. M. T., & Fernandes, J. B. K. (2019). Dietary protein quality and proper protein to energy ratios: A bio-economic approach in aquaculture feeding practices. *Latin American Journal of Aquatic Research, 47*(2), 232–239. https://doi.org/10.3856/vol47-issue2-fulltext-3.

Khan, M. A., Jafri, A. K., & Chadha, N. K. (2005). Effects of varying dietary protein levels on growth, reproductive performance, body and egg composition of rohu, *Labeo rohita* (Hamilton). *Aquaculture Nutrition, 11*(1), 11–17. https://doi.org/10.1111/j.1365-2095.2004.00318.x.

Kim, K., Wang, X., & Bai, S. C. (2003). Re-evaluation of the dietary protein requirement of Japanese flounder (*Paralichthys olivaceus*). *Journal of the World Aquaculture Society, 34*(2), 133–139. https://doi.org/10.1111/j.1749-7345.2003.tb00049.x.

Kim, L. O., & Lee, S. M. (2005). Effects of dietary protein and lipid levels on growth and body composition of bagrid catfish, *Pseudobagrus fulvidraco*. *Aquaculture, 243*(1–4), 323–329. https://doi.org/10.1016/j.aquaculture.2004.11.003.

Kim, S., & Lee, K. (2009). Dietary protein requirement of juvenile tiger puffer (*Takifugu rubripes*). *Aquaculture, 287*(1, 2), 219–222. https://doi.org/10.1016/j.aquaculture.2008.10.021.

Klein, S., Lorenz, E. K., Bueno, G. W., Signor, A., Feiden, A., & Boscolo, W. R. (2014). Levels of crude protein in diets for pacu (*Piaractus mesopotamicus*) from 150 to 400 g reared in cages. *Archivos de Zootecnia, 63*(244), 599–610. https://doi.org/10.4321/S0004-05922014000400004.

Köprücü, K. (2012). Effects of dietary protein and lipid levels on growth, feed utilization and body composition of juvenile grass carp (*Ctenopharyngodon idella*). *Journal of Fisheries Sciences, 6*(3), 243–251. https://doi.org/10.3153/jfscom.2012028.

Kpogue, D., Gangbazo, H., & Fiogbe, E. (2013). A preliminary study on the dietary protein requirement of *Parachanna obscura* (Günther, 1861) larvae. *Turkish Journal of Fisheries and Aquatic Sciences, 13*(1), 111–117. https://doi.org/10.4194/1303-2712-v13_1_14.

Lall, S. P., & Anderson, S. (2005). Amino acid nutrition of salmonids: Dietary requirements and bioavailability. In D. Montero, B. Basurco, I. Nengas, M. Alexis, & M. Izquierdo

(Eds.), *Mediterranean Fish Nutrition* (Cahiers Options Méditerranéennes, No. 63, pp. 73–90). CIHEAM.

Lee, J. K., Cho, S. H., Park, S. U., Kim, K., & Lee, S. (2003). Dietary protein requirement for young turbot (*Scophthalmus maximus* L.). *Aquaculture Nutrition, 9*(4), 283–286. https://doi.org/10.1046/j.1365-2095.2003.00255.x.

Lee, S., Kim, K., Park, H. G., Kim, C. H., & Hong, K. E. (2001). Protein requirement of juvenile Manchurian trout *Brachymystax lenok*. *Fisheries Science, 67*(1), 46–51. https://doi.org/10.1046/j.1444-2906.2001.00197.x.

Lee, S., Masagounder, K., Hardy, R. W., & Small, B. C. (2019). Effects of lowering dietary fishmeal and crude protein levels on growth performance, body composition, muscle metabolic gene expression, and chronic stress response of rainbow trout (*Oncorhynchus mykiss*). *Aquaculture, 513*, 734435. https://doi.org/10.1016/j.aquaculture.2019.734435.

Lemme, A. (2010). Availability and effectiveness of free amino acids in aquaculture. In L. E. Cruz-Suarez, D. Ricque-Marie, M. Tapia-Salazar, M. G. Nieto-López, D. A. Villarreal-Cavazos, & J. Gamboa-Delgado (Eds.), *Avances en Nutrición Acuícola X–Memorias del Décimo Simposio Internacional de Nutrición Acuícola* (pp. 264–275). Universidad Autónoma de Nuevo León.

Lemos, M. V. A. D., Arantes, T. Q., Souto, C. N., Martins, G. P., Araújo, J. G., & Guimarães, I. G. (2014). Effects of digestible protein to energy ratios on growth and carcass chemical composition of Siamese fighting fish (*Betta splendens*). *Ciência e Agrotecnologia, 38*(1). https://doi.org/10.1590/S1413-70542014000100009.

Li, X. F., Liu, W. B., Jiang, Y. Y., Zhu, H., & Ge, X. P. (2010). Effects of dietary protein and lipid levels in practical diets on growth performance and body composition of blunt snout bream (*Megalobrama amblycephala*) fingerlings. *Aquaculture, 303*(1–4), 65–70. https://doi.org/10.1016/j.aquaculture.2010.03.014.

Liebert, F., & Benkendorff, K. (2007). Modeling lysine requirements of *Oreochromis niloticus* due to principles of the diet dilution technique. *Aquaculture, 267*(1–4), 100–110. https://doi.org/10.1016/j.aquaculture.2007.02.022.

López, S. R., Lazo, J. P., Reyes, G. C., & Viana, M. T. (2011). Effect of dietary protein and energy levels on growth, survival and body composition of juvenile *Totoaba macdonaldi*. *Aquaculture, 319*(3), 385–390. https://doi.org/10.1016/j.aquaculture.2011.07.007.

Luo, Z., Liu, Y. J., Mai, K. S., Tian, L. X., Liu, D. H., & Tan, X. Y. (2004). Optimal dietary protein requirement of grouper, *Epinephelus coioides* juveniles fed iso-energetic diets in floating net cages. *Aquaculture Nutrition, 10*(4), 247–252. https://doi.org/10.1111/j.1365-2095.2004.00296.x.

Ma, R., Liu, X., Meng, Y., Wu, J., Zhang, L., Han, B., Qian, K., Luo, Z., Wei, Y., & Li, C. (2019). Protein nutrition on sub-adult triploid rainbow trout (1): Dietary requirement and effect on anti-oxidative capacity, protein digestion and absorption. *Aquaculture, 507*, 428–434. https://doi.org/10.1016/j.aquaculture.2019.03.069.

Ma, X., Hu, Y., Wang, X. Q., Ai, Q. H., He, Z. G., Feng, F. X., & Lu, X. Y. (2014). Effects of practical dietary protein to lipid levels on growth, digestive enzyme activities and body composition of juvenile rice field eel (*Monopterus albus*). *Aquaculture International, 22*, 749–760. https://doi.org/10.1007/s10499-013-9703-0.

Mack, S., Bercovici, D., De Groote, G., Leclercq, B., Lippens, M., Pack, M., Schutte, J. B., & Van Cauwenberghe, S. (1999). Ideal amino acid profile and dietary lysine specification for broiler chickens of 20 to 40 days of age. *British Poultry Science, 40*(2), 257–265. https://doi.org/10.1080/00071669987683.

Mansano, C. F. M., da Silva, E. P., Khan, K. U., Macente, B. I., Nascimento, T. M. T., Sakomura, N. K., Fernandes, J. B. K., & Takahashi, L. S. (2021). Digestible protein requirements for maintenance, growth, and efficiency of protein utilization in pacu (*Piaractus mesopotamicus*) juveniles: An exponential nitrogen utilization model. *Latin American Journal of Aquatic Research, 49*(4), 608–619. https://doi.org/10.3856/vol49-issue4-fulltext-2678.

Marammazi, J. G., & Kahkesh, F. (2011). Effects of dietary protein and energy levels on growth performance, feed utilization and body composition of juvenile shirbot *Barbus grypus* (Heckle, 1843). *Iranian Journal of Fisheries Sciences, 10*(3), 461–474. http://hdl.handle.net/1834/11394 (accessed on 3 June 2025).

McGoogan, B. B., & Gatlin, III. D. M. (1999). Dietary manipulations affecting growth and nitrogenous waste production of red drum, *Sciaenops ocellatus*. I. Effects of dietary protein and energy levels. *Aquaculture, 178*(3, 4), 333–348. https://doi.org/10.1016/S0044-8486(99)00137-4.

McGoogan, B. B., & Gatlin, III. D. M. (2000). Dietary manipulations affecting growth and nitrogenous waste production of red drum, *Sciaenops ocellatus* II. Effects of energy level and nutrient density at various feeding rates. *Aquaculture, 182*(3, 4), 271–285. https://doi.org/10.1016/S0044-8486(99)00260-4.

Meer, M. B. V. D., Machiels, M. A. M., & Verdegem, M. C. J. (1995). The effect of dietary protein level on growth, protein utilization and body composition of *Colossoma macropomum* (Cuvier). *Aquaculture Research, 26*(12), 901–909. https://doi.org/10.1111/j.1365-2109.1995.tb00885.x.

Melo, J. F. B., Lundstedt, L. M., Metón, I., Baanante, I. V., & Moraes, G. (2006). Effects of dietary levels of protein on nitrogenous metabolism of *Rhamdia quelen* (Teleostei: Pimelodidae). *Comparative Biochemistry and Physiology Part A: Molecular & Integrative Physiology, 145*(2), 181–187. https://doi.org/10.1016/j.cbpa.2006.06.007.

Meyer, G., & Fracalossi, D. M. (2004). Protein requirement of jundia fingerlings, *Rhmdia quelen*, at two dietary energy concentrations. *Aquaculture, 240*(1–4), 341–343. https://doi.org/10.1016/j.aquaculture.2004.01.034.

Miao, S., Han, B., Li, J., Hu, J., Wan, W., Sun, L., & An, Z. (2020). Effects of dietary protein level on the growth performance, feed utilization and immunity of red swamp crayfish *Procambarus clarkia. Aquaculture Reports, 18*, 100540. https://doi.org/10.1016/j.aqrep.2020.100540.

Mohanta, K. N., Mohanty, S. N., & Jena, J. K. (2007). Protein-sparing effect of carbohydrate in silver barb, *Puntius gonionotus* fry. *Aquaculture Nutrition, 13*(4), 311–317. https://doi.org/10.1111/j.1365-2095.2007.00482.x.

Monentcham, S. E., Pouomigne, V., & Kestemont, P. (2010). Influence of dietary protein levels on growth performance and body composition of African bony tongue fingerlings *Heterotis niloticus* (Cuvier, 1829). *Aquaculture Nutrition, 16*(2), 144–152. https://doi.org/10.1111/j.1365-2095.2008.00646.x.

Moraes, G., & Bidinotto, P. M. (2004). Digestive proteases of pacu, *Piaractus mesopotamicus*, fed on distinct protein-starch diets. *Journal of Applied Aquaculture, 15*(3–4), 197–207. https://doi.org/10.1300/J028v15n03_17.

National Research Council. (2011). *Nutrient Requirements of Fish and Shrimp*. National Academy Press.

Nelson, D. L., & Cox, M. M. (2005). Lipids. In A. L. Lehninger, D. L. Nelson, & M. M. Cox (Eds.), *Lehninger Principles of Biochemistry* (4th ed.). W. H. Freeman and Company. http://aulanni.lecture.ub.ac.id/files/2012/01/15616949-Lehninger-Principles-of-Biochemistry-1-copy.pdf (accessed on 3 June 2025).

Nguyen, T. N., Davis, D. A., & Saoud, I. P. (2009). Evaluation of alternative protein sources to replace fish meal in practical diets for juvenile tilapia, *Oreochromis* spp. *Journal of the World Aquaculture Society, 40*(1), 113–121. https://doi.org/10.1111/j.1749-7345.2008.00230.x.

Oesman, Rachmansyah, Laining, A., Ahmad, T., & Williams, K. C. (2005). Optimum dietary protein and lipid specifications for grow-out of humpback grouper *Cromileptes altivelis* (Valenciennes). *Aquaculture Research, 36*(13), 1285–1292. https://doi.org/10.1111/j.1365-2109.2005.01341.x.

Okorie, O. E., Kim, Y., Lee, S., Bae, J., Yoo, J. H., Han, K., & Bai, S. C. (2007). Reevaluation of the dietary protein requirements and optimum dietary protein to energy ratios in Japanese eel, *Anguilla japonica*. *Journal of the World Aquaculture Society, 38*(3), 418–426. https://doi.org/10.1111/j.1749-7345.2007.00113.x.

Ozório, R. O., Valente, L. M., Pousão-Ferreira, P., & Oliva-Teles, A. (2006). Growth performance and body composition of white seabream (*Diplodus sargus*) juveniles fed diets with different protein and lipid levels. *Aquaculture Research, 37*(3), 255–263. https://doi.org/10.1111/j.1365-2109.2005.01427.x.

Pelletier, N., Klinger, D. H., Sims, N. A., Yoshioka, J. R., & Kittinger, J. N. (2018). Nutritional attributes, substitutability, scalability, and environmental intensity of an illustrative subset of current and future protein sources for aquaculture feeds: Joint consideration of potential synergies and trade-offs [Review article]. *Environmental Science & Technology, 52*(10), 5532–5544. https://doi.org/10.1021/acs.est.7b05468.

Pilbrow, P. J., & Morris, T. R. (1974). Comparison of lysine requirements amongst eight stocks of laying fowls. *British Poultry Science, 15*(1), 51–73. https://doi.org/10.1080/00071667408416081.

Portz, L., & Cyrino, J. E. P. (2004). Digestibility of nutrients and amino acids of different protein sources in practical diets by largemouth bass *Micropterus salmoides* (Lacépède, 1802). *Aquaculture Research, 35*(4), 312–320. https://doi.org/10.1111/j.1365-2109.2004.00984.x.

Qiang, J., Yang, H., Wang, H., Kpundeh, M. D., & Xu, P. (2012). Interactive effects of temperature–dietary protein level on somatotropic gene expression and its interrelationship with growth in juvenile GIFT tilapia *Oreochromis niloticus*. *Aquaculture, 364, 365*, 263–271. https://doi.org/10.1016/j.aquaculture.2012.08.021.

Rahimnejad, S., Bang, I. C., Park, J. Y., Sade, A., Choi, J., & Lee, S. M. (2015). Effects of dietary protein and lipid levels on growth performance, feed utilization and body composition of juvenile hybrid grouper, *Epinephelus fuscoguttatus* × *E. lanceolatus*. *Aquaculture, 446*, 283–289. https://doi.org/10.1016/j.aquaculture.2015.05.019.

Robb, D. H. F., & Crampton, V. O. (2013). On-farm feeding and feed management: perspectives from the fish feed industry. In M. R. Hasan & M. B. New (Eds.), *On-Farm Feeding and Feed Management in Aquaculture* (FAO Fisheries and Aquaculture

Technical Paper No. 583, pp. 489–518). Rome: FAO. https://www.fao.org/fishery/docs/CDrom/T583/root/19.pdf (accessed on 3 June 2025).

Sa, R., Gavilán, M., Rioseco, M.J., Llancabure, A., Vargas-Chacoff, L., Augsburger, A., Bas, F. (2014). Dietary protein requirement of *Patagonian blennie* (*Eleginops maclovinus*, Cuvier 1830) juveniles, *Aquaculture*, *428–429*, 125–134, https://doi.org/10.1016/j.aquaculture.2014.02.017.

Sá, R., Pousão-Ferreira, P., & Oliva-Teles, A. (2008). Dietary protein requirement of white sea bream (*Diplodus sargus*) juveniles. *Aquaculture Nutrition, 14*(4), 309–317. https://doi.org/10.1111/j.1365-2095.2007.00532.x.

Sadasivam, J., Kaushik, S., & Seiliez, I. (2010). Protein and amino acid nutrition and metabolism in fish: Current knowledge and future needs. *Aquaculture Research, 41*(3), 322–332. https://doi.org/10.1111/j.1365-2109.2009.02174.x.

Sagada, G., Chen, J., Shen, B., Huang, A., Sun, L., Jiang, J., & Jin, C. (2017). Optimizing protein and lipid levels in practical diet for juvenile northern snakehead fish (*Channa argus*). *Animal Nutrition, 3*(2), 156–163. https://doi.org/10.1016/j.aninu.2017.03.003.

Sankian, Z., Khosravi, S., Kim, Y., & Lee, S. (2017). Effect of dietary protein and lipid level on growth, feed utilization, and muscle composition in golden mandarin fish *Siniperca scherzeri*. *Fisheries and Aquatic Sciences, 20*(7). https://doi.org/10.1186/s41240-017-0053-0.

Sargent, J. R., & Tacon, A. G. J. (1999). Development of farmed fish: A nutritionally necessary alternative to meat. *Proceedings of the Nutrition Society, 58*(2), 377–383. https://doi.org/10.1017/S0029665199001366.

Satpathy, B. B., & Ray, A. K. (2009). Effect of dietary protein and carbohydrate levels on growth, nutrient utilization and body composition in fingerling rohu, *Labeo rohita* (Hamilton). *Journal of Applied Ichthyology, 25*(6), 728–733. https://doi.org/10.1111/j.1439-0426.2009.01326.x.

Satpathy, B. B., Mukherjee, D., & Ray, A. K. (2003). Effects of dietary protein and lipid level on growth, feed conversion and body composition in rohu, *Labeo rohita* (Hamilton), fingerlings. *Aquaculture Nutrition, 9*(1), 17–24. https://doi.org/10.1046/j.1365-2095.2003.00223.x.

Schuhmacher, A., Wax, C., & Gropp, J. M. (1997). Plasma amino acids in rainbow trout (*Oncorhynchus mykiss*) fed intact protein or a crystalline amino acid diet. *Aquaculture, 151*(1–4), 15–28. https://doi.org/10.1016/S0044-8486(96)01502-5.

Seenappa, D., & Devaraj, K. V. (1995). Effect of different levels of protein, fat and carbohydrate on growth, feed utilization and body carcass composition of fingerlings in *Catla catla* (Ham.). *Aquaculture, 129*(1–4), 243–249. https://doi.org/10.1016/0044-8486(94)00273-Q.

Shapawi, R., Ebi, I., Yong, A. S. K., & Ng, W. K. (2014). Optimizing the growth performance of brown-marbled grouper, *Epinephelus fuscoguttatus* (Forskal), by varying the proportion of dietary protein and lipid levels. *Animal Feed Science and Technology, 191*, 98–105. https://doi.org/10.1016/j.anifeedsci.2014.01.020.

Shiau, S. Y., & Lan, C. W. (1996). Optimum dietary protein level and protein to energy ratio for growth of grouper (*Epinephelus malabaricus*). *Aquaculture, 145*(1–4), 259–266. https://doi.org/10.1016/S0044-8486(96)01324-5.

Shipton, T. A., & Hecht, T. (2013). Economic, regulatory and legal review of feed management practices. In M. R. Hasan & M. B. New (Eds.), *On-Farm Feeding and Feed*

Management in Aquaculture (FAO Fisheries and Aquaculture Technical Paper No. 583, pp. 565–585). Rome: FAO. https://www.fao.org/fishery/docs/CDrom/T583/root/22.pdf (accessed on 3 June 2025).

Signor, A. A., Boscolo, W. R., Feiden, A., Bittencourt, F., Coldebella, A., & Reidel, A. (2010). Proteína e energia na alimentação de pacus criados em tanques-rede. *Revita Brasileira de Zootecnia, 39*(11), 2336–2341. https://doi.org/10.1590/S1516-35982010001100004.

Singh, R. K., Balange, A. K., & Ghughuskar, M. M. (2006). Protein sparing effect of carbohydrates in the diet of *Cirrhinus mrigala* (Hamilton, 1822) fry. *Aquaculture, 258*(1–4), 680–684. https://doi.org/10.1016/j.aquaculture.2006.03.049.

Sweilum, M. A., Abdella, M. M., & El-Din, S. S. (2005). Effect of dietary protein-energy levels and fish initial sizes on growth rate, development and production of Nile tilapia, *Oreochromis niloticus* L. *Aquaculture Research, 36*(14), 1414–1421. https://doi.org/10.1111/j.1365-2109.2005.01362.x.

Tacon, A. G. J. (1992). *Nutrición y Alimentación De Peces Y Camarones Cultivados: Manual De Capacitación*. FAO. https://www.fao.org/3/af286s/af286s.pdf (accessed on 3 June 2025).

Tacon, A. G. J., & Metian, M. (2015). Feed matters: Satisfying the feed demand of aquaculture. *Reviews in Fisheries Science & Aquaculture, 23*(1), 1–10. https://doi.org/10.1080/23308249.2014.987209.

Tacon, A. G. J., Lemos, D., & Metian, M. (2020). Fish for health: improved nutritional quality of cultured fish for human consumption. *Reviews in Fisheries Science & Aquaculture*, 1–10. https://doi.org/10.1080/23308249.2020.1762163.

Thoman, E. S., Davis, D. A., & Arnold, C. R. (1999). Evaluation of grow-out diets with varying protein and energy levels for red drum (*Sciaenops ocellatus*). *Aquaculture, 176*(3, 4), 343–353. https://doi.org/10.1016/S0044-8486(99)00118-0.

Tibbetts, S. M., Lall, S. P., & Anderson, D. M. (2000). Dietary protein requirement of juvenile American eel (*Anguilla rostrata*) fed practical diets. *Aquaculture, 186*(1, 2), 145–155. https://doi.org/10.1016/S0044-8486(99)00363-4.

Tidwell, J. H., & Allan, G. L. (2001). Fish as food: aquaculture's contribution ecological and economic impacts and contributions of fish farming and capture fisheries. *EMBO Reports, 2*(11), 958–963. https://doi.org/10.1093/embo-reports/kve236.

Toan, D. H., Mi, L. T., & Phuong, N. T. (2011). Effects of restricted and alternative feeding methods on the growth of the striped catfish (*Pangasianodon hypophthalmus*) fingerlings. In *Proceedings of the 4th Workshop in Aquaculture Science at Can Tho University* (pp. 178–190). https://caf.ctu.edu.vn/en/publication/scientific-paper-2005-2013.html (accessed on 3 June 2025).

Tuan, L. A., & Williams, K. C. (2007). Optimum dietary protein and lipid specifications for juvenile malabar grouper (*Epinephelus malabaricus*). *Aquaculture, 267*(1), 129–138. https://doi.org/10.1016/j.aquaculture.2007.03.007.

Vandergeest, P. (2007). Certification and communities: Alternatives for regulating the environmental and social impacts of shrimp farming. *World Development, 35*(7), 1152–1171. https://doi.org/10.1016/j.worlddev.2006.12.002.

Wang, J., Jiang, Y., Li, X., Han, T., Yang, Y., Hu, S., & Yang, M. (2016). Dietary protein requirement of juvenile red spotted grouper (*Epinephelus akaara*). *Aquaculture, 450*, 289–294. https://doi.org/10.1016/j.aquaculture.2015.08.007.

Wang, X., Liu, G., Xie, S., Pan, L., & Tan, Q. (2022). Growth and meat quality of grass carp (*Ctenopharyngodon idella*) responded to dietary protein (soybean meal) level through the muscle metabolism and gene expression of myosin heavy chains. *Frontiers in Nutrition, 9*, 833924. https://doi.org/10.3389/fnut.2022.833924.

Wang, Y. Y., Ma, G. J., Shi, Y., Liu, D. S., Guo, J. X., Yang, Y. H., & Che, C. D. (2013). Effects of dietary protein and lipid levels on growth, feed utilization and body composition in *Pseudobagrus ussuriensis* fingerlings. *Aquaculture Nutrition, 19*(3), 390–398. https://doi.org/10.1111/j.1365-2095.2012.00972.x.

Webb, K. A., Jr., & Gatlin III, D. M. (2003). Effects of dietary protein level and form on production characteristics and ammonia excretion of red drum *Sciaenops ocellatus*. *Aquaculture, 225*(1–4), 17–26. https://doi.org/10.1016/S0044-8486(03)00274-6.

Wethli, E., Morris, T. R., & Shresta, T. P. (1975). The effect of feeding high levels of low-quality proteins to growing chicks. *British Journal of Nutrition, 34*(3), 363–373. https://doi.org/10.1017/S0007114575000426.

White, P. (2013). Environmental consequences of poor feed quality and feed management. In M. R. Hasan & M. B. New (Eds.), *On-Farm Feeding and Feed Management in Aquaculture* (FAO Fisheries and Aquaculture Technical Paper No. 583, pp. 553–564). Rome: FAO. https://www.fao.org/fishery/docs/CDrom/T583/root/21.pdf (accessed on 3 June 2025).

Wilson, R. P. (2002). Amino acids and protein. In J. E. Halver & R. W. Hardy (Eds.), *Fish Nutrition* (3rd ed., pp. 143–179). Academic Press.

Wilson, R. P., & Poe, W. E. (1985). Apparent digestible protein and energy coefficients of common feed ingredients for channel catfish. *Progressive Fish-Culturist, 47*(3), 154–158. https://doi.org/10.1577/1548-8640(1985)47<154:ADPAEC>2.0.CO;2.

Wood, C. M. (1993). Ammonia and urea metabolism and excretion. In D. H. Ewans (Ed.), *Physiology of Fishes* (pp. 379–425). CRC Press.

Wu, W., Ji, H., Yu, H., Sun, J., & Zhou, J. (2020). Effect of refeeding dietary containing different protein and lipid levels on growth performance, body composition, digestive enzyme activities and metabolic related gene expression of grass carp (*Ctenopharyngodon idellus*) after overwinter starvation. *Aquaculture, 523*, 735196. https://doi.org/10.1016/j.aquaculture.2020.735196.

Wu, X., & Gatlin III, D. M. (2014). Effects of altering dietary protein content in morning and evening feedings on growth and ammonia excretion of red drum (*Sciaenops ocellatus*). *Aquaculture, 434*, 33–37. https://doi.org/10.1016/j.aquaculture.2014.07.019.

Xu, J., Wu, P., Jiang, W. D., Liu, Y., Jiang, J., Kuang, S. Y., Tang, L., Tang, W. N., Zhang, Y. A., Zhou, X. Q., & Feng, L. (2016). Optimal dietary protein level improved growth, disease resistance, intestinal immune and physical barrier function of young grass carp (*Ctenopharyngodon idella*). *Fish & Shellfish Immunology, 55*, 64–87. https://doi.org/10.1016/j.fsi.2016.05.021.

Yadav, A. K., Mandal, S. C., Patel, A. B., & Maurya, P. K. (2019). Evaluation of dietary protein requirement for the growth performance of minor carp, *Cirrhinus reba* (Hamilton, 1822) fingerlings. *Aquaculture Research, 50*(11), 3343–3349. https://doi.org/10.1111/are.14292.

Yang, S. D., Lin, T. S., Liou, C. H., & Peng, H. K. (2003). Influence of dietary protein level on growth performance, carcass composition and liver lipid classes of juvenile

Spinibarbus hollandi (Oshima). *Aquaculture Research, 34*(8), 661–666. https://doi.org/10.1046/j.1365-2109.2003.00880.x.

Yang, S. D., Liou, C., & Liu, F. (2002). Effects of dietary protein level on growth performance, carcass composition and ammonia excretion in juvenile silver perch, *Bidyanus bidyanus*. *Aquaculture, 213*(1–4), 363–372. https://doi.org/10.1016/S0044-8486(02)00120-5.

Ye, W., Han, D., Zhu, X., Yang, Y., Jin, J., & Xie, S. (2017). Comparative study on dietary protein requirements for juvenile and pre-adult gibel carp (*Carassius auratus* gibelio var. CAS III). *Aquaculture Nutrition, 23*(4), 755–765. https://doi.org/10.1111/anu.12442.

Zarate, D. D., & Lovell, R. T. (1997). Free lysine is utilized for growth less efficiently than protein-bound lysine (soybean meal) in practical diets by young channel catfish (*Ictalurus punctatus*). *Aquaculture, 159*(1, 2), 87–100. https://doi.org/10.1016/S0044-8486(97)00184-1.

Zeitoun, I. H., Hullrey, D. E., Magee, W. T., Gill, J. L., & Bergen, W. G. (1976). Quantifying nutrient requirements in fish. *Journal of the Fisheries Research Board of Canada, 33*(1), 167–172. https://doi.org/10.1139/f76-019.

Zhang, J., Zhou, F., Wang, L., Shao, Q., & Xu, Z. (2010). Dietary protein requirement of juvenile black sea bream, Sparus macrocephalus. *Journal of the World Aquaculture Society, 41*(s2), 151–154. https://doi.org/10.1111/j.1749-7345.2010.00356.x.

Zhang, Y., Sun, Z., Wang, A., Ye, C., & Zhu, X. (2017). Effects of dietary protein and lipid levels on growth, body and plasma biochemical composition and selective gene expression in liver of hybrid snakehead (*Channa maculata* ♀ × *Channa argus* ♂) fingerlings. *Aquaculture, 468*(Part-1), 1–9. https://doi.org/10.1016/j.aquaculture.2016.09.052.ss

CHAPTER 4

Lysine Requirement of Different Fish Species

KRISHNA PADA SINGHA[1,*] and MANAS KUMAR MAITI[2]

[1]*Department of Animal, Veterinary, and Food Sciences, Aquaculture Research Institute, University of Idaho, Moscow, Idaho, USA*

[2]*Fish Nutrition, Biochemistry and Physiology Division, ICAR–Central Institute of Fisheries Education, Versova, Mumbai, Maharashtra, India*

*Corresponding author

ABSTRACT

Lysine, an indispensable amino acid (AA), is crucial for fish growth, health, and feed efficiency. The shift towards sustainable aquaculture practices and the increasing inclusion of plant-based proteins in diets highlight the importance of lysine supplementation. This chapter provides an synthesized review of lysine's metabolic roles, emphasizing the need for species-specific lysine requirements to optimize aquafeed formulations and support the productivity of farmed fish. Advances in understanding the molecular impact of lysine on nutrient-sensing pathways have profound implications for improving the utilization of alternative protein sources. The chapter also addresses lysine's interactions with other AAs, noting the potential for antagonism that can affect metabolic processes. Additionally, it considers the environmental and economic aspects of sourcing high-quality lysine for aquafeed, underscoring the need for innovative solutions to meet nutritional demands sustainably. Collectively, the chapter sets a foundation for future research directions, focusing on refining lysine supplementation strategies in line with the evolving landscape of global aquaculture.

Fish Nutrition: Protein and Amino Acid Requirements. Imtiaz Ahmed, Ishtiyaq Ahmad, and Shabihul Fatma Sayed (Eds.)

DOI: 10.1201/9781998511570-4

4.1 INTRODUCTION

The growing field of aquaculture nutrition continually seeks to improve the efficiency of feed utilization, the growth performance of cultured species, and the sustainability of production practices (Singha et al., 2020, 2021). Central to these objectives is the understanding of essential amino acids (EAAs), particularly lysine, which is vital for the optimal growth and health of fish. With the shift towards sustainable aquaculture practices, lysine's significance is amplified as plant-based proteins often replace lysine-rich fishmeal in aquafeeds (Maiti et al., 2022). This change has led to a heightened focus on accurately determining lysine requirements, influenced by various factors, including species-specific needs, life stages, and dietary compositions.

In fish metabolism, lysine is linked to several key pathways, such as the target of rapamycin (TOR) and amino acid response (AAR) pathways, which are fundamental for growth and energy metabolism (Wang et al., 2020). The methodologies to ascertain lysine requirements have progressed significantly, reflecting the depth of fish nutritional science. From growth assays to factorial models and beyond, each method has contributed to a nuanced understanding of lysine's role (NRC, 2011). The expression of lysine requirements has been subject to debate, considering whether to base it on the percentage of diet, dietary crude protein, or digestible energy. Contemporary research suggests that lysine requirements are more accurately expressed as a percentage of dietary protein (DP), which aligns better with the physiological effects of protein on lysine needs, as opposed to the minimal impact of dietary energy.

This chapter provides a synthesized overview of the current knowledge of lysine requirements in fish, addressing the controversies and offering insights into feed formulation practices. As aquaculture grows, understanding lysine's role is paramount in ensuring industry sustainability, fish welfare, and the ability to meet the global seafood demand.

4.2 LYSINE METABOLISM

In the metabolic processes of fish, lysine emerges as an indispensable amino acid (AA), crucial for protein architecture and metabolic pathways. The overview of lysine metabolism is provided in Figure 4.1.

Lysine metabolism in fish is intricately linked to several key metabolic pathways, which are essential for maintaining growth, health, and overall physiological function.

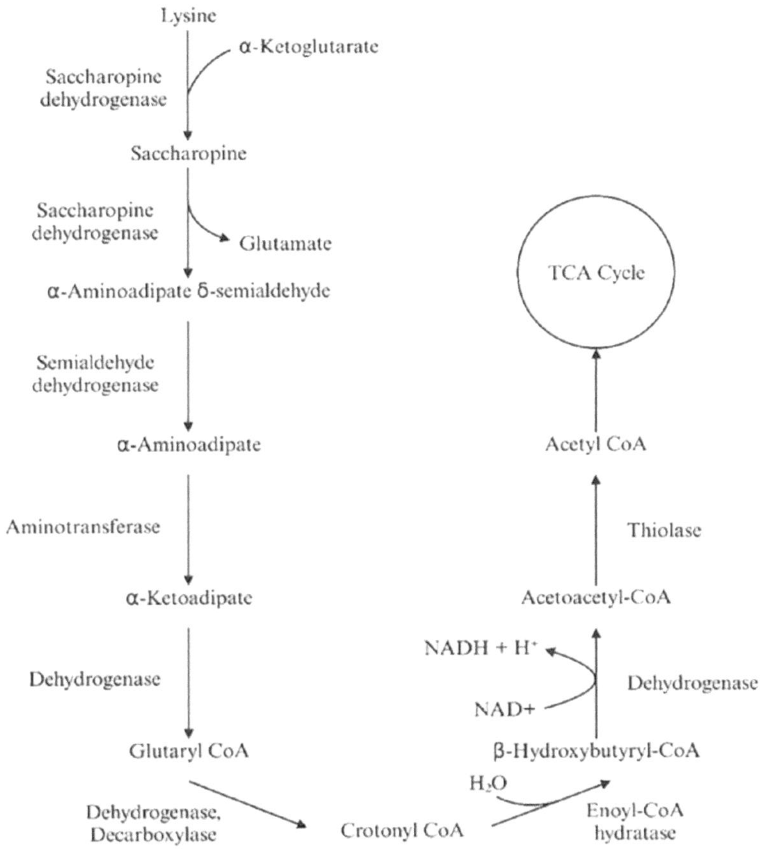

FIGURE 4.1 Overview of lysine metabolism pathway.

4.2.1 TARGET OF RAPAMYCIN (TOR) PATHWAY

The TOR pathway is a central component of cellular nutrient sensing and is directly influenced by AA availability, particularly lysine. It governs growth by modulating protein generation and cell growth. Abundant lysine activates the TOR pathway, encouraging anabolic activities and muscle development, while its deficiency can suppress these processes

(Wang et al., 2020). This pathway's significance is evident in lysine's ability to regulate intermediary metabolism, which includes lipid and carbohydrate metabolism, thus affecting the overall energy balance within the fish body (Wang et al., 2020).

4.2.2 AMINO ACID RESPONSE (AAR) PATHWAY

Complementing the TOR pathway, the AAR pathway is sensitive to the availability of EAAs and is triggered by a lack of lysine. Activation of the AAR pathway by uncharged tRNAs, which accumulate due to insufficient lysine, leads to the inhibition of global protein synthesis. This response helps the organism to conserve resources under conditions of nutrient stress (Wang et al., 2020).

4.2.3 CARNITINE BIOSYNTHESIS

Lysine is also a precursor for synthesizing carnitine (Walton et al., 1984), a compound vital for transporting long-chain fatty acids into the mitochondria for β-oxidation. This process is critical for energy production in muscle and other tissues. After being released during proteolysis, the methylated forms of lysine residues in proteins are converted into carnitine through a series of enzymatic reactions. This conversion is essential for maintaining energy balance and is especially important in fish, which often rely on fat as a major energy source.

Lysine metabolism in fish is a multifaceted process, encompassing various biochemical pathways fundamental for growth, energy production, and overall metabolic health. Understanding these pathways and the role of lysine within them is crucial for developing nutritional strategies to enhance fish growth and aquaculture productivity. The ongoing research in this field aims to elucidate the complex interactions between dietary lysine, metabolic regulation, and fish physiology, to improve aquaculture efficiency and fish health.

4.3 EMPIRICAL METHODS USED FOR LYSINE REQUIREMENT

For over five decades, researchers have explored the EAA requirements of fish. The initial investigations, primarily focused on Chinook salmon, were conducted by Halver and colleagues in the late 1950s. Most quantitative

EAA requirement estimates have been derived from dose-response studies, complemented by various analytical techniques such as evaluating plasma or tissue AA concentrations and rates of AA oxidation. The discussion below outlines various methods used to quantify EAAs, including lysine, as detailed in the NRC (2011) report, presented comprehensively and concisely for better comprehension of each method.

4.3.1 GROWTH RESPONSE ASSAYS

The backbone of EAA requirement estimations in fish lies in conventional growth response assays. These experiments involve offering a basal diet deficient in a single EAA but meet all other known nutrient requirements of the fish. Gradually, the diet is supplemented with increasing levels of the specific EAA being studied. Various diet formulations, including chemically defined, purified, and natural ingredients, have provided incremental amounts of the target AA. In many instances, young, fast-growing fish are used for these tests, though the methodology can also be applied to other life stages, depending on the study's objectives.

Quantification of EAA requirements primarily relies on dose-response curves, with weight gain as the primary response criterion. The lowest level of the EAA that maximizes live weight gain (LWG) is identified as the minimum dietary requirement. In addition to weight gain, protein and EAA depositions are sometimes used as response parameters. It's worth mentioning that protein deposition-based estimates are higher than those based on weight gain in certain studies (Encarnação et al., 2004).

4.3.2 WHOLE-BODY AMINO ACID PROFILE

For many high-quality diets, AA deposition into body protein constitutes a significant portion of total AAs consumed. This makes the composition of the deposited body protein an indicator of the EAA profile required by the fish. Early estimates of AA requirements in fish were drawn from the AA profiles of fish, eggs, and whole-body proteins (Wilson, 1989). Despite the simplicity of this approach, it's important to recognize that it might not fully mirror the EAA requirements because some AAs are preferentially retained in tissues while others play more active metabolic roles and are less efficiently retained.

In most cases, there is good alignment between the body EAA profile and EAA requirements, although minor discrepancies can exist. This alignment allows for simple quantitative estimates of EAA requirements, with whole-body AA profiles and muscle protein AA content ratios often serving as a foundation for determining the "ideal protein" pattern of AAs for fish (Kaushik, 1998; Kaushik and Seiliez, 2010). However, it's acknowledged that such estimates might overstate the requirements for AAs preferentially deposited in body protein, like leucine and lysine, while underestimating the requirements for AAs with critical metabolic roles, such as methionine, threonine, histidine, and arginine.

4.3.3 FACTORIAL MODELS

Recent methodologies for estimating EAA requirements in fish involve mathematical models integrating biological processes governing EAA utilization for body protein deposition (BPD) (Richard et al., 2010). These models, often extending existing models for protein and energy requirements, combine information from various sources, including whole-body EAA profiles, maintenance requirements, and catabolism of AAs, obtained through experimentation or meta-analyses of published studies, along with growth and bioenergetic modeling.

4.3.4 AA LEVELS IN BLOOD AND MUSCLE

At sub-requirement intake levels, the serum or tissue content of the tested EAA should remain low until the requirement is met, subsequently increasing when excessive amounts of the AA are fed (Wilson, 1989). This technique has occasionally confirmed EAA requirements, notably in the case of channel catfish (Wilson et al., 1977). Nevertheless, it's not always reliable due to factors such as the nature of the tested EAA, interactions between AAs, and the timing between meal and blood sampling.

4.3.5 AMINO ACID OXIDATION STUDIES

Amino acid oxidation studies provide insights into the partitioning of EAAs between protein synthesis and oxidation. Direct and indirect oxidation studies are based on tissue-free AA concentrations and offer measures of EAA utilization. The direct method suggests that, at limiting levels, EAA oxidation rates should be low until the requirement is reached and then increase sharply, serving as an indicator of the requirement (Kim et al., 1983). Indirect oxidation studies measure the oxidation of an EAA other than the one under study, with its incorporation into tissue protein being constrained by the intake level of the EAA being examined. This results in high rates of oxidation for the other AAs. However, these techniques have varying degrees of success and reliability in estimating specific EAA requirements.

In summary, a wide range of methodologies and approaches are used to estimate the EAA requirements in fish, each with its advantages and limitations. These methods have evolved over time and are continuously refined to provide more accurate and reliable data for the nutrition and aquaculture industries.

4.4 DIETARY LYSINE REQUIREMENT

The dietary lysine requirement of fish is a critical aspect of aquaculture management, as it plays a fundamental role in ensuring the optimal growth, health, and overall performance of these aquatic organisms. Lysine is one of the EAAs that fish cannot synthesize *de novo*, making it necessary to obtain this nutrient through their diet. It's often used as a reference AA for formulating fish diets because it's one of the limiting AAs in many plant-based feed ingredients commonly used in aquaculture (Small and Soares, 1998). Further, the dietary lysine requirement of fish can vary depending on the species, life stage, environmental conditions, and overall diet composition (Thu et al., 2009; Webb and Gatlin, 2003). For example, carnivorous species might have higher lysine requirements than omnivorous or herbivorous species. Additionally, the growth stage of the fish also influences their lysine needs, with younger and faster-growing fish generally requiring more lysine than older fish. Understanding the specific lysine needs of different fish species is crucial for formulating nutritionally balanced feeds that promote efficient protein utilization, minimize waste production, and support sustainable aquaculture practices (NRC, 2011). In this context, elucidating the dietary lysine requirements of various fish species is not only a matter of economic significance but also contributes to the broader goals of conserving natural fish populations and meeting the growing global demand for quality fish. For that necessity, accurate lysine requirement determination often involves conducting growth trials where different lysine levels are tested in controlled environments to determine the optimal inclusion level for maximizing growth and performance. Lysine requirements are typically expressed as a percentage of the total diet or as a ratio to another EAA and crude protein. Common ranges for lysine inclusion in fish diets might fall between 10 to 30 g/kg of the total diet or around 50 to 100 g/kg of the total crude protein content. Still, these values can vary widely depending on the factors mentioned earlier. The quantitative lysine requirement of some commercially important finfish species is summarized in Table 4.1.

4.5 SOURCES OF DIETARY LYSINE

Dietary lysine in aquafeed or fish feed can come from a variety of natural sources of ingredients, as well as synthetic origin. Lysine is an EAA that needs to be provided in the diet since fish cannot synthesize it in sufficient amounts (Peres and Oliva-Teles, 2008). When formulating fish diets, the lysine content of the selected ingredients must be considered carefully, along with complete AA profiles, digestibility, and availability. The goal is to create a balanced diet that meets the lysine and other nutritional requirements of the target fish species, supporting optimal growth and overall health.

Aquatic feeds for farmed fish differ from livestock feeds primarily in their demand for protein-rich ingredients due to the higher crude protein requirements of fish (Henry et al., 2015). Historically, fish meal has been the preferred protein source due to its nutritional value and balanced EAA profile with high lysine levels. However, the rapid growth of aquaculture has led to a plateau in global fish meal supplies, making it less available and expensive. Consequently, alternative protein sources like blood meal, meat and bone meal, insect meal, soybean meal (SB), canola meal, feather meal, corn gluten meal, ground nut oil cake, cotton seed meal, and different plant-based protein concentrates (Kim et al., 2019) are being explored. However, lysine and methionine are often limiting AAs in these alternatives (Galili and Amir, 2013). Traditional methods of increasing lysine levels through fish meal or marine proteins are costly due to market constraints.

TABLE 4.1 Dietary Lysine Requirement of some Commercially Important Finfish

Species	Average Initial Weight (g)	Studied Lysine Range (g/kg Diet)	Salinity (ppt)	Temperature (°C)	Optimum Lysine Requirement (g/kg Diet)	References
Greater amberjack *(Seriola dumerili)*	32.8 ± 3.0	19.3–22.9	35	19.8 ± 1.7	20.3–21.1	Kotzamanis et al. (2021)
Yellowtail (*Seriola quinqueradiata*)	68.2	10.5–30.5	30.7 ± 1.8	25.8 ± 1.6	16.6	Ruchimat et al. (1997)
Milkfish (*Chanos chanos*)	5.92 ± 0.14	7–27	30.7	29	20	Borlongan and Benitez (1990)
Hybrid grouper (*Epinephelus fuscoguttatus* × *Epinephelus lanceolatus)*	11.44 ± 0.07	17.4–39.9	Not recorded (NR)	27	21.6	Li et al. (2019)
Orange-spotted grouper (*Epinephelus coioides*)	15.84 ± 0.23	19.2–39.5	24–28	28–30.5	28.3	Luo et al. (2006)
European seabass (*Dicentrarchus Labrax*)	0.85 ± 0.03	12–34.8	34.6 ± 1.6	25.5 ± 1.2	21.7–22.2	Tibaldi and Lanari (1991)
Gilthead bream (*Sparus aurata*)	3.5	36.3–79.7	32 ± 0.5	24 ± 0.5	50.4	Marcouli et al. (2006)
Japanese sea bass (*Lateolabrax japonicus*)	5.50 ± 1.63	12.8–42.5	25–28	27–30	24.9–26.1	Mai et al. (2006)
Large yellow croaker (*Pseudosciaena crocea*)	2.75 ±0.11 mg	24.8–41.0	23 ±1	23 ± 2	33.4–33.7	Xie et al. (2012)
	1.23 ± 0.02	12.7–42.2	25–28	27–30	24.3–24.8	Zhang et al. (2008)
Rainbow trout (*Oncorhynchus mykiss*)	8.54 ± 0.19	18.3–35.2	Fresh Water (FW)	20–22	22.3–27.5	Lee et al. (2020)
	512 ± 6.8	7.2–35.2	FW	16 ± 0.4	22–23.4	Yun et al. (2016)
Red drum (*Sciuenops ocellatus*)	6–7	10–25	5 ± 1	NR	15.5	Craig and Gatlin (1992)
	2.7	16–24	4–6	26–30	20	Brown et al. (1988)
Atlantic salmon (*Salmo salar)*	376 ± 46	5.9–21.1	33	8	16–18	Berge et al. (1998)
	62.8 ± 1.3	19.7–31.6	32.7 ± 0.4	11.3 ± 0.4	28.8	Grisdale-Helland et al. (2011)
	4.7 ± 0.2	12.4–29.4	10	15 ± 0.9	18.4 ± 1.6	Anderson et al. (1993)
Asian sea bass (*Lates calcarifer*)	13.12 ± 0.12	10–25	31	27	20.6	Murillo-Gurrea et al. (2001)
Silver pompano (*Trachinotus blochii)*	6.28	15.2–29.8	17–20	27	24–24.5	Ebeneezar et al. (2019)
Florida pompano (*Trachinotus carolinus*)	13.07 ± 0.46	14.2–24.2	35.48 ± 3.43	27.97 ± 1.59	16.7	Stites et al. (2022)

TABLE 4.1 *(Continued)*

Species	Average Initial Weight (g)	Studied Lysine Range (g/kg Diet)	Salinity (ppt)	Temperature (°C)	Optimum Lysine Requirement (g/kg Diet)	References
Catla (*Catla catla*)	0.58 ± 0.02	12.5–25	FW	26.8 ± 1.6	17–18	Zehra and Khan (2013)
	1.24	0–35	FW	NR	25–30	Ravi and Devaraj (1991)
Rohu (*Labeo rohita*)	0.66 ± 0.081	12–22	FW	25–26.5	18–19.1	Abidi and Khan (2010)
Mrigal (*Cirrhinus mrigala*)	0.63 ± 0.02	15–27.5	FW	26.5 ± 1.80	23	Ahmed and Khan (2004)
Common carp (*Cyprinus carpio*)	7.89 ± 0.04	7–22	FW	21 ± 1	18.88	Zhou et al. (2008)
	22.77	12.1–22.1	FW	20 ± 1	14.5	D'Mello et al. (1989)
	21.92	12.1–22.1	FW	25 ± 1	22.4	
Grass carp (*Ctenopharyngodon idella*)	3.15 ± 0.01	6.9–30.8	FW	28.7 ± 2.5	20.7	Wang et al. (2005)
	164.85 ± 0.79	5.6–20.7	FW	28.3 ± 3.2	13.51–14.56	Hu et al. (2021)
	0.36	14.4–28.7	FW	27 ± 2	24.4	Huang et al. (2021)
	460 ± 1.5	6.6–16.7	FW	25 ± 2	10.9	Li et al. (2014)
Nile tilapia (*Oreochromis niloticus*)	274.9 ± 3.3	11.1–16.4	FW	27.5 ± 2.6	14.6	Michelato et al. (2016)
	10	12.4–20.4	FW	NR	17.4–18.4	Hussein et al. (2016)
	1.44	11.3–20.8	FW	27.1 ± 1.45	15.21	Furuya et al. (2012)
	519 ± 27.23	9.3–21.3	FW	24.89 ± 1.07	13.1	Junior et al. (2023)
GIFT tilapia	5.2 ± 0.1	14.3–20.5	FW	28 ± 1	18–19.5	Prabu et al. (2019)
	3.16 ± 0.01	12.2–27.2	9.97–10.27	27.7–28.3	19.3–20.7	Garg et al. (2022)
Sutchi catfish (*Pangasius hypophthalmus*)	8.21 g ± 0.32	14.8–26.7	FW	24.6 ± 0.39	22.2–22.4	Rachmawati and Nurhayati (2022)
Clarias gariepinus	15.4 ± 1.6	16–26	FW	28	22.8	Fagbenro et al. (1999)
Largemouth Bass (*Micropterus salmoides*)	17.34 ± 0.02	25.6–40.9	FW	28 ± 2	30.3–30.7	Huang et al. (2022)
Blunt snout bream (*Megalobrama amblycephala*)	3.34 ± 0.03	12.9–32.7	NR	NR	22.2–23.6	Liao et al. (2013)

Thus, synthetic sources are commonly used to supplement deficient AAs. To reduce excessive dietary crude protein (CP) in commercial aquatic feeds, the ideal protein concept is gaining popularity, emphasizing EAA requirements met through synthetic AAs. In addition, the synthetic sources of lysine are more useful for requirement studies carried out with purified or semi-purified diets.

4.5.1 NATURAL SOURCES OF DIETARY LYSINE

1. **Fish Meal:** Fish meal is considered a high-quality protein source in aqua feed. It is mainly derived from various small marine fish species and contains a well-balanced AA profile, including a significant amount of lysine. Fish meal is particularly valuable for its digestibility and EAA content.
2. **Soybean Meal:** SB is a common plant-based protein source used in fish feed, especially for omnivorous and herbivorous fish. While SB is not as rich in lysine as animal-based proteins, it still provides a reasonable lysine content. It's important to note that the lysine content can vary depending on the quality of the SB and its processing.
3. **Rapeseed (Canola) Meal:** Rapeseed meal is another plant-based protein source used in aqua feed formulations. It contains a moderate amount of lysine, although its overall AA profile may be influenced by processing methods. Proper processing, such as heat treatment, is essential to reduce the levels of antinutritional factors.
4. **Pea Protein:** Pea protein is gaining popularity as a sustainable protein source in aqua feed. While not as commonly used as fish meal or SB, pea protein contains a decent amount of lysine and other EAAs. Pea protein is particularly suitable for omnivorous and herbivorous fish species.
5. **Corn Gluten Meal:** Corn gluten meal is a byproduct of corn processing and is used in fish feed to provide both protein and energy. It contains some lysine, but its overall AA profile can be imbalanced for certain fish species, so it's often used in combination with other protein source.

6. **Poultry Byproduct Meal:** This ingredient is derived from poultry processing and includes various poultry tissues. Poultry byproduct meal can contribute to the lysine content in fish feed formulations, along with other EAAs and nutrients.
7. **Wheat Gluten:** Wheat gluten is obtained from wheat processing and is used as a protein source in some aqua feed formulations. It contains a moderate amount of lysine and can be valuable for its protein content and binding properties.
8. **Krill Meal:** Krill meal is a premium marine ingredient derived from krill, a small crustacean. It is highly valued for its excellent nutritional profile, including lysine, omega-3 fatty acids, and other essential nutrients. Krill meal is particularly used in high-end feed formulations.
9. **Algae and Seaweed:** Some algae and seaweed species can provide lysine and other EAAs. They are particularly useful for formulating diets for herbivorous and omnivorous fish, as these ingredients also offer vitamins, minerals, and pigments.
10. **Insects:** Insect-based ingredients, such as mealworms and black soldier fly larvae, are gaining attention due to their sustainable nature and potential to provide lysine and other essential nutrients. These ingredients are being explored as alternatives to traditional protein sources.

4.5.2 SYNTHETIC SOURCES OF DIETARY LYSINE

1. **L-lysine Monohydrochloride:** Lysine monohydrochloride is a crystalline form of lysine that contains a high concentration of lysine molecules. It is produced through a fermentation process that involves bacterial (mainly using *Corynebacterium glutamicum* and *Escherichia coli*) cultures converting sugars into lysine (Pandey et al., 2020). This form of lysine is easily soluble in water and can be efficiently mixed into fish feed formulations (Rodehutscord et al., 2000). Its high purity and concentration make it a cost-effective and popular choice for supplementing lysine in fish diets.

2. **L-lysine Sulfate:** Like lysine monohydrochloride, lysine sulfate is another synthetic form of lysine. It is produced by reacting lysine with sulfuric acid (Powell et al., 2015). Lysine sulfate provides a concentrated source of lysine and can be included in aquafeed formulations to meet the AA requirements of fish.

4.6 INTERACTION OF LYSINE WITH OTHER AMINO ACIDS

AA interaction (antagonism) in fish is an area of nutritional biochemistry that explores the intricate dynamics between EAAs and their impact on fish growth and metabolism. EAAs like lysine can interact with arginine, which thereby can be critical for various physiological processes, including protein synthesis, immune function, and nutrient metabolism. However, their interaction can be complex; imbalances, particularly in diets high in plant proteins, may lead to suboptimal growth rates and health outcomes in aquaculture species. Understanding these interactions is crucial for formulating diets that support the optimal growth and well-being of cultured fish.

A study by Berge et al. (1999) on the uptake of lysine and arginine in Atlantic salmon's mid-intestine highlighted the complex interaction between these AAs, indicating they may share a common transport carrier across the brush border membrane. The research demonstrated that lysine could both stimulate and inhibit the uptake of arginine, depending on their relative concentrations. In contrast, arginine consistently inhibited lysine uptake. Moreover, lysine acts as a potent inhibitor of arginase, which converts arginine into ornithine; its elevated levels could potentially decrease the synthesis of polyamines, which are crucial for RNA and DNA synthesis and associated with growth regulation (Berge et al., 2002). Consequently, high dietary lysine might result in reduced growth rates due to its interference with arginine metabolism and polyamine synthesis.

Nguyen et al. (2013) conducted an experiment on cobia to evaluate the effects of varying dietary lysine-to-arginine ratios. The study concluded that cobia fed a diet with a balanced lysine to arginine ratio (BL/A; 1.1) showed better growth performance compared to diets with high (HL/A; 1.8) or low (LL/A; 0.8) lysine to arginine ratios. The findings underscore the importance of maintaining a balanced ratio of these AAs in diets to optimize growth and health in cobia, suggesting that an imbalance can lead to reduced performance.

Contrary to the previous reports, Robinson et al. (1981) reported that fingerling channel catfish showed no evidence of arginine-lysine antagonism when excess lysine was included in their diets that were adequate or marginal in arginine. Growth and feed efficiency data suggested that channel catfish are not as sensitive to disproportionate levels of lysine and arginine as other animals.

Beyond its well-documented antagonism with arginine, lysine competes with other large neutral AAs like leucine, isoleucine, and tryptophan. These AAs share similar transport mechanisms in the body, particularly at the intestinal level and across the blood-brain barrier, leading to potential competition for absorption. Such interactions can affect the overall nutritional status, necessitating a carefully balanced AA profile in dietary formulations to prevent imbalances that could compromise growth, health, and function, especially in animals and aquaculture species where precise nutritional management is critical for optimal productivity.

The interaction between arginine and lysine is a complex issue, where excess dietary lysine can negatively impact arginine utilization, potentially hindering growth. Findings across different fish species have been inconsistent, suggesting that the effect of dietary lysine on arginine and vice versa may vary (Hoseini et al., 2020). This inconsistency in results underscores the need for further research to clarify the nature of these interactions and their implications for fish nutrition.

4.7 LYSINE AND IDEAL PROTEIN CONCEPT

The term "ideal protein" is described as a protein with a balanced AA profile to perfectly fulfill the animal's needs without any excess or deficiency (NRC, 2011). Lysine is crucial in this "ideal protein" concept when formulating fish feed. Lysine is a key component in this concept due to its importance as a limiting AA in many feed ingredients and its role in protein synthesis, growth, and overall metabolism. The EAA needs are typically calculated in relation to the lysine requirement. Consequently, if you have information about the dietary lysine requirement and the overall AA composition of the fish, it should be possible to approximate the dietary needs for the other EAAs in relation to the lysine requirement for the fish (Kaushik and Seiliez, 2010).

Here's how lysine contributes to the ideal protein concept in fish feed formulation:

1. **Limiting Amino Acid:** Lysine is often a limiting AA in many plant-based feed ingredients commonly used in aquafeed, like SB or corn gluten meal. This means that even if other AAs are present in excess, growth and performance will be constrained by the availability of lysine. By ensuring an adequate level of lysine, the growth potential of fish can be maximized, and other AAs can be better utilized.
2. **Optimal Amino Acid Balance:** The ideal protein concept seeks to balance the AA profile in feed to match the AA profile required by the fish for optimal growth. Lysine serves as a reference AA for this balancing act. The ratio of lysine to other AAs, particularly sulfur-containing AAs like methionine, is important for efficient protein synthesis and growth. Achieving the right lysine-methionine ratio helps prevent AA imbalances that could hinder growth.
3. **Protein Synthesis and Growth:** Lysine is an essential component of proteins and is crucial for the synthesis of muscle proteins and enzymes that support growth and metabolic functions. It participates in various biochemical processes, including cell division, tissue repair, and hormone production. Providing adequate lysine ensures that the fish can build and maintain their body tissues and achieve their growth potential.
4. **Feed Efficiency:** Including sufficient lysine in fish feed can improve feed efficiency, as it reduces the need for the fish to catabolize existing proteins to obtain lysine. This contributes to better utilization of the energy in the feed for growth rather than protein breakdown.
5. **Minimizing Environmental Impact:** By formulating diets with the ideal balance of AAs, including lysine, aquaculture practitioners can reduce the overuse of protein sources and limit the nitrogen waste released into the environment. This is especially important for sustainable aquaculture practices and minimizing environmental impacts.

6. **Customized Formulation:** Different fish species have varying lysine requirements. The ideal protein concept allows nutritionists to tailor feed formulations to the specific needs of each species and their growth stages. This customization ensures that lysine and other EAAs are provided in the right amounts for optimal performance.

4.8 PHYSIOLOGICAL AND NUTRITIONAL FUNCTIONS OF LYSINE

Lysine, an indispensable AA, plays a pivotal role in the physiological and nutritional landscape of fish, profoundly influencing their growth, health, and overall performance. This AA, unable to be synthesized by fish, must be supplied through their diet to meet their multifaceted requirements. Lysine's multifunctional significance is evident across a spectrum of physiological and nutritional functions, rendering it a cornerstone in the formulation of aquafeeds and the optimization of aquaculture practice.

At the core of lysine's role is its involvement in protein synthesis, an intricate process by which AAs are assembled into proteins vital for cellular function. Lysine acts as a fundamental building block in this process, contributing to the formation of peptide bonds that string AAs together. Consequently, proteins synthesized with lysine serve as the backbone of various biological structures, including muscle tissues, enzymes, hormones, and antibodies (Li et al., 2011). This close connection between lysine and growth underscores its significance in aquaculture. Lysine's participation in the synthesis of muscle proteins directly impacts fish growth rates and body composition (Valente et al., 2013). As lysine availability increases, so does the potential for robust lean body mass development, a fundamental objective in modern aquaculture striving to produce fish with optimal flesh quality and marketability. Energy metabolism, a vital component of fish vitality, also relies on lysine. In times of energy demand, lysine can be metabolized through gluconeogenesis, where AAs are converted into glucose for energy production (Encarnação et al., 2006). This adaptation is particularly valuable during fasting or stressful events when other energy sources may be limited, sustaining energy levels and maintaining physiological equilibrium.

Lysine's contributions extend into the realm of immunity and defense mechanisms. Through its role in immune-related protein synthesis, including antibodies and enzymes, lysine bolsters the immune response, defending fish against potential pathogens and diseases (Sandell and Daniel, 1988). By supporting the production of immune components, lysine enhances the ability of fish to mount a robust defense and preserve their health status. Enzymatic activity, vital for the execution of metabolic processes, also hinges on lysine's presence. Serving as a cofactor or structural component, lysine aids in the functionality of enzymes that govern digestion, nutrient absorption, detoxification, and hormonal regulation. Without lysine, these processes could be compromised, leading to suboptimal nutrient utilization and overall physiological function. Structurally, lysine contributes to the integrity of connective tissues. Collagen, a structural protein that supports tissues such as skin, scales, bones, and cartilage, relies on lysine for its formation (Hema and Mathew, 2015). Lysine's involvement ensures the resilience of these tissues, allowing fish to withstand environmental stressors and maintain their structural soundness.

Reproductive performance, a fundamental aspect of aquaculture, is also under the influence of lysine. This AA plays a role in the synthesis of reproductive hormones and proteins crucial for successful reproduction. Lysine supports the development of gametes, such as eggs and sperm, as well as the embryonic stages of growth. Beyond these specific roles, lysine's interactions with other AAs are essential for achieving AA balance and synergy (Li et al., 2021). AAs like methionine, rich in sulfur, collaborate with lysine to maintain an optimal ratio that supports protein synthesis and growth. This intricate balance underscores the importance of formulating diets that not only provide adequate lysine but also ensure the harmonious interplay of AAs. Furthermore, lysine's availability influences nutrient utilization and feed efficiency. In lysine-deficient diets, fish may resort to breaking down existing proteins to obtain lysine, diverting resources from growth-related processes. By supplying sufficient lysine, aquaculture practitioners can ensure efficient nutrient utilization, reducing wastage and optimizing growth. Lysine's significance extends even into the management of stress. Stressors, such as environmental fluctuations or disease challenges, can disrupt metabolic processes. Adequate lysine levels support stress adaptation mechanisms, aiding fish in coping with challenges and maintaining overall health in the face of adversity (Ciji and Akhtar, 2021).

In the context of modern aquaculture, where precision and sustainability are paramount, lysine's functions become essential considerations. Its role in protein synthesis, growth, energy metabolism, immunity, and structural integrity underscores its impact on fish health and productivity. Formulating diets that encompass optimal lysine levels, customized to specific species and growth stages, is pivotal for realizing these functions in practice. Ultimately, the physiological and nutritional functions of lysine in fish are intricately intertwined with every facet of their biology. By acknowledging lysine's multifaceted roles, aquaculture practitioners can craft diets that harness its potential, resulting in enhanced growth rates, improved immune responses, and overall sustainability. As the aquaculture industry continues to evolve, lysine remains a beacon of scientific understanding, guiding the formulation of feeds that enable fish to thrive in their aquatic habitats.

4.9 NUTRITIONAL DISORDERS OF LYSINE DEFICIENCY

Lysine deficiency in fish has been studied across various species with consistent observations that it adversely affects growth, health, and metabolism. Moreover, lysine deficiency in fish has been associated with specific diseases and symptoms, including dorsal and caudal fin erosion in rainbow trout, increased mortality in rainbow trout and common carp, and incidence of lordosis (a curvature of the spine) in common carp (Tacon, 1992).

Deficiencies prompt fish to adapt by altering their feeding patterns or metabolic pathways, such as increased nitrogen turnover and selective plankton consumption. Growth performance is typically suppressed, with effects manifesting as increased susceptibility to diseases, changes in muscle tissue composition, and impaired immune responses. Adequate lysine is critical for maintaining optimal growth rates, feed efficiency, and overall health, as demonstrated by improved growth metrics when dietary lysine is sufficient. The studies indicate a species-specific requirement for lysine, underscoring the need for precise formulation of fish diets to include this EAA for maintaining robust aquaculture practices. Table 4.2 covers the summarized information on lysine deficiency across different species.

TABLE 4.2 Deficiency Symptoms of Lysine in Various Fish Species

Fish Species	Initial Body Weight (g)	Deficiency Symptoms	References
Grass carp (*Ctenopharyngodon idella*)	165	Multiple adverse effects, such as suppressed growth performance, increased susceptibility to skin lesions, and impaired immune response; reduces the expression of immune-related genes and antibacterial peptides while exacerbating inflammatory responses	Hu et al. (2021)
Nile tilapia (*Oreochromis niloticus*)	17.2	Adverse effect on intestinal morphology by increasing the distal villi length, submucosal layer, muscular layer and mid villi width, proximal / mid goblet cell counts and proximal lamina propria	Khalil et al. (2022)
Pacu (*Piaractus mesopotamicus*)	10.2	Increased the incorporation rate of nitrogen-free compounds like lipids and glycogen into muscle tissue, higher metabolic turnover of nitrogen leads to greater breakdown of nitrogen compounds	Abimorad et al. (2017)
Gilthead seabream (*Sparus aurata*)	3.5	Reduced growth performance	Marcouli et al. (2006)
	–	Downregulated IGF-I, IGF-II, and IGF-IRb mRNA levels, as well as the gene expression of various MRFs, including Myf5, Myogenin, and MyoD2; a significant decrease in proliferation and differentiation markers	Azizi et al. (2016)
Rainbow trout (*Oncorhynchus mykiss)*	100	Decreased growth attributed to reduced protein synthesis and voluntary feed intake, and increased lipid content in the liver	Gatrell et al. (2016)
	–	Reduced weight, high mortality, and fin erosion	Ketola et al. (1983)
Turbot (*Scophthalmus maximus*)	–	Increased plasma total cholesterol (T-CHO) and low-density lipoprotein cholesterol (LDL-C) levels	Huang et al. (2022)
Zebrafish (*Danio rerio*)	–	Reduced growth, higher protein turnover for basal metabolism rather than muscle protein accretion which indicates sarcomeric degradation	de Vareilles et al. (2012)

4.10 FACTORS AFFECTING LYSINE REQUIREMENT

The lysine requirement in fish is influenced by a variety of factors that collectively shape their nutritional needs for optimal growth, health, and performance (Hauler and Carter, 2001; Kaushik and Seiliez, 2010). These factors underscore the complexity of determining lysine requirements for fish. They interact dynamically, necessitating a comprehensive approach to aquafeed formulation. Through careful consideration of these variables and conducting thorough research and trials, aquaculture practitioners can ensure that lysine requirements are met effectively, optimizing growth, health, and performance in their fish populations. Here are the key factors that affect the lysine requirement in fish:

1. **Species and Life Stage:** Different fish species exhibit variations in their physiological characteristics, metabolic rates, and growth patterns. These differences contribute to varying lysine requirements among species. Furthermore, the life stage of fish is a critical determinant. Juvenile fish experience rapid growth and development, necessitating higher lysine levels to support protein synthesis. Mature fish might have lower lysine requirements as growth slows down.

2. **Dietary Protein Level:** The overall protein content of the diet can impact lysine requirements. Diets with higher protein levels might need proportionally more lysine to maintain a balanced AA profile. If other AAs are provided in abundance, lysine can become the limiting AA, affecting overall growth potential.

3. **Dietary Protein Source:** The protein source used in aquafeeds significantly affects the availability of lysine. Animal-based protein sources like fish meal are generally well-balanced in AAs, including lysine. However, plant-based protein sources such as SB might be deficient in lysine and require supplementation to meet fish requirements adequately.

4. **Amino Acid Interactions:** Lysine's interactions with other AAs, particularly sulfur-containing AAs like methionine and cysteine, are critical. An improper balance between these AAs can affect protein synthesis and growth. Maintaining an optimal ratio is essential for ensuring that lysine is effectively utilized.

5. **Dietary Lipid Content:** Dietary lipid levels can influence lysine requirements. High-lipid diets can alter lysine utilization due to changes in nutrient absorption and metabolism. Consequently, fish on high-lipid diets might require higher lysine levels to counteract potential imbalances.

6. **Energy-to-Protein Ratio:** The energy-to-protein ratio in the diet can impact lysine needs. Diets with imbalanced energy and protein levels can affect AA utilization. A mismatch in energy and protein can lead to lysine becoming the limiting AA, necessitating increased supplementation.

7. **Water Temperature:** Water temperature profoundly affects fish metabolism. Higher temperatures result in increased metabolic rates and higher energy requirements. This can lead to greater demand for AAs, including lysine, during periods of elevated growth, such as warm seasons.

8. **Environmental Stress:** Stressors like handling, transport, changes in water quality, and disease outbreaks can elevate lysine requirements. Stress increases metabolic demands and disrupts normal physiological processes, necessitating higher lysine levels to support immune responses and stress adaptation.

9. **Feed Formulation and Processing:** The composition and processing of feeds can affect lysine availability. Certain processing methods, such as heat treatment, can reduce lysine content due to degradation. Ensuring minimal lysine losses during feed processing is vital for meeting fish requirements.

10. **Nutrient Digestibility:** The digestibility of feed ingredients influences the actual intake of lysine. Poorly digestible ingredients might limit lysine absorption, affecting its availability for metabolic processes.

11. **Genetic Variation:** Genetic diversity among fish species and even within a species can result in varying lysine requirements. Selective breeding for growth performance can impact lysine needs, as genetic traits influencing metabolism and growth come into play.

12. **Dietary Anti-Nutritional Factors:** Plant-based protein sources may contain antinutritional factors that impede lysine utilization.

These factors can inhibit digestion and nutrient absorption, necessitating higher lysine levels to compensate.

13. **Water Quality:** Water quality directly impacts fish health and stress levels. Poor water quality can lead to stress and impaired nutrient utilization, potentially increasing lysine requirements as fish allocate resources to cope with adverse conditions.

4.11 FUTURE PERSPECTIVES

The future of lysine research in aquaculture is set to pivot around the customization of lysine requirements to the unique biological profiles of diverse fish species. This precision approach is expected to improve not only growth and feed efficiency but also overall health outcomes for different aquaculture stocks. Additionally, as the industry leans towards plant-based diets, there will be an increased emphasis on understanding how lysine from these sources interacts within the metabolic pathways of fish.

Advancements in understanding the molecular mechanisms of lysine will likely inform the next generation of feed formulations. For example, recent studies, such as those investigating the nutrient-sensing signaling pathways in largemouth bass, have begun to elucidate how lysine supplementation can significantly boost growth performance, feed efficiency, and nutrient retention. Delving deeper into such molecular effects will pave the way for optimizing the utilization of alternative protein sources in aquafeeds, such as poultry by-product meals.

The search for cost-effective and high-quality lysine sources is a concurrent trend, which may benefit from biotechnological innovations. The industry is poised to explore new methods of enhancing lysine content in feed ingredients or uncovering alternative sustainable sources that can meet nutritional demands without straining the environment. These efforts are integral to maintaining the delicate balance between economic viability and ecological responsibility in aquaculture.

4.12 CONCLUSION

Essentially, lysine is a cornerstone in the ideal protein concept for fish feed formulation. Its fundamental role as a limiting AA, critical for protein synthesis and growth, underscores the necessity of creating nutritionally balanced feeds tailored to the specific needs of farmed fish. The impact of lysine on the balance of AAs highlights the importance of precise determination and supplementation strategies to support the health and productivity of aquatic species. As aquaculture advances, integrating these insights into feed development is paramount for enhancing industry sustainability and meeting the nutritional demands of a growing global population.

KEYWORDS

- **amino acid metabolism in fish**
- **essential amino acids**
- **feed formulation strategies**
- **fish growth and nutrition**
- **lysine supplementation in aquafeed**
- **sustainable aquaculture practices**

REFERENCES

Abimorad, E. G., Ducatti, C., Castellani, D., Jomori, R. K., Portella, M. C., Sartori, M. M. P., & Carneiro, D. J. (2017). Dietary lysine deficiency increases the incorporation rate of nitrogen-free compounds in muscle of pacu (*Piaractus mesopotamicus*). *Aquaculture, 479*, 522–527.

Ahmed, I., & Khan, M. A. (2004). Dietary lysine requirement of fingerling Indian major carp, *Cirrhinus mrigala* (Hamilton). *Aquaculture, 235*(1–4), 499–511.

Anderson, J. S., Lall, S. P., Anderson, D. M., & McNiven, M. A. (1993). Quantitative dietary lysine requirement of Atlantic salmon (*Salmo salar*) fingerlings. *Canadian Journal of Fisheries and Aquatic Sciences, 50*(2), 316–322.

Azizi, S., Nematollahi, M. A., Mojazi Amiri, B., Velez, E. J., Lutfi, E., Navarro, I., Capilla, E., & Gutierrez, J. (2016). Lysine and leucine deficiencies affect myocytes development and IGF signaling in gilthead sea bream (*Sparus aurata*). *PLOS ONE, 11*(1), e0147618.

Berge, G. E., Bakke-McKellep, A. M., & Lied, E. (1999). *In vitro* uptake and interaction between arginine and lysine in the intestine of Atlantic salmon (*Salmo salar*). *Aquaculture, 179*(1–4), 181–193.

Berge, G. E., Sveier, H., & Lied, E. (1998). Nutrition of Atlantic salmon (*Salmo salar*); the requirement and metabolic effect of lysine. *Comparative Biochemistry and Physiology Part A: Molecular & Integrative Physiology, 120*(3), 477–485.

Berge, G. E., Sveier, H., & Lied, E. (2002). Effects of feeding Atlantic salmon (*Salmo salar* L.) imbalanced levels of lysine and arginine. *Aquaculture Nutrition, 8*(4), 239–248.

Borlongan, I. G., & Benitez, L. V. (1990). Quantitative lysine requirement of milkfish (*Chanos chanos*) juveniles. *Aquaculture, 87*(3, 4), 341–347.

Brown, P. B., Davis, D. A., & Robinson, E. H. (1988). An estimate of the dietary lysine requirement of juvenile red drum *Sciaenops ocellatus*. *Journal of the World Aquaculture Society, 19*(2), 109–112.

Ciji, A., & Akhtar, M. S. (2021). Stress management in aquaculture: A review of dietary interventions. *Reviews in Aquaculture, 13*(4), 2190–2247.

Coloso, R. M., Borlongan, I. G., & Serrano, A. E. (2001). Lysine and arginine requirements of juvenile Asian sea bass (*Lates calcarifer*). *Journal of Applied Ichthyology, 17*(2), 49–53.

Craig, S. R., & Gatlin, D. M., III. (1992). Dietary lysine requirement of juvenile red drum *Sciaenops ocellatus*. *Journal of the World Aquaculture Society, 23*(2), 133–137.

de Vareilles, M., Conceição, L. E., Gómez-Requeni, P., Kousoulaki, K., Richard, N., Rodrigues, P. M., Fladmark, K. E., & Rønnestad, I. (2012). Dietary lysine imbalance affects muscle proteome in zebrafish (*Danio rerio*): A comparative 2D-DIGE study. *Marine Biotechnology, 14*, 643–654.

D'Mello, J. P. F., Al-Salman, M. H., & Mills, D. H. (1989). Responses of fingerling carp, *Cyprinus carpio* L., to dietary lysine. *Aquaculture Research, 20*(4), 417–426.

Ebeneezar, S., Vijayagopal, P., Srivastava, P. P., Gupta, S., Varghese, T., Prabu, D. L., Chandrasekar, S., Varghese, E., Sayooj, P., Tejpal, C. S., & Wilson, L. (2019). Dietary lysine requirement of juvenile silver pompano, *Trachinotus blochii* (Lacepède, 1801). *Aquaculture, 511*, 734234.

Encarnação, P., de Lange, C., & Bureau, D. P. (2006). Diet energy source affects lysine utilization for protein deposition in rainbow trout (*Oncorhynchus mykiss*). *Aquaculture, 261*(4), 1371–1381.

Encarnação, P., de Lange, C., Rodehutscord, M., Hoehler, D., Bureau, W., & Bureau, D. P. (2004). Diet digestible energy content affects lysine utilization, but not dietary lysine requirements of rainbow trout (*Oncorhynchus mykiss*) for maximum growth. *Aquaculture, 235*(1–4), 569–586.

Fagbenro, O. A., Balogun, A. M., & Fasakin, E. A. (1999). Dietary methionine requirement of the African catfish, *Clarias gariepinus*. *Journal of Applied Aquaculture, 8*(4), 47–54.

Fatma Abidi, S., & Khan, M. A. (2010). Growth, protein retention, and body composition of fingerling Indian major carp, rohu, *Labeo rohita* (Hamilton), fed diets with various levels of lysine. *Journal of the World Aquaculture Society, 41*(5), 791–799.

Furuya, W. M., Graciano, T. S., Vidal, L. V. O., Xavier, T. O., Gongora, L. D., Righetti, J. S., & Furuya, V. R. B. (2012). Digestible lysine requirement of Nile tilapia fingerlings fed arginine-to-lysine-balanced diets. *Revista Brasileira de Zootecnia, 41*, 485–490.

Galili, G., & Amir, R. (2013). Fortifying plants with the essential amino acids lysine and methionine to improve nutritional quality. *Plant Biotechnology Journal, 11*(2), 211–222.

Garg, C. K., Sardar, P., Sahu, N. P., Maiti, M. K., Shamna, N., Varghese, T., Deo, A. D., & Harikrishna, V. (2022). Dietary lysine requirement of genetically improved farmed tilapia (GIFT) juvenile reared in inland saline water of 10 ppt salinity. *Aquaculture, 555*, 738223.

Gatrell, S. K., Silverstein, J. T., Barrows, F. T., Grimmett, J. G., Cleveland, B. M., & Blemings, K. P. (2017). Effect of dietary lysine and genetics on growth and indices of lysine catabolism in rainbow trout (*Oncorhynchus mykiss*). *Aquaculture Nutrition, 23*(5), 917–925.

Grisdale-Helland, B., Gatlin, D. M., III, Corrent, E., & Helland, S. J. (2011). The minimum dietary lysine requirement, maintenance requirement and efficiency of lysine utilization for growth of Atlantic salmon smolts. *Aquaculture Research, 42*(10), 1509–1529.

Hema, G. S., & Mathew, S. (2015). Biochemical Characterization and Bio-Evaluation of Collagen and Collagen Peptides Extracted and Purified from Fish Skin: *In Vitro* and *In Vivo* Studies on Antiarthritic and Wound Healing Properties. Doctoral dissertation, Cochin University of Science and Technology.

Halver, J. E., & Shanks, W. E. (1960). Nutrition of salmonoid fishes: VIII. Indispensable amino acids for sockeye salmon. *The Journal of Nutrition, 72*(3), 340–346.

Hauler, R. C., & Carter, C. G. (2001). Reevaluation of the quantitative dietary lysine requirements of fish. *Reviews in Fisheries Science, 9*(3), 133–163.

Henry, M., Gasco, L., Piccolo, G., & Fountoulaki, E. (2015). Review on the use of insects in the diet of farmed fish: Past and future. *Animal Feed Science and Technology, 203*, 1–22.

Hoseini, S. M., Ahmad Khan, M., Yousefi, M., & Costas, B. (2020). Roles of arginine in fish nutrition and health: Insights for future researches. *Reviews in Aquaculture, 12*(4), 2091–2108.

Hu, Y., Feng, L., Jiang, W., Wu, P., Liu, Y., Kuang, S., Tang, L., & Zhou, X. (2021). Lysine deficiency impaired growth performance and immune response and aggravated inflammatory response of the skin, spleen and head kidney in grown-up grass carp (*Ctenopharyngodon idella*). *Animal Nutrition, 7*(2), 556–568.

Huang, D., Liang, H., Ge, X., Zhu, J., Li, S., Wang, Y., Ren, M., & Chen, X. (2022). Effects of dietary lysine levels on growth performance and glycolipid metabolism via the AKT/FoxO1 pathway in juvenile largemouth bass (*Micropterus salmoides*). *Aquaculture Nutrition*, Article e14170.

Huang, D., Liang, H., Ren, M., Ge, X., Ji, K., Yu, H., & Maulu, S. (2021). Effects of dietary lysine levels on growth performance, whole body composition and gene expression related to glycometabolism and lipid metabolism in grass carp (*Ctenopharyngodon idella*) fry. *Aquaculture, 530*, 735806.

Huang, X., Song, X., Wang, X., Zhou, H., Liu, C., Mai, K., & He, G. (2022). Dietary lysine level affects digestive enzyme, amino acid transport, and hepatic intermediary metabolism in turbot (*Scophthalmus maximus*). *Fish Physiology and Biochemistry, 48*(4), 1091–1103.

Hussein, M. S., Zaghloul, A. M., Tolan, A. Z., & Ahmed, M. H. I. (2016). Studies on lysine requirement for growing mono-sex Nile tilapia (*Oreochromis niloticus*). *Journal of Animal and Poultry Production, 7*(9), 355–361.

Kaushik, S. J. (1998). Nutritional bioenergetics and estimation of waste production in non-salmonids. *Aquatic Living Resources, 11*(4), 211–217.

Kaushik, S. J., & Seiliez, I. (2010). Protein and amino acid nutrition and metabolism in fish: Current knowledge and future needs. *Aquaculture Research, 41*(3), 322–332.

Ketola, H. G. (1983). Requirement for dietary lysine and arginine by fry of rainbow trout. *Journal of Animal Science, 56*(1), 101–107.

Khalil, H. S., Momoh, T., Al-Kenawy, D., Yossa, R., BadrEldin, A. M., Roem, A., Schrama, J., & Verdegem, M. (2022). Metabolic growth, plankton selectivity, haemato-biochemical and intestinal morphometry of Nile tilapia (*Oreochromis niloticus*) fed a lysine-deficient diet in earthen ponds. *Aquaculture Reports, 24*, 101122.

Kim, K. I., McMillan, I., & Bayley, H. S. (1983). Determination of amino acid requirements of young pigs using an indicator amino acid. *British Journal of Nutrition, 50*(2), 369–382.

Kim, S. W., Less, J. F., Wang, L., Yan, T., Kiron, V., Kaushik, S. J., & Lei, X. G. (2019). Meeting global feed protein demand: Challenge, opportunity, and strategy. *Annual Review of Animal Biosciences, 7*, 221–243.

Kotzamanis, Y., Fawole, F. J., Brezas, A., Kumar, V., Fontanillas, R., Antonopoulou, E., Kouroupakis, E., & Ilia, V. (2021). Dietary lysine requirement of greater amberjack juvenile (*Seriola dumerili*, Risso, 1810). *Aquaculture Nutrition, 27*(6), 2107–2118.

Lee, S., Small, B. C., Patro, B., Overturf, K., & Hardy, R. W. (2020). The dietary lysine requirement for optimum protein retention differs with rainbow trout (*Oncorhynchus mykiss* Walbaum) strain. *Aquaculture, 514*, 734483.

Li, X. Y., Tang, L., Hu, K., Liu, Y., Jiang, W. D., Jiang, J., Wu, P., Chen, G. F., Li, S. H., Kuang, S. Y., & Feng, L. (2014). Effect of dietary lysine on growth, intestinal enzymes activities and antioxidant status of sub-adult grass carp (*Ctenopharyngodon idella*). *Fish Physiology and Biochemistry, 40*, 659–671.

Li, X., Wu, X., Dong, Y., Gao, Y., Yao, W., & Zhou, Z. (2019). Effects of dietary lysine levels on growth, feed utilization and related gene expression of juvenile hybrid grouper (*Epinephelus fuscoguttatus*♀ × *Epinephelus lanceolatus*♂). *Aquaculture, 502*, 153–161.

Li, X., Zheng, S., & Wu, G. (2021). Nutrition and functions of amino acids in fish. In *Amino Acids in Nutrition and Health: Amino Acids in the Nutrition of Companion, Zoo and Farm Animals* (pp. 133–168).

Liao, Y., Liu, B., Ren, M., Cui, H., Xie, J., & Ge, X. (2013). Effects of lysine on growth, physiological and biochemical indexes of blood and essential amino acids of serum in juvenile blunt snout bream (*Megalobrama amblycephala*). *Journal of Fisheries of China, 37*(11), 1716–1724.

Luo, Z., Liu, Y. J., Mai, K. S., Tian, L. X., Tan, X. Y., Yang, H. J., Liang, G. Y., & Liu, D. H. (2006). Quantitative L-lysine requirement of juvenile grouper *Epinephelus coioides*. *Aquaculture Nutrition, 12*(3), 165–172.

Mai, K., Zhang, L., Ai, Q., Duan, Q., Zhang, C., Li, H., Wan, J., & Liufu, Z. (2006). Dietary lysine requirement of juvenile Japanese seabass, *Lateolabrax japonicus*. *Aquaculture, 258*(1–4), 535–542.

Maiti, M. K., Sahu, N. P., Sardar, P., Garg, C. K., Varghese, T., Shamna, N., Deo, A. D., & Harikrishna, V. (2022). Dietary lysine requirement of juvenile Pacific white shrimp, *Litopenaeus vannamei* (Boone, 1931) reared in inland saline water of 10 ppt salinity. *Animal Feed Science and Technology, 291*, 115378.

Marcouli, P. A., Alexis, M. N., Andriopoulou, A., & Iliopoulou-Georgudaki, J. (2006). Dietary lysine requirement of juvenile gilthead seabream *Sparus aurata* L. *Aquaculture Nutrition, 12*(1), 25–33.

Michelato, M., de Oliveira Vidal, L. V., Xavier, T. O., de Moura, L. B., de Almeida, F. L. A., Pedrosa, V. B., Furuya, V. R. B., & Furuya, W. M. (2016). Dietary lysine requirement to enhance muscle development and fillet yield of finishing Nile tilapia. *Aquaculture, 457*, 124–130.

Muniz Junior, J. C. B., Haidamus, S. L., Machado, N. D. J. B., Dilelis, F., Brasil, R. J. M., Quaresma, D. V., & Lima, C. A. R. D. (2023). Requirement for digestible lysine in Nile tilapias (*Oreochromis niloticus*) with live weight between 500 and 600 g. *Acta Scientiarum. Animal Sciences*, *45*, e61547.

National Research Council. (2011). *Nutrient Requirements of Fish and Shrimp* (pp. 57–92). The National Academies Press. https://doi.org/10.17226/13039.

Nunes, A. J., Sá, M. V., Browdy, C. L., & Vazquez-Anon, M. (2014). Practical supplementation of shrimp and fish feeds with crystalline amino acids. *Aquaculture*, *431*, 20–27.

Pandey, A. K., Pandey, K., & Singh, L. K. (2020). Microbial production and applications of L-lysine. In *Innovations in Food Technology: Current Perspectives and Future Goals* (pp. 211–229).

Peres, H., & Oliva-Teles, A. (2008). Lysine requirement and efficiency of lysine utilization in turbot (*Scophthalmus maximus*) juveniles. *Aquaculture*, *275*(1–4), 283–290.

Powell, C. D., Chowdhury, M. K., & Bureau, D. P. (2015). Assessing the bioavailability of L-lysine sulfate compared to L-lysine HCl in rainbow trout (*Oncorhynchus mykiss*). *Aquaculture*, *448*, 327–333.

Prabu, E., Felix, N., Uma, A., & Praveenraj, J. (2020). Effects of dietary L-lysine supplementation on growth, body composition and muscle-growth-related gene expression with an estimation of lysine requirement of GIFT tilapia. *Aquaculture Nutrition*, *26*(2), 568–578.

Rachmawati, D., & Nurhayati, D. (2022). Effect of dietary lysine on the growth performance of *Pangasius hypophthalmus*. *Depik*, *11*(2), 111–116.

Ravi, J., & Devaraj, K. V. (1991). Quantitative essential amino acid requirements for growth of catla, *Catla catla* (Hamilton). *Aquaculture*, *96*(3, 4), 281–291.

Richard, L., Blanc, P. P., Rigolet, V., Kaushik, S. J., & Geurden, I. (2010). Maintenance and growth requirements for nitrogen, lysine and methionine and their utilization efficiencies in juvenile black tiger shrimp, *Penaeus monodon*, using a factorial approach. *British Journal of Nutrition*, *103*(7), 984–995.

Robinson, E. H., Wilson, R. P., & Poe, W. E. (1981). Arginine requirement and apparent absence of a lysine-arginine antagonist in fingerling channel catfish. *The Journal of Nutrition*, *111*(1), 46–52.

Rodehutscord, M., Borchert, F., Gregus, Z., & Pfeffer, E. (2000). Availability and utilisation of free lysine in rainbow trout (*Oncorhynchus mykiss*): 2. Comparison of L-lysine·HCl and L-lysine sulphate. *Aquaculture*, *187*(1, 2), 177–183.

Ruchimat, T., Masumoto, T., Hosokawa, H., & Shimeno, S. (1997). Quantitative methionine requirement of yellowtail (*Seriola quinqueradiata*). *Aquaculture*, *150*(1, 2), 113–122.

Singha, K. P., Shamna, N., Sahu, N. P., Sardar, P., HariKrishna, V., Thirunavukkarasar, R., Kumar, M., & Krishna, G. (2020). Feeding graded levels of protein to genetically improved farmed tilapia (GIFT) juveniles reared in inland saline water: Effects on growth and gene expression of IGFI, IGF-IR, and IGF-BPI. *Aquaculture*, *525*, 735306.

Singha, K. P., Shamna, N., Sahu, N. P., Sardar, P., Harikrishna, V., Thirunavukkarasar, R., Chowdhury, D. K., Maiti, M. K., & Krishna, G. (2021). Optimum dietary crude protein for culture of genetically improved farmed tilapia (GIFT), *Oreochromis niloticus* (Linnaeus, 1758) juveniles in low inland saline water: Effects on growth, metabolism, and gene expression. *Animal Feed Science and Technology*, *271*, 114713.

Small, B. C., & Soares, J. H. (1998). Estimating the quantitative essential amino acid requirements of striped bass *Morone saxatilis*, using fillet A/E ratios. *Aquaculture Nutrition*, *4*(4), 225–232.

Stites, W., Weldon, A., Reis, J., Walsh, S., Davis, R., Rhodes, M., & Davis, D. A. (2022). Quantitative lysine requirement for juvenile Florida pompano, *Trachinotus carolinus* fed plant-based diets. *Aquaculture*, *547*, 737548.

Tacon, A. G. J. (1992). *Nutritional Fish Pathology: Morphological Signs of Nutrient Deficiency and Toxicity in Farmed Fish* (FAO Fish Technical Paper No. 330). FAO.

Thu, T. T. N., Parkouda, C., De Saeger, S., Larondelle, Y., & Rollin, X. (2009). Protein level does not affect lysine utilization efficiency at marginal lysine intake in growing rainbow trout (*Oncorhynchus mykiss*) fry. *Aquaculture*, *288*(3, 4), 312–320.

Tibaldi, E., & Lanari, D. (1991). Optimal dietary lysine levels for growth and protein utilisation of fingerling sea bass (*Dicentrarchus labrax* L.) fed semipurified diets. *Aquaculture*, *95*(3, 4), 297–304.

Van Nguyen, M., Rønnestad, I., Buttle, L., Van Lai, H., & Espe, M. (2014). Imbalanced lysine to arginine ratios reduced performance in juvenile cobia (*Rachycentron canadum*) fed high plant protein diets. *Aquaculture Nutrition*, *20*(1), 25–35.

Walton, M. J., Cowey, C. B., & Adron, J. (1984). The effect of dietary lysine levels on growth and metabolism of rainbow trout (*Salmo gairdneri*). *British Journal of Nutrition*, *52*(1), 115–122.

Wang, L., Ye, L., Hua, Y., Zhang, G., Li, Y., Zhang, J., He, J., Liu, M., & Shao, Q. (2019). Effects of dietary DL-methionyl-DL-methionine (Met-Met) on growth performance, body composition and hematological parameters of white shrimp (*Litopenaeus vannamei*) fed with plant protein–based diets. *Aquaculture Research*, *50*(6), 1718–1730.

Wang, S., Liu, Y. J., Tian, L. X., Xie, M. Q., Yang, H. J., Wang, Y., & Liang, G. Y. (2005). Quantitative dietary lysine requirement of juvenile grass carp (*Ctenopharyngodon idella*). *Aquaculture*, *249*(1–4), 419–429.

Wang, W., Xu, Y., Chi, S., Yang, P., Mai, K., & Song, F. (2020). Dietary lysine regulates body growth performance via the nutrient-sensing signaling pathways in largemouth bass (*Micropterus salmoides*). *Frontiers in Marine Science*, *7*, 595682. https://doi.org/10.3389/fmars.2020.595682.

Webb, K. A., & Gatlin III, D. M. (2003). Effects of dietary protein level and form on production characteristics and ammonia excretion of red drum (*Sciaenops ocellatus*). *Aquaculture*, *225*(1–4), 17–26.

Wilson, R. P. (2003). Amino acid requirements of finfish and crustaceans. In J. P. F. D'Mello (Ed.), *Amino Acids in Animal Nutrition* (pp. 427–447). CABI Publishing.

Wilson, R. P., Harding, D. E., & Garling Jr, D. L. (1977). Effect of dietary pH on amino acid utilization and the lysine requirement of fingerling channel catfish. *The Journal of Nutrition*, *107*(1), 166–170.

Xie, F., Ai, Q., Mai, K., Xu, W., & Wang, X. (2012). Dietary lysine requirement of large yellow croaker (*Pseudosciaena crocea*, Richardson 1846) larvae. *Aquaculture Research*, *43*(6), 917–928.

Yun, H., Park, G., Ok, I., Katya, K., Hung, S., & Bai, S. C. (2016). Determination of the dietary lysine requirement by measuring plasma free lysine concentrations in rainbow trout (*Oncorhynchus mykiss*) after dorsal aorta cannulation. *Fisheries and Aquatic Sciences*, *19*, 1–7.

Zehra, S., & Khan, M. A. (2013). Dietary lysine requirement of fingerling *Catla catla* (Hamilton) based on growth, protein deposition, lysine retention efficiency, RNA/DNA ratio and carcass composition. *Fish Physiology and Biochemistry*, *39*, 503–512.

Zhang, C., Ai, Q., Mai, K., Tan, B., Li, H., & Zhang, L. (2008). Dietary lysine requirement of large yellow croaker, *Pseudosciaena crocea* R. *Aquaculture*, *283*(1–4), 123–127.

Zhou, X. Q., Zhao, C. R., Jiang, J., Feng, L., & Liu, Y. (2008). Dietary lysine requirement of juvenile Jian carp (*Cyprinus carpio* var. Jian). *Aquaculture Nutrition*, *14*(5), 381–386.

CHAPTER 5

Methionine Requirement of Different Fish Species

MANAS KUMAR MAITI[1] and KRISHNA PADA SINGHA[2*]

[1]*Fish Nutrition, Biochemistry and Physiology Division, ICAR–Central Institute of Fisheries Education, Versova, Mumbai, Maharashtra, India*

[2]*Department of Animal, Veterinary, and Food Sciences, Aquaculture Research Institute, University of Idaho, Moscow, Idaho, USA*

*Corresponding author

ABSTRACT

Methionine, an essential amino acid (EAA), is a linchpin in the complex web of metabolic pathways that govern fish nutrition. This chapter of the book delves into the multifaceted nature of methionine, shedding light on its vital functions, dietary requirements, and its profound implications in the context of sustainable aquaculture practices. Methionine is pivotal in protein synthesis, methylation reactions, and numerous other critical processes. This chapter emphasizes the importance of comprehending these roles in the context of maximizing growth and health in farmed fish and reducing environmental impacts. The chapter provides insights into the diverse dietary sources of methionine and explores the intricate interactions that methionine has with other amino acids (AAs). Moreover, the chapter discusses the exciting prospects in methionine research within aquaculture to enhance its efficiency and sustainability and minimize its ecological footprint. Conclusively, this chapter underscores the indispensable role of methionine in fish nutrition, influencing growth, health, and the sustainability of aquaculture practices. It aims to provide valuable insights

Fish Nutrition: Protein and Amino Acid Requirements. Imtiaz Ahmed, Ishtiyaq Ahmad, and Shabihul Fatma Sayed (Eds.)

DOI: 10.1201/9781998511570-5

and guidance to researchers, aquaculturists, and nutritionists operating in the dynamic and evolving field of aquaculture.

5.1 INTRODUCTION

Methionine, an essential amino acid (EAA), stands at the crossroads of growth, health, and sustainability in fish nutrition. Far more than a fundamental building block of proteins, methionine is a linchpin in the intricate metabolic pathways that govern a fish's well-being and productivity (Belghit et al., 2014). The journey of methionine in aquatic species traverses complex biochemical processes, from its vital role in methyl group transfers to its contributions to synthesizing critical molecules. In the rapidly evolving field of aquaculture, comprehending the multifaceted nature of methionine's functions and its intricate dietary requirements is paramount. This introduction invites us to explore methionine's significance in the context of fish nutrition, unraveled through various factors and framed against sustainable aquaculture practices.

The narrative of methionine in fish nutrition unfolds as a tale of metabolic intricacy and precision. As an indispensable amino acid (AA), methionine plays a multifaceted role beyond serving as a constituent of proteins. Its function as a catalyst in methyl group transfers, fundamental for regulating protein activity and essential biological processes, sets the stage for intricate metabolic cascades (Wang et al., 2023). The metabolic journey continues as methionine metamorphoses into cysteine and cystine, forming the building blocks of vital molecules. This, in turn, reflects the versatility of methionine, as it contributes its sulfur atom to cysteine's structure, fostering the synthesis of compounds like taurine and glutathione (GSH).

Numerous chapters remain to be written in the complex and interconnected world of methionine in fish nutrition. As we journey through this exploration, we will delve deeper into the empirical methods used to determine methionine requirements, diverse dietary methionine sources, intricate interactions between methionine and other AAs, and the potential ramifications of methionine deficiency in fish populations. Moreover, we will navigate the waters of various factors that influence methionine requirements, from species-specific nuances to environmental dynamics, thereby shaping the practices and future of aquaculture. This multifaceted narrative beckons researchers, aquaculturists, and nutritionists to understand methionine's pivotal role in shaping the world beneath the surface, where growth, health, and sustainability converge.

5.2 METHIONINE METABOLISM

The metabolism of methionine in fish, as in other animals, is a complex biochemical process that involves several steps and is crucial for various metabolic functions (Wang et al., 2023). The metabolism of methionine is categorized into three parts: the use of methionine in methyl group transfers, the transformation of methionine into cysteine and cystine, and finally, the breakdown of cysteine and its subsequent conversion into various specialized compounds. In transmethylation, methionine is first activated to S-adenosylmethionine (SAM), which then donates a methyl group to a receiving molecule. This step is key for various biological functions, including regulating protein activity and lifespan, as well as the synthesis of important molecules such as polyamines and plant hormones like ethylene. The conversion of homocysteine to cysteine involves a reaction with serine, producing cystathionine, which is then further processed to generate cysteine. This sequence showcases the versatility of methionine as it donates its sulfur atom to form cysteine, while serine contributes the remainder of the cysteine structure. Figure 5.1 illustrates how the methionine metabolism pathway produces cysteine, taurine, and GSH.

Furthermore, methionine works with vitamin B12 in the one-carbon metabolism pathway (Adam et al., 2022). Methionine acts as a methyl group donor, and after releasing this group, it is transformed into homocysteine. Vitamin B12 is vital in converting homocysteine back to methionine, illustrating the interconnected nature of these metabolic pathways and the importance of both methionine and vitamin B12. Methionine metabolism in fish is a vital process that affects not only growth and development but also the regulation of various metabolic pathways. It involves the interplay of enzymes, vitamins, and other cofactors to balance methionine's utilization, conversion to other sulfur-containing AAs, and its role in one-carbon metabolism.

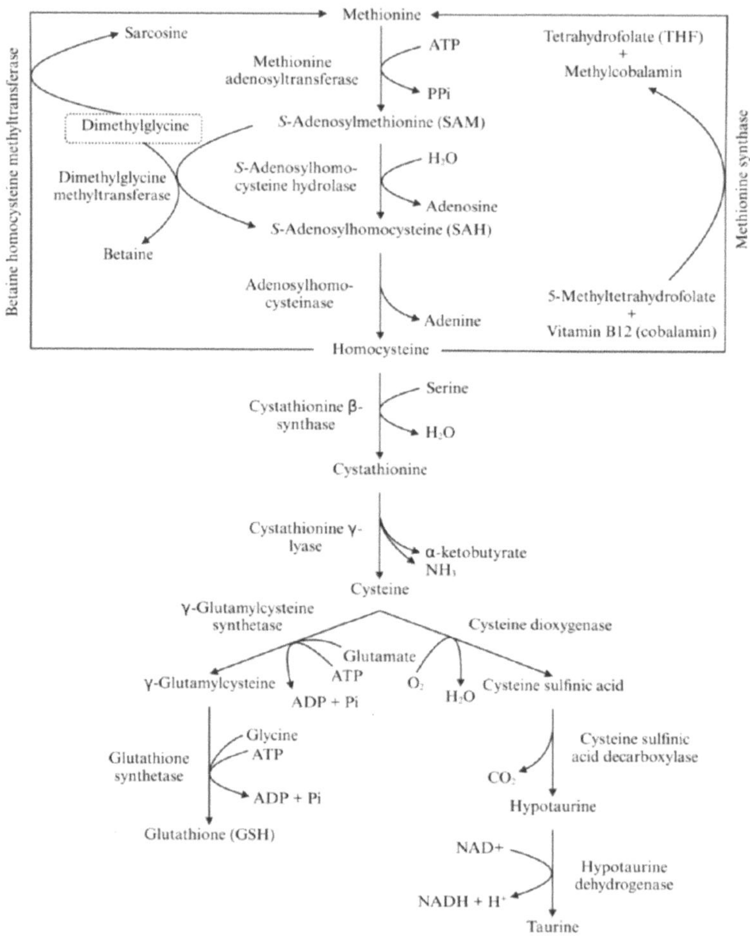

FIGURE 5.1 Overview of methionine metabolism pathway.

5.3 EMPIRICAL METHODS USED FOR METHIONINE REQUIREMENT

In fish nutrition, determining EAA requirements shares a common set of methodologies, which are applied consistently across various AAs, including lysine and methionine. These methodologies encompass a range of techniques, as discussed in detail in the "Lysine requirement of different fish species" (Chapter 4). Researchers use growth response assays, whole-body AA profiles, factorial models, blood and muscle AA levels, and AA oxidation studies to estimate EAA requirements for fish (NRC, 2011). This uniform approach for lysine and methionine requirements underscores the overarching methodology employed in fish nutrition research. Because of the notable similarity in these methodologies, we encourage readers to refer to the chapter "*Lysine requirement of different fish species*" for an in-depth understanding of how these techniques are applied to estimate EAA requirements for fish nutrition. Each method offers its own unique strengths and limitations, and ongoing refinements aim to enhance the precision and reliability of data for the aquaculture and nutrition industries. The unified use of these methodologies across different EAAs highlights their fundamental importance in the science of fish nutrition and underscores their consistency in the field.

5.4 DIETARY METHIONINE REQUIREMENT

Methionine is an EAA that plays a critical role in the growth, health, and overall performance of fish. Like humans and other animals, fish cannot synthesize methionine within their bodies and must obtain it from their diet (Pascual and Kanazawa, 1986). Besides, methionine participates in protein synthesis and other crucial physiological and metabolic functions (Lovell, 1989). Methionine is the precursor of SAM, which aids as a major methyl donor for various molecules such as amine, choline, polyamines, nucleic acids, and other metabolic intermediates (Bender, 2003). Methionine is involved in carnitine synthesis through methylation of L-lysine to trimethyl-lysine (Zhou et al., 2015). Understanding the methionine requirements of fish is paramount for aquaculturists and fish nutritionists as it directly impacts the success of aquaculture operations.

The dietary methionine requirement for fish varies depending on several factors, including the species, age, and specific nutritional needs. For different fish species, recommendations are typically expressed as a percentage of the total DP. Tilapia, a widely farmed freshwater fish, generally requires approximately 1.0–1.2% of the total DP to maintain optimal growth and health. On the other hand, salmonids, such as rainbow trout and Atlantic salmon, tend to have a higher methionine requirement, ranging from 1.0% to 1.4% of the total DP. Catfish, a popular species in aquaculture, typically require around 0.8–1.0% of the total DP for methionine. Similarly, common carp's methionine requirement is within the 0.8–1.0% range. Wang et al. (2023) reviewed that the dietary methionine requirement of fish ranges from 0.49% to 2.5% of the diet (1.49% to 4.7% of DP). It is important to note that various factors, including water temperature, life stage, methionine forms, dietary cystine and taurine levels, and the quality of protein sources in the fish feed can influence these figures. As a result, fish farmers and nutritionists must tailor their feeding strategies to meet the specific methionine requirements of their fish species. This ensures not only the promotion of growth but also the overall health and well-being of the fish population. Additionally, ongoing research and advancements in aquaculture nutrition continue to refine our understanding of the methionine requirements of different fish species. The quantitative methionine requirements of some commercially important finfish species are summarized in Table 5.1.

5.5 SOURCES OF DIETARY METHIONINE

Dietary methionine is a critical component of fish nutrition, impacting their growth, health, and overall performance. Both natural and synthetic methionine sources are utilized in fish feeds to meet their specific dietary requirements. However, the choice of methionine source in fish feeds depends on factors such as the farmed species, the feed formulation, and cost considerations (Nunes et al., 2014). Balancing natural and synthetic sources allows aquaculturists to create nutritionally balanced diets that promote healthy growth and efficient feed utilization while considering sustainability and cost-effectiveness. It's crucial for fish nutritionists and farmers to plan and monitor the methionine content in fish feeds carefully to ensure the health and productivity of their fish populations while minimizing environmental impact.

5.5.1 NATURAL SOURCES OF DIETARY METHIONINE

1. **Fishmeal:** Fishmeal, produced from the rendering of fish and fishery byproducts, is a highly digestible and valuable source of methionine in fish diets. It offers a balanced profile of EAAs and is particularly effective for carnivorous fish species.

TABLE 5.1 Dietary Methionine Requirement of Some Commercially Important Finfish

Species	Average Initial Weight (g)	Studied Methionine Range (g/kg Diet)	Salinity (ppt)	Temperature (°C)	Dietary Cystine Level (g/kg diet)	Optimum Methionine Requirement (g/kg Diet)	References
Yellowtail (*Seriola quinqueradiata*)	23.3	5.7–15.8	30.5 ± 1.9	24.8 ± 1.4	3.1	11.1	Ruchimat et al. (1997)
Milkfish (*Chanos chanos*)	8	5.9–16	32	28	7.5	10	Borlongan and Coloso (1993)
Hybrid grouper (*Epinephelus fuscoguttatus* × *Epinephelus lanceolatus*)	10.61 ± 0.05	5.6–26.2	Not recorded (NR)	27–29	6.9	14.5	Li et al. (2020)
Orange-spotted grouper (*Epinephelus coioides*)	13.25 ± 0.19	5.5–18.1	24–28	28–30.5	2.6	13.1	Lue et al. (2005)
European seabass (*Dicentrarchus Labrax*)	13.4	1.6–16.2	35	23 ± 0.5	4	8–9.1	Tulli et al. (2010)
Cobia (*Rachycentron canadum*)	11.61 ± 0.16	6.1–16.8	22–24	28–30	6.7	11.9	Zhou et al. (2006)
	150.9 ± 0.8	6.2–14.2	26–31	28–31	4.2	10.4–11.5	Wang et al. (2016)
Gilthead bream (*Sparus aurata*)	17	6.6–13.6	32 ± 1.6	22 ± 1.2	5.4	10.5	Guerreiro et al. (2023)
Large yellow croaker (*Pseudosciaena crocea*)	1.23±0.02	6.6–18.9	25–28	27–30	2.9	13.9–14.4	Mai et al. (2006)
Rainbow trout (*Oncorhynchus mykiss*)	11.2	2.3–8	Fresh Water (FW)	15	5	5.2	Kim et al. (1992)
	8.8	5.5–9.5	FW	14	3	5.5–7.5	Rumsey et al. (1983)
	505±6.5	2.5–9.5	FW	15±0.3	5	5.9–6.7	Bae et al. (2011)
Red drum (*Sciuenops ocellatus*)	0.9	3.5–18.5	6 ± 1	27 ± 2	1.2	9.4	Moon and Gatlin (1991)
Atlantic salmon (*Salmo salar)*	493	6.9–12.9	33	8–9	6	9–9.9	Espe et al. (2008)
	1066	6.4–12.3	30	9.8±0.4	NR	9.2	Sveier et al. (2001)
Rock fish (*Sebastes schlegeli*)	43.61	5.8–30.8	30–34	19.5	1.2	13.7	Yan et al. (2007)

TABLE 5.1 *(Continued)*

Species	Average Initial Weight (g)	Studied Methionine Range (g/kg Diet)	Salinity (ppt)	Temperature (°C)	Dietary Cystine Level (g/kg diet)	Optimum Methionine Requirement (g/kg Diet)	References
Asian sea bass (*Lates calcarifer*)	2.59 ± 0.08	6.2–12.6	31	26	3.1	13.4	Coloso et al. (1999)
Silver pompano (*Trachinotus blochii*)	5.47	5.1–15.2	17–20	27	5.3	11.6–11.8	Ebeneezar et al. (2020)
Golden pompano (*Trachinotus ovatus*)	12.4	8.6–14.5	28	28–31	2	10.6–12.7	Niu et al. (2013)
Catla (*Catla catla*)	0.65 ± 0.02	5–17.6	FW	25.3–28.5	0.6	12.2	Zehra and Khan (2016)
	1.2 ± 0.15	5–25	FW	27 ± 2	0.8	14.2	Ravi and Devaraj (1991)
Rohu (*Labeo rohita*)	0.66±0.1	4–14	FW	26–27	0.4	9.1–11.5	Abidi and Khan (2011)
	2.4 ± 0.05	3.2–16.5	FW	27.5–29	1.4	11.5	Murthy and Varghese (1998)
Mrigal (*Cirrhinus mrigala*)	0.50 ± 0.02	5–20	FW	26.0–27.5	1	12	Ahmed et al. (2003)
	0.19 ± 0.02	11–21	FW	27–28	8.5	16–16.9	Khan and Abidi (2013)
Common carp (*Cyprinus carpio*)	569 ± 65	6.1–13.4	FW	23	4.2	8.55	Schwarz et al. (1998)
Grass carp (*Ctenopharyngodon idella*)	10.10 ± 0.03	7.9–28.5	FW	27 ± 2	3.7	5.87	Su et al. (2018)
	178.47±0.36	2.54–15.11	FW	28.6±3.1	1.5	9.28–9.56	Fang et al. (2021)
	0.36 ± 0.01	3.3–12	FW	28	3.4	10.3	Ji et al. (2022)
Nile tilapia (*Oreochromis niloticus*)	4.3±0.01	5–25	FW	29.21±0.52	1.1	14–15	Mohammed et al. (2020)
	1.28	3.1–10.1	FW	28	0.4	8.1	Nguyen and Davis (2009)
GIFT tilapia	3.16 ± 0.01	4.9–19.8	10	27.67–28.23	3.2	12.1–14.1	Garg et al. (2023)
	17.51–19.55	5.2–12.7	FW	28	5	7.5–8	Yossa et al. (2023)
Pangasius catfish (*Pangasius bocourti*)	161.74±9.59	4.6–8.1	FW	25–29	4.3	6.3	Yuangsoi et al. (2016)

TABLE 5.1 *(Continued)*

Species	Average Initial Weight (g)	Studied Methionine Range (g/kg Diet)	Salinity (ppt)	Temperature (°C)	Dietary Cystine Level (g/kg diet)	Optimum Methionine Requirement (g/kg Diet)	References
Indian catfish (*Heteropneustes fossilis*)	4.33 ± 0.52	3–15.5	FW	24.20 ± 1.85	4	10.5	Ahmed (2014)
–	6.9±0.1	6.5–14	FW	27±0.8	4	10.7	Khan (2014)
Largemouth Bass (*Micropterus salmoides*)	37.88±0.40	6.1–15.9	FW	27±1	3	12.2	Chen et al. (2010)
Blunt snout bream (*Megalobrama amblycephala*)	3.34 ± 0.03	3.9–15.4	FW	26–31	2.2	8.5	Liao et al. (2014)

2. **Fish and Seafood Byproducts:** Processing waste and byproducts from fish and seafood, such as fish trimmings and shrimp heads, can be used as natural sources of methionine in fish feeds. These materials help reduce waste and enhance the sustainability of aquaculture operations.

3. **Poultry Byproducts:** Byproducts from poultry processing, including feather meal and poultry meal, are rich in methionine and can be included in fish feeds. These materials are processed to concentrate the AA content, making them valuable components of fish diet formulations.

4. **Crustacean Meal:** Meal derived from crustaceans like shrimp and crab shells contains methionine and is often used in feeds for fish species that require high-quality protein sources.

5. **Soybean Meal:** While primarily a plant-based ingredient, soybean meal (SB) is a valuable source of methionine in fish feeds. It is often fortified with synthetic methionine to ensure that it meets the specific dietary requirements of fish.

6. **Algae and Seaweed:** Certain species of algae and seaweed contain methionine and can be used as a natural source of this AA in feeds for herbivorous and omnivorous fish.

7. **Insect Meal:** Insect meal, derived from insects such as black soldier fly larvae, is gaining popularity as a sustainable source of protein and methionine in fish feeds, especially for species like tilapia and catfish.

5.5.2 SYNTHETIC SOURCES OF DIETARY METHIONINE

DL-methionine is the most prevalent crystalline methionine utilized in aqua-feed, composed of a mixture of D- and L-isomers of methionine. In contrast, newer crystalline forms of methionine have emerged, including DL-2-hydroxy-4-methylthiobutanoic acid (HMTBa), methionine hydroxy analogue (MHA), and DL-methionyl-DL-methionine (Met-Met). Within these compounds, the D-isomer of DL-methionine and both D- and L-isomers of HMTBa can be enzymatically converted into L-methionine (Nunes et al., 2014). Notably, MHA distinguishes itself by featuring a hydroxyl group on the alpha carbon, unlike DL-methionine, which exhibits an amino group resembling commonly employed organic acids. Additionally, Met-Met, a novel dipeptide with a methionine substitution, encompasses four different methionine stereoisomers, that is, LL-methionine-methionine, LD-methionine-methionine, DD-methionine-methionine, and DL-methionine-methionine (Wang et al., 2019).

5.6 INTERACTION OF METHIONINE WITH OTHER AMINO ACIDS

AA antagonism refers to the competitive relationship between different AAs when they are absorbed or utilized by cells or organisms. It often occurs in the context of nutrition and can have important implications for the health and functioning of living organisms. Although there is limited study on the aspect of methionine antagonism in fish, and most of the studies are about methionine and cysteine interaction in terms of requirement (Feng et al., 2020; Walton et al., 1982; Zehra et al., 2016), brief understanding can be drawn from these reported studies as follows:

1. **Methionine and cysteine metabolism:** Methionine is an EAA that serves as a precursor for cysteine synthesis. Methionine is converted into cysteine through a series of enzymatic reactions in the body. Cysteine, in turn, is used to synthesize proteins, GSH (an important antioxidant), and other bioactive molecules.
2. **Cysteine competition:** Methionine and cysteine share the same metabolic pathways and enzymes, so they can compete for utilization within the body. When there is an excess of methionine in the diet or body, it may lead to reduced uptake and utilization of cysteine. This competition can impact the synthesis of important molecules like GSH, which plays a crucial role in cellular defense against oxidative stress.

Besides methionine and cysteine interaction, methionine, taurine, and choline are interconnected through various metabolic pathways. Methionine provides the methyl groups needed for many biochemical reactions, including those requiring taurine and choline. Taurine synthesis in fish depends on the presence of sufficient methionine. In the form of betaine, choline can also act as a methyl donor, contributing to methionine metabolism. Supplementation of dietary methionine can reduce the dietary requirement for taurine, as reported in the case of juvenile Yellowtail Kingfish (*Seriola lalandi*) (Candebat et al., 2020). Whereas choline synthesis or transulfuration processes, which depend on methionine-derived SAM, might not be affected by taurine supplementation in the presence of low methionine in the case of Atlantic salmon (Espe et al., 2016). The interactions between these nutrients are complex and crucial for maintaining fish health and metabolism.

Although methionine AA antagonism is primarily associated with cysteine due to their shared metabolic pathways, there can be interactions or competition between methionine and other AAs to a lesser extent, depending on specific metabolic processes or dietary factors. For example, methionine and lysine are EAAs and can interact in certain metabolic pathways. Methionine and tryptophan are involved in serotonin synthesis, an important neurotransmitter. There can be competition between these two AAs for utilization in serotonin production. Methionine and threonine are both involved in protein synthesis. An imbalance in the intake of these AAs could affect overall protein production, although this antagonism is generally less pronounced than methionine and cysteine.

5.7 PHYSIOLOGICAL AND NUTRITIONAL FUNCTIONS OF METHIONINE

Methionine is among the first limiting AAs in many plant protein sources. Methionine participates in protein synthesis and other crucial physiological and metabolic functions. Again, being an EAA, dietary methionine at an optimum level can decrease the oxidation of other AAs by improving their use in protein synthesis. Consecutively, methionine is involved in carnitine synthesis through methylation of L-lysine to trimethyl-lysine, essential in transporting long-chain fatty acids for β-oxidation (Zhou et al., 2016). These roles of methionine are also supported by the lower levels of body lipid content in the higher methionine supplement groups *Heteropneustes fossilis* (Khan, 2014), *Oreochromis niloticus* (Michelato et al., 2013), *Pseudobagrus ussuriensis* (Wang et al., 2016), and *Clarias gariepinus* (Elesho et al., 2021). Methionine is the precursor of S-adenosylmethionine, which aids as a major methyl donor for various molecules such as amine, choline, polyamines, nucleic acids, and other metabolic intermediates

(Bender, 2003). Previous studies have shown that inadequate dietary methionine can lead to poor growth (Ahmad, 2014; He et al., 2017), reduced feed utilization (Yuan et al., 2011), decreased digestive and metabolic enzyme activities (Wang et al., 2016; Wu et al., 2017), altered immune system (Machado et al., 2021), and bilateral cataract in fish (Zhou et al., 2011). Methionine deficiency also causes a rapid reduction of different antioxidants, such as ascorbic acid, GSH, and vitamin E, in different fish tissues, which may result in irreversible oxidative stress (Li et al., 2009). It was also observed that excess dietary methionine would be stored in fish tissue and produce ketones and other toxic metabolites, thus undesirably affecting growth performance (Murthy and Varghese, 1997). Therefore, methionine should be in adequate amounts in the diet to smoothen different metabolic functions of the body, including protein synthesis, trans-methylation, and re-methylation reactions (Buono et al., 2001).

5.8 NUTRITIONAL DISORDERS OF METHIONINE DEFICIENCY

Methionine deficiency in fish can lead to various deficiency disorders and health issues. Methionine is an EAA, and its deficiency can have widespread effects on fish growth, metabolism, and overall health. A well-documented disorder due to methionine deficiency in the case of salmonids was found to be the occurrence of cataracts (Tacon et al., 1992). Based on the report of Séité et al. (2018), here are some of the potential deficiency disorders and problems that can occur in fish due to methionine deficiency:

1. **Impaired Growth and Protein Synthesis:** Methionine is a critical building block for protein synthesis. A deficiency can lead to modulated intracellular signaling pathways, resulting in increased protein degradation and inhibition of protein synthesis, impacting growth negatively.
2. **Increased Oxidative Stress:** Methionine plays a role in anti-oxidant defense by contributing to GSH and taurine formation. A deficiency can reduce GSH levels, increasing oxidative damage in tissues and impacting the overall oxidative status of the liver and other organs.

3. **Weakened Immune Function:** AAs, including methionine, are vital for the immune response. Deficiency can impair immune function and increase susceptibility to infectious diseases.

Methionine deficiency in fish can lead to many disorders and health issues affecting their growth, metabolism, and overall well-being. To ensure the health and productivity of farmed fish, it is imperative to provide diets that contain adequate levels of methionine, either through natural dietary sources or supplementation.

5.9 FACTORS AFFECTING METHIONINE REQUIREMENT

A variety of factors influence the methionine requirement of fish. Nutritionists and fish farmers must consider these variables when formulating fish feeds, accounting for species-specific needs, life stages, and changing environmental conditions to promote healthy growth and minimize waste in aquaculture practices (Arechavala-Lopez et al., 2022; Eriegha and Ekokotu, 2017). The following are some of the major factors that affect the dietary methionine requirement of fish.

1. **Fish Species:** Different fish species have varying methionine requirements due to differences in their metabolic and physiological processes. Carnivorous fish, for example, may have higher methionine requirements than herbivorous or omnivorous species. For instance, rainbow trout and Atlantic salmon have higher methionine requirements than species like tilapia or catfish (Lovell, 1991).
2. **Life Stage:** The methionine requirement of fish can change throughout their life stages. Juvenile fish, actively growing, typically have higher methionine requirements than adult fish. Broodstock fish, used for breeding purposes, may also have different methionine needs. Therefore, feed formulations must be adjusted to meet the specific requirements of fish at different life stages.
3. **Water Temperature:** Water temperature can influence the metabolic rate of fish, which, in turn, affects their methionine requirements. Fish in warmer water may have higher metabolic rates and

thus increased dietary methionine needs. Cold-water species may have different requirements from warm-water species.

4. **Dietary Protein Level:** The DP level plays a significant role in the methionine requirement of fish. Diets with high protein content may affect the optimal methionine-to-protein ratio, requiring higher methionine supplementation. Conversely, diets with lower protein levels may require a more balanced AA profile to optimize growth.
5. **Protein Source:** The source of protein in fish feeds can impact methionine availability. Ingredients like fishmeal, which is rich in methionine, can contribute to meeting fish's methionine requirements effectively. Diets primarily based on plant proteins may necessitate additional methionine supplementation.
6. **Water Quality:** The water quality in aquaculture systems, including factors such as pH, dissolved oxygen levels, and ammonia concentration, can affect the overall health of fish. Poor water quality may lead to stress and increased methionine requirements for maintaining health and immune function.
7. **Stress and Disease:** Stressful conditions, such as overcrowding, transportation, or exposure to diseases, can increase the methionine requirement of fish. Under such circumstances, methionine plays a vital role in supporting the immune system and maintaining fish health.
8. **Genetics:** Genetic differences among individual fish or fish strains can impact their methionine requirements. Some fish may have inherent variations in AA metabolism and utilization, affecting their response to dietary methionine levels.
9. **Dietary Antinutritional Factors:** Certain plant-based ingredients in fish feeds may contain antinutritional factors that can reduce the bioavailability of methionine. These factors may interfere with AA absorption, further necessitating adjustments in methionine supplementation.
10. **Environmental Sustainability:** As the aquaculture industry seeks to reduce its reliance on finite and ecologically sensitive resources like fishmeal, there is a growing interest in developing sustainable

feed ingredients. This shift can influence the methionine requirements of fish, as alternative protein sources, such as insect meal or single-cell proteins, may have different AA profiles.

5.10 FUTURE PERSPECTIVES

The future of methionine research in aquaculture is expected to revolve around customizing methionine requirements to meet the unique metabolic profiles of diverse fish species. This precision approach will likely enhance the growth, feed efficiency, and overall health outcomes for aquaculture systems. Advancements in understanding the molecular mechanisms of methionine's role in fish metabolism will inform the next generation of feed formulations. Recent studies delving into the interplay between methionine and other AAs have provided valuable insights into its multifaceted effects. This deeper understanding will facilitate the formulation of well-balanced diets, particularly as aquaculture adapts to new protein sources. The search for cost-effective and eco-friendly methionine sources is another ongoing trend in aquaculture nutrition. The industry is poised to explore new methods for enhancing methionine content in feed ingredients and discovering sustainable alternatives that can meet nutritional demands without straining the environment. These efforts are essential for balancing economic viability and ecological responsibility in aquaculture.

In the coming years, the methionine requirements for various fish species are expected to be fine-tuned further, considering species-specific differences and the diverse nutritional needs of fish at different life stages. Molecular insights will continue to drive the development of tailored diets that optimize growth, health, and feed utilization.

5.11 CONCLUSION

Methionine serves various vital functions in fish. It's a precursor for proteins and peptides, contributing to muscle and tissue growth. Additionally, it's a component of important molecules like *S*-adenosylmethionine, which is crucial for various metabolic processes, including DNA methylation and neurotransmitter synthesis. Fish feed formulations are carefully designed to balance natural and synthetic methionine sources to meet the specific needs of different fish species and their developmental stages. Inadequate methionine intake can result in growth retardation, poor feed conversion, and reduced immune function, ultimately impacting the overall health and performance of the fish. Ongoing research in aquaculture aims to refine our understanding of methionine's role and optimize its inclusion in fish diets. This research is essential for promoting better fish health and sustainable aquaculture practices, contributing to the world's growing demand for high-quality seafood.

KEYWORDS

- **amino acid metabolism**
- **essential amino acids**
- **fish feed formulation**
- **growth and health in aquaculture**
- **methionine in fish nutrition**
- **sustainable aquaculture**

REFERENCES

Abidi, S. F., & Khan, M. A. (2011). Total sulphur amino acid requirement and cystine replacement value for fingerling rohu, *Labeo rohita*: Effects on growth, nutrient retention and body composition. *Aquaculture Nutrition, 17*(2), e583–e594.

Adam, A. C., Saito, T., Espe, M., Whatmore, P., Fernandes, J. M. D. O., Vikeså, V., & Skjærven, K. H. (2022). Metabolic and molecular signatures of improved growth in Atlantic salmon (*Salmo salar*) fed surplus levels of methionine, folic acid, vitamin B6 and B12 throughout smoltification. *British Journal of Nutrition, 127*(9), 1289–1302.

Ahmed, I. (2014). Dietary amino acid L-methionine requirement of fingerling Indian catfish, *Heteropneustes fossilis* (Bloch-1974) estimated by growth and haemato-biochemical parameters. *Aquaculture Research, 45*(2), 243–258.

Ahmed, I., Khan, M. A., & Jafri, A. K. (2003). Dietary methionine requirement of fingerling Indian major carp, *Cirrhinus mrigala* (Hamilton). *Aquaculture International, 11*, 449–462.

Arechavala-Lopez, P., Cabrera-Álvarez, M. J., Maia, C. M., & Saraiva, J. L. (2022). Environmental enrichment in fish aquaculture: A review of fundamental and practical aspects. *Reviews in Aquaculture, 14*(2), 704–728.

Belghit, I., Skiba-Cassy, S., Geurden, I., Dias, K., Surget, A., Kaushik, S., Panserat, S., & Seiliez, I. (2014). Dietary methionine availability affects the main factors involved in muscle protein turnover in rainbow trout (*Oncorhynchus mykiss*). *British Journal of Nutrition, 112*(4), 493–503.

Bender, D. A. (2003). *Nutritional Biochemistry of the Vitamins* (pp. 5–50). Cambridge University Press.

Borlongan, I. G., & Coloso, R. M. (1993). Requirements of juvenile milkfish (*Chanos chanos* Forsskal) for essential amino acids. *The Journal of Nutrition, 123*(1), 125–132.

Buono, D. M., Wykes, L. J., Ball, R. O., & Pencharz, P. B. (2001). Dietary cysteine reduces the methionine requirement in men. *The American Journal of Clinical Nutrition, 74*(6), 761–766.

Candebat, C. L., Booth, M., Codabaccus, M. B., & Pirozzi, I. (2020). Dietary methionine spares the requirement for taurine in juvenile Yellowtail Kingfish (*Seriola lalandi*). *Aquaculture, 522*, 735090.

Chen, N., Ma, J., Zhou, H., Zhou, J., Qiu, X., Jin, L., & Lin, L. (2010). Assessment of dietary methionine requirement in largemouth bass, *Micropterus salmoides. Journal of Fisheries of China, 34*(8), 1244–1253.

Coloso, R. M., Murillo-Gurrea, D. P., Borlongan, I. G., & Catacutan, M. R. (1999). Sulphur amino acid requirement of juvenile Asian sea bass *Lates calcarifer. Journal of Applied Ichthyology, 15*(2), 54–58.

Ebeneezar, S., Vijayagopal, P., Srivastava, P. P., Gupta, S., Varghese, T., Prabu, D. L., Chandrasekar, S., Varghese, E., Tejpal, C. S., Anikuttan, K. K., & Sayooj, P. (2020). Optimum dietary methionine requirement of juvenile silver pompano, *Trachinotus blochii* (Lacepede, 1801). *Animal Feed Science and Technology, 268*, 114592.

Elesho, F. E., Sutter, D. A. H., Swinkels, M. A. C., Verreth, J. A. J., Kröckel, S., & Schrama, J. W. (2021). Quantifying methionine requirement of juvenile African catfish (*Clarias gariepinus*). *Aquaculture, 532*, 736020.

Eriegha, O. J., & Ekokotu, P. A. (2017). Factors affecting feed intake in cultured fish species: A review. *Animal Research International, 14*(2), 2697–2709.

Espe, M., Hevrøy, E. M., Liaset, B., Lemme, A., & El-Mowafi, A. (2008). Methionine intake affects hepatic Sulphur metabolism in Atlantic salmon, *Salmo salar. Aquaculture, 274*(1), 132–141.

Espe, M., Zerrahn, J. E., Holen, E., Rønnestad, I., Veiseth-Kent, E., & Aksnes, A. (2016). Choline supplementation to low methionine diets increases phospholipids in Atlantic salmon, while taurine supplementation had no effects on phospholipid status, but improved taurine status. *Aquaculture Nutrition, 22*(4), 776–785.

Fang, C. C., Feng, L., Jiang, W. D., Wu, P., Liu, Y., Kuang, S. Y., Tang, L., Liu, X. A., & Zhou, X. Q. (2021). Effects of dietary methionine on growth performance, muscle nutritive deposition, muscle fiber growth and type I collagen synthesis of on-growing grass carp (*Ctenopharyngodon idella*). *British Journal of Nutrition, 126*(3), 321–336.

Feng, H., Yi, K., Qian, X., Niu, X., Sun, Y., & Ye, J. (2020). Growth and metabolic responses of juvenile grouper (*Epinephelus coioides*) to dietary methionine/cystine ratio at constant sulfur amino acid levels. *Aquaculture, 518*, 734869.

Garg, C. K., Sardar, P., Sahu, N. P., Maiti, M. K., Shamna, N., Varghese, T., & Deo, A. D. (2023). Effect of graded levels of dietary methionine on growth performance, carcass composition and physio-metabolic responses of genetically improved farmed tilapia (GIFT) juveniles reared in inland saline water of 10 ppt. *Animal Feed Science and Technology, 298*, 115602.

Guerreiro, I., Basto-Silva, C., Magalhães, R., Martins, N., Fontinha, F., Peres, H., Mercier, Y., Nuez-Ortín, W. G., Mahmood, T., & Oliva-Teles, A. (2023). Reevaluation of the dietary methionine requirement of gilthead seabream (*Sparus aurata*) juveniles: A comparison of the efficiency of hydroxy methionine with DL-methionine as a dietary methionine source. *Aquaculture*, *576*, 739853.

He, J. Y., Tian, L. X., Lemme, A., Figueiredo-Silva, C., Gao, W., Yang, H. J., Han, B., Zeng, S. L., & Liu, Y. J. (2017). The effect of dietary methionine concentrations on growth performance of juvenile Nile tilapia (*Oreochromis niloticus*) fed diets with two different digestible energy levels. *Aquaculture Nutrition, 23*(1), 76–89.

Ji, K., Liang, H., Ge, X., Ren, M., Pan, L., & Huang, D. (2022). Optimal methionine supplementation improved the growth, hepatic protein synthesis and lipolysis of grass carp fry (Ctenopharyngodon idella). *Aquaculture, 554*, 738125.

Khan, M. A. (2014). Total sulfur amino acid requirement and cystine replacement value for fingerling stinging catfish, *Heteropneustes fossilis* (Bloch). *Aquaculture, 426*, 270–281.

Khan, M. A., & Abidi, S. F. (2013). Dietary methionine requirement of Indian major carp fry, *Cirrhinus mrigala* (Hamilton) based on growth, feed conversion and nitrogen retention efficiency. *Aquaculture Research, 44*(2), 268–281.

Li, P., Mai, K., Trushenski, J., & Wu, G. (2009). New developments in fish amino acid nutrition: Towards functional and environmentally oriented aquafeeds. *Amino Acids, 37*(1), 43–53.

Li, X., Mu, W., Wu, X., Dong, Y., Zhou, Z., Wang, X., Ma, L., Ye, B., & Geng, L. (2020). The optimum methionine requirement in diets of juvenile hybrid grouper (*Epinephelus fuscoguttatus* ♀× *Epinephelus lanceolatus* ♂): Effects on survival, growth performance, gut micromorphology and immunity. *Aquaculture, 520*, 735014.

Liao, Y. J., Ren, M. C., Liu, B., Sun, S. M., Cui, H. H., Xie, J., Zhou, Q. L., Pan, L. K., Chen, R. L., & Ge, X. P. (2014). Dietary methionine requirement of juvenile blunt snout bream (*Megalobrama amblycephala*) at a constant dietary cystine level. *Aquaculture Nutrition, 20*(6), 741–752.

Lovell, R. T. (1991). Nutrition of aquaculture species. *Journal of Animal Science, 69*(10), 4193–4200.

Lovell, T. (1989). *Nutrition and Feeding of Fish*. Springer New York, 205–222.

Luo, Z., Liu, Y. J., Mai, K. S., Tian, L. X., Yang, H. J., Tan, X. Y., & Liu, D. H. (2005). Dietary L-methionine requirement of juvenile grouper *Epinephelus coioides* at a constant dietary cystine level. *Aquaculture, 249*(1–4), 409–418.

Machado, M., Moura, J., Peixoto, D., Castro-Cunha, M., Conceição, L. E., Dias, J., & Costas, B. (2021). Dietary methionine as a strategy to improve innate immunity in rainbow trout (*Oncorhynchus mykiss*) juveniles. *General and Comparative Endocrinology, 302*, 683–690.

Mai, K., Wan, J., Ai, Q., Xu, W., Liufu, Z., Zhang, L., Zhang, C., & Li, H. (2006). Dietary methionine requirement of large yellow croaker, *Pseudosciaena crocea* R. *Aquaculture, 253*(1–4), 564–572.

Michelato, M., Furuya, W. M., Graciano, T. S., Vidal, L. V. O., Xavier, T. O., Moura, L. B. D., & Furuya, V. R. B. (2013). Digestible methionine + cystine requirement for Nile tilapia from 550 to 700 g. *Revista Brasileira de Zootecnia, 42*(1), 7–12.

Mohammed, F. A., Yousif, R. A., Hilal, F. M., Adam, R. A., & Ahmed, T. K. (2020). The effect of dietary methionine levels on growth, feed conversion and protein retention efficiency of Nile Tilapia (*Oreochromis niloticus*) fingerlings. *Nusantara Bioscience, 12*, 21–27.

Moon, H. Y., & Gatlin III, D. M. (1991). Total sulphur amino acid requirement of juvenile red drum, *Sciaenops ocellatus*. *Aquaculture, 95*(1, 2), 97–106.

Murthy, H. S., & Varghese, T. J. (1997). Dietary requirements of juveniles of the Indian major carp, *Labeo rohita*, for the essential amino acid lysine. *Israeli Journal of Aquaculture–Bamidgeh, 49*, 19–24.

Murthy, H. S., & Varghese, T. J. (1998). Total Sulphur amino acid requirement of the Indian major carp, *Labeo rohita* (Hamilton). *Aquaculture Nutrition, 4*(1), 61–65.

National Research Council (NRC). (2011). *Nutrient Requirements of Fish and Shrimp*. The National Academies Press. https://doi.org/10.17226/13039.

Nguyen, T. N., & Davis, D. A. (2009). Re-evaluation of total sulphur amino acid requirement and determination of replacement value of cystine for methionine in semi-purified diets of juvenile Nile tilapia, *Oreochromis niloticus*. *Aquaculture Nutrition, 15*(3), 247–253.

Niu, J., Du, Q., Lin, H. Z., Cheng, Y. Q., Huang, Z., Wang, Y., Wang, J., & Chen, Y. F. (2013). Quantitative dietary methionine requirement of juvenile golden pompano *Trachinotus ovatus* at a constant dietary cystine level. *Aquaculture Nutrition, 19*(5), 677–686.

Nunes, A. J., Sá, M. V., Browdy, C. L., & Vazquez-Anon, M. (2014). Practical supplementation of shrimp and fish feeds with crystalline amino acids. *Aquaculture, 431*, 20–27.

Pascual, F. P., & Kanazawa, A. (1986). Specific amino acid-free semi purified diets for *Penaeus monodon* juveniles. Kagoshima University Research Center for the Pacific Islands, Japan, 65–72.

Ravi, J., & Devaraj, K. V. (1991). Quantitative essential amino acid requirements for growth of catla, *Catla catla* (Hamilton). *Aquaculture, 96*(3, 4), 281–291.

Ruchimat, T., Masumoto, T., Hosokawa, H., & Shimeno, S. (1997). Quantitative methionine requirement of yellowtail (*Seriola quinqueradiata*). *Aquaculture, 150*(1, 2), 113–122.

Schwarz, F. J., Kirchgessner, M., & Deuringer, U. (1998). Studies on the methionine requirement of carp (*Cyprinus carpio* L.). *Aquaculture, 161*(1–4), 121–129.

Séité, S., Mourier, A., Camougrand, N., Salin, B., Figueiredo-Silva, A. C., Fontagné-Dicharry, S., Panserat, S., & Seiliez, I. (2018). Dietary methionine deficiency affects oxidative status, mitochondrial integrity and mitophagy in the liver of rainbow trout (*Oncorhynchus mykiss*). *Scientific Reports, 8*(1), 10151.

Su, Y. N., Wu, P., Feng, L., Jiang, W. D., Jiang, J., Zhang, Y. A., Figueiredo-Silva, C., Zhou, X. Q., & Liu, Y. (2018). The improved growth performance and enhanced immune function by DL-methionyl-DL-methionine are associated with NF-κB and TOR signaling in intestine of juvenile grass carp (*Ctenopharyngodon idella*). *Fish & Shellfish Immunology, 74*, 101–118.

Sveier, H., Nordås, H., Berge, G. M., & Lied, E. (2001). Dietary inclusion of crystalline D-and L-methionine: Effects on growth, feed and protein utilization, and digestibility in small and large Atlantic salmon (*Salmo salar* L.). *Aquaculture Nutrition, 7*(3), 169–181.

Tacon, A. G. J. (1992). *Nutritional Fish Pathology. Morphological Signs of Nutrient Deficiency and Toxicity in Farmed Fish* (FAO Fish Technical Paper No. 330). Rome: FAO.

Tulli, F., Messina, M., Calligaris, M., & Tibaldi, E. (2010). Response of European sea bass (*Dicentrarchus labrax*) to graded levels of methionine (total sulfur amino acids) in soya protein-based semi-purified diets. *British Journal of Nutrition, 104*(5), 664–673.

Walton, M. J., Cowey, C. B., & Adron, J. W. (1982). Methionine metabolism in rainbow trout fed diets of differing methionine and cystine content. *The Journal of Nutrition, 112*(8), 1525–1535.

Wang, L., Gao, C., Wang, B., Wang, C., Sagada, G., & Yan, Y. (2023). Methionine in fish health and nutrition: Potential mechanisms, affecting factors, and future perspectives. *Aquaculture*, 739310.

Wang, L., Ye, L., Hua, Y., Zhang, G., Li, Y., Zhang, J., He, J., Liu, M., & Shao, Q. (2019). Effects of dietary dl-methionyl-dl-methionine (Met-Met) on growth performance, body composition and hematological parameters of white shrimp (*Litopenaeus vannamei*) fed with plant protein–based diets. *Aquaculture Research, 50*(6), 1718–1730.

Wang, Z., Mai, K., Xu, W., Zhang, Y., Liu, Y., & Ai, Q. (2016). Dietary methionine level influences growth and lipid metabolism via GCN2 pathway in cobia (*Rachycentron canadum*). *Aquaculture, 454*, 148–156.

Wu, P., Tang, L., Jiang, W., Hu, K., Liu, Y., Jiang, J., Kuang, S., Tang, W., Zhang, Y., Zhou, X., & Feng, L. (2017). The relationship between dietary methionine and growth, digestion, absorption, and antioxidant status in intestinal and hepatopancreatic tissues of sub-adult grass carp (*Ctenopharyngodon idella*). *Journal of Animal Science and Biotechnology, 8*(1), 1–14.

Yan, Q., Xie, S., Zhu, X., Lei, W., & Yang, Y. (2007). Dietary methionine requirement for juvenile rockfish, *Sebastes schlegeli*. *Aquaculture Nutrition, 13*(3), 163–169.

Yossa, R., Fatan, N. A., & Palanivelu, K. (2023). Revaluating the dietary methionine requirements in an improved strain of Nile tilapia (*Oreochromis niloticus*). *Heliyon, 9*(3).

Yuan, Y. C., Gong, S. Y., Yang, H. J., Lin, Y. C., Yu, D. H., & Luo, Z. (2011). Effects of supplementation of crystalline or coated lysine and/or methionine on growth performance and feed utilization of the Chinese sucker, *Myxocyprinus asiaticus*. *Aquaculture, 316*(1–4), 31–36.

Yuangsoi, B., Wongmaneeprateep, S., & Sangsue, D. (2016). The optimal dietary DL-methionine on growth performance, body composition and amino acids profile of Pangasius catfish (*Pangasius bocourti*). *Aquaculture, Aquarium, Conservation & Legislation, 9*(2), 369–378.

Zehra, S., & Khan, M. A. (2016). Total Sulphur amino acid requirement and maximum cysteine replacement value for methionine for fingerling *Catla catla* (Hamilton). *Aquaculture Research, 47*(1), 304–317.

Zhou, F., Xiao, J. X., Hua, Y., Ngandzali, B. O., & Shao, Q. J. (2011). Dietary l-methionine requirement of juvenile black sea bream (*Sparus macrocephalus*) at a constant dietary cystine level. *Aquaculture Nutrition, 17*(5), 469–481.

Zhou, L. L., Dan, X. M., Li, Y. W., Li, X. X., Chen, X. L., Li, Y., & Gan, L. (2015). Effect of reducing 3.2% dietary protein level on the growth performance and immunity of Nile tilapia (*Oreochromis niloticus*) with supplementation of multi amino acids. *AquaDocs, 27*, 43–50.

Zhou, Q. C., Wu, Z. H., Tan, B. P., Chi, S. Y., & Yang, Q. H. (2006). Optimal dietary methionine requirement for juvenile cobia (*Rachycentron canadum*). *Aquaculture, 258*(1–4), 551–557.

Zhou, X., He, L., Wan, D., Yang, H., Yao, K., Wu, G., Wu, X., & Yin, Y. (2016). Methionine restriction on lipid metabolism and its possible mechanisms. *Amino Acids, 48*(7), 1533–1540.

CHAPTER 6

Valine Requirement of Different Fish Species

MUSTAPHA ABA*

Independent Researcher and Aquaculture Expert, Fish Nutrition, Rabat, Morocco

*Corresponding author

ABSTRACT

With the expansion of aquaculture to ensure global food security over the last few years, research has been able to unveil the functional roles of essential amino acids (EAAs) contained in fishmeal, which is becoming expensive and scarce, in optimizing the effective nutrition of farmed fish species, which has become a major concern for the commercial success of aquaculture. Valine is considered one of the EAAs in fish, and its content in aquaculture feeds depends on the type of ingredients used. Valine is a glucogen, while the synthesis of participating proteins is the most important physiological function of this AA, so it is closely linked to fish growth and plays an important role in regulating growth. It also plays an important role in fish immunity as it acts on intestinal absorption. Several studies have shown that a deficient or excessive level of valine in diets negatively affects the growth performance of fish. Valine cannot be synthesized *de novo* by fish and other animals and must be obtained by breaking down proteins in the diet, particularly fishmeal. To overcome this problem, dietary supplementation with synthetic L-valine is used as an alternative to reduce dependence on fishmeal in fish diets. Given the importance of valine in animal and fish production, its production has become industrialized through synthesis using microorganisms.

Fish Nutrition: Protein and Amino Acid Requirements. Imtiaz Ahmed, Ishtiyaq Ahmad, and Shabihul Fatma Sayed (Eds.)

DOI: 10.1201/9781998511570-6

6.1 INTRODUCTION

Amino acids (AAs) are essential sources of nitrogen for live animals; therefore, they are of prime interest in the field of aquaculture nutrition, especially in view of the limitation of fisheries that provides fish meal whose profile is very well balanced in EAAs (Glencross, 2020). In general, AAs can be grouped according to the chemical nature of their side chains, transport affinities, role in animal nutrition, or the catabolic fate of their carbon skeletons (Falco et al., 2020). The success of aquaculture, particularly in intensive aquaculture systems, depends primarily on food, as proteins and AAs are essential and important components of fish feed due to their significant role in the structure, the formation, and metabolism of all living organisms (Lee et al., 2001, 2003; Ma et al., 2019; National Research Council, 2011).

Fish feed accounts for a significant share of the production cost, about 50–70% (Pereira et al., 2022), and is a powerful lever to control the level of performance and the development of nutritionally efficient and cost-effective diets for farmed fish species, and plays a fundamental role in the commercial success of aquaculture (Ahmad and Khan, 2006; Ahmad et al., 2021; Peres and Oliva-Teles, 2009). The quality of aquaculture feed ensures the sustainability of the aquaculture sector through its three societal, environmental, and economic pillars (Le Gouvello and Simard, 2017); therefore, to optimize the nutrient utilization efficiency of aquaculture feeds, it is important to feed each fish with appropriate levels of protein and AAs, especially as aquaculture faces a strategic challenge of improving feed adequacy and supporting the overall increase in fish production.

Combining a rational diet with a balanced AA composition for fish is one of the main problems of global aquaculture (Li and Wu, 2020). Like other vertebrate animals, fish do not have the ability to synthesize certain AAs, which must be provided by food and are said to be EAAs (Guillaume et al., 2001). For fish, the following AAs are classified as essential: arginine, phenylalanine, histidine, isoleucine, leucine, lysine, methionine, threonine, tryptophan, and valine (NRC, 2011). Valine, like leucine and isoleucine, belongs to the branched-chain AA group, which contains approximately 35% of the AAs essential to muscle proteins (Harper et al., 1984). Valine is an EAA and potentially limiting in conventional fish feeds (Rodrigues et al., 2020), and it plays a significant role in regulating fish growth (Ahmed et al., 2019; Mokrani et al., 2020). In 1901, valine was first isolated from casein by Emil Fischer. Its name comes from valerian acid, which is named after the plant "valerian" *Valeriana officinalis*, because of the presence of valeric acid in the roots of the plant, which allowed him to receive the Nobel Prize in Chemistry in 1902 (Yang et al., 2023). Valine has the following chemical formula: $C_5H_{11}NO_2$, and the following figure shows its chemical structure.

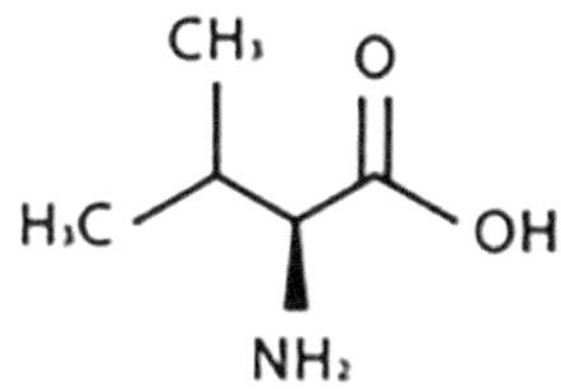

Valine

Since valine belongs to the group of branched-chain AAs, it is an essential hydrophobic AA, making it an important substrate for the formation of protein structure of cell membranes (Brosnan and Brosnan, 2006). Valine also plays a major role in the formation of various metabolites and is usually concentrated in skeletal muscle tissue (Khan and Abidi, 2007). The main role of valine is to participate in the synthesis of proteins in the body. However, its important physiological functions are the maintenance and growth of muscle tissue (Ahmad et al., 2021; Suryawan et al., 2011), tissue repair with nitrogen balance in fish (Ahmad et al., 2021; Brosnan and Brosnan, 2006; Rahimnejad and Lee, 2013), as it is directly involved in the synthesis of glutamine and alanine (Wu, 2009).

6.2 VALINE METABOLISM

Fish nutrition is one of the main areas studied by aquaculture researchers. In fish farming, protein and AAs are the major and most expensive macronutrients in diets for fish, knowing that aquaculture feeds account for about 50% of the cost of production (Craig 2017; Rawles et al., 2018). Fish, like other vertebrate animals, are unable to synthesize the EAAs; these latter must be provided by the diet (Guillaume et al., 2001; NRC, 2011) and include: arginine, phenylalanine, histidine, isoleucine, leucine, lysine, methionine, threonine, tryptophan, and valine. This group of EAAs has multiple and important physiological functions beyond their roles as basic components of proteins (Wu et al., 2014). In addition, branched-chain amino acids (BCAAs), including leucine (Leu), isoleucine (Ile), and valine (Val), play an essential role in regulating energy homeostasis, nutritional metabolism, gut health, immunity, and disease in humans and animals (Nie et al., 2018), and assessment of the need for BCAAs, of which valine in fish is essential because of its central role in promoting muscle growth and maintaining various physiological and metabolic functions (Doi et al., 2003; Lei et al., 2012; Li et al., 2009; Negro et al., 2008; Nishimura et al., 2010; Teodoro et al., 2012). Unlike most other oxidized AAs in the liver, BCAAs are mainly oxidized in skeletal muscle but not in the liver (Ahmed and Khan 2006; Ahmad et al., 2021; Lynch and Adams, 2014).

The main role of valine is to participate in the synthesis of body proteins, as it performs other important physiological functions, including the maintenance and growth of muscle tissue (Suryawan et al., 2011). Therefore, valine plays an important role in the formation of various metabolites and is generally concentrated in skeletal muscle tissue (Khan and Abidi, 2007), and acts in the synthesis of glutamine and alanine (Wu, 2009). In addition, it is important to meet the requirements of valine, a branched-chain AA, as it is involved in the synthesis of other EAAs. AAs play a key role in metabolic pathways for body maintenance, growth, dietary intake, nutrient utilization, immunity, behavior, larval metamorphosis, reproduction, and resistance to environmental stress and pathogenic organisms in fish (Baki et al., 2017; Li, Mai, Trushenski, and Wu, 2008; Namulawa, Rutaisire, and Britz, 2012).

Recent research on the 10 EAAs, including valine, has shown the quantitative dietary requirements and their effects on fish nutrition and metabolism (Kan et al., 2023; Li et al., 2023), mainly valine, whose metabolism performs different functions, producing multiple necessary products needed by animals, including aquatic animals (Wang et al., 2014). In aquaculture, studies on the metabolism and functions of AAs are important for the development of low-fishmeal diets by including plant-derived ingredients and crystalline AAs (Li et al., 2021), which will help in the manufacture of more sustainable aquaculture feeds. It should be noted that the most active enzymatic system for the oxidation of branched-chain AAs, including valine, is mainly found in skeletal muscle, while for other AAs, liver tissue is their main oxidation site (Neinast et al., 2019).

The branched-chain AA valine is derived from the degradation of dietary proteins (DPs) contained in aquaculture feeds and undergoes metabolism comprising two complementary processes: catabolism (degradation) and anabolism, using metabolic intermediates as substrates for AA valine biosynthesis. There is a similarity between the metabolism of BCAA in fish and that of mammals (Ahmad et al., 2021). Knowing that during metabolism, valine undergoes two important processes: anabolism, which leads to the synthesis of valine, and catabolism, which leads to the catabolism of valine.

6.2.1 VALINE CATABOLISM

The metabolism of valine, leucine, and isoleucine begins with the degradation of these branched-chain AAs, undergoing transamination and oxidative decarboxylation by shared enzymes (Biswas et al., 2019; Mann et al., 2021; Manoli and Venditti, 2016). Valine (as leucine and isoleucine) metabolism (Figure 6.1) is initiated by the degradation of this branched-chain AA, undergoing transamination catalyzed by branched-chain aminotransferase (BCAT) and oxidative decarboxylation by the branched-chain α-ketoacid dehydrogenase complex (BCKD) (Holeček, 2020). The reversible transamination of valine by BCAT yields 2-keto-isovalerate (KIV) (Ahmed and Khan, 2006; Ferrando et al., 1995; Kainulainen et al., 2013), which undergoes irreversible oxidative decarboxylation by BCKD to produce isobutyryl-CoA, which then undergoes dehydrogenation by the enzyme isobutyryl-CoA dehydrogenase to produce 3-methacrylyl-CoA, which undergoes enoyl CoA hydratase transformation to 3-hydroxyisobutyryl CoA, which in turn converts to 3-hydroxyisobutyrate under the reaction of the enzyme 3-hydroxyisobutyryl CoA hydrolase. The latter, under the action of 3-hydroxyisobutyrate dehydrogenase, gives methylmalonate semialdehyde (methylmalonate semialdehyde dehydrogenase generates propionyl-CoA from methylmalonate semialdehyde) (Adeva-Andany, 2017). Valine is catabolized into propionyl-CoA that is converted into d-methylmalonyl-CoA by propionyl-CoA carboxylase via fatty acids β-oxidation. D-methylmalonyl-CoA is subsequently racemized into l-methylmalonyl-CoA by methylmalonyl-CoA racemase (Wongkittichote et al., 2017). The reaction catalyzed by methylmalonyl mutase eventually produces succinyl-CoA, which finally is integrated into the tricarboxylic acid (TCA) cycle (Ahmad et al., 2021; Zhang et al., 2017), demonstrating the glucogenic role of valine (Dimou et al., 2022).

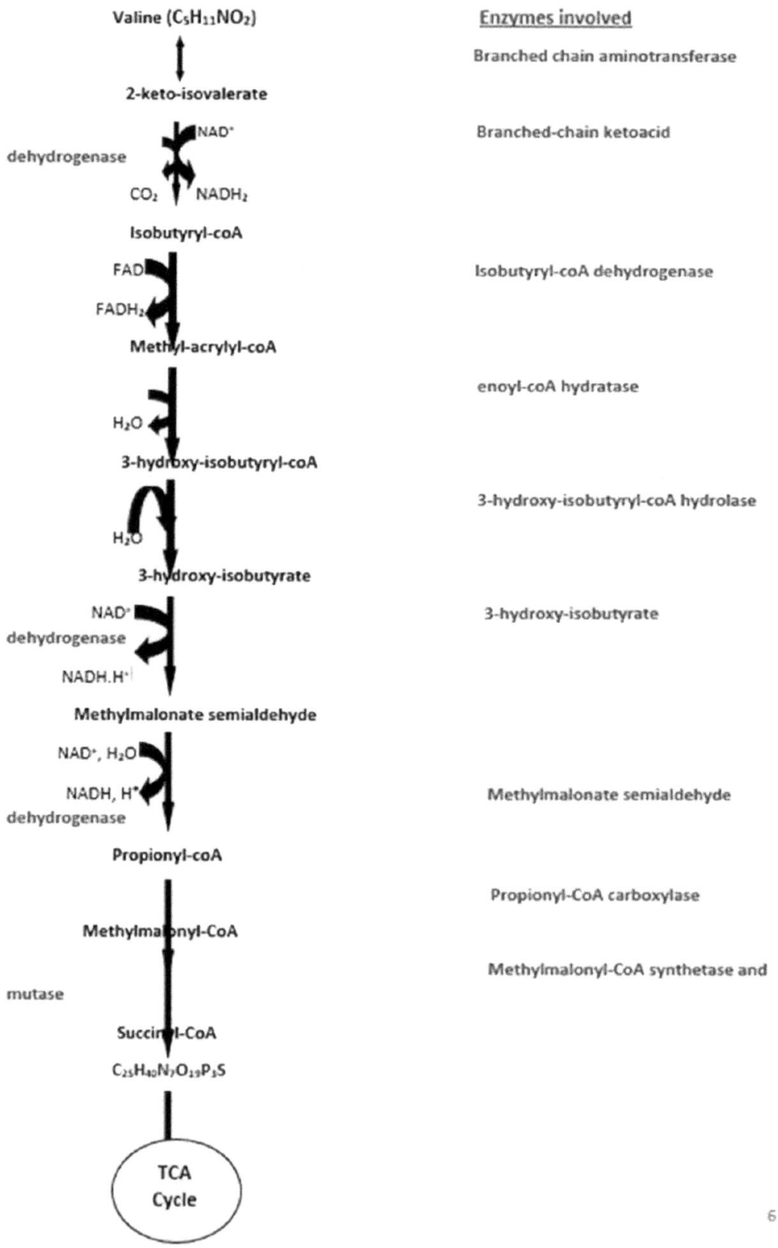

FIGURE 6.1 The main steps of valine catabolism, conversion to succinyl-CoA.

Source: Adapted from Wongkittichote et al. (2017); Jakubowski and Flatt (2023); Adeva-Andany et al. (2017).

The branched-chain hydrophobic AA, valine, can be converted to succinyl-CoA, which can be converted to α-ketoglutarate in the Krebs cycle in net fashion, and hence is a glucogenic AA (Prabhu and Shetty, 2014).

6.2.2 VALINE ANABOLISM

Anabolism, often called biosynthesis, is the production of more complex molecules used in the structure and function of the cell from simpler organic molecules (Jakubowski and Flatt, 2023). However, the branched-chain AAs are essentially anabolic, aiding in regulating the translation and initiation of protein synthesis in various tissues, and are thus necessary for protein synthesis and muscle growth (Shimomura; Yamamoto; Bajotto, 2006). Valine has important physiological functions in muscle tissue growth and protein synthesis in the body (Suryawan et al., 2011). In addition, Chung and Baker (1992) describe that valine deficiency works by reducing the use of other limiting AAs for protein deposition. This has been proven by Rodrigues et al. (2018), who have been able to demonstrate in their research that valine has an inducing effect on protein synthesis because the increase in valine concentrations in experimental diets contributed to the development of muscle fibers.

Dietary AA are required by animals primarily for maintenance (including the synthesis of nonprotein metabolites) and protein accretion. Like the other branched-chain AAs, Val has anabolic effects on skeletal muscle; however, a lack of Val, as with all EAAs, slows protein synthesis (Anthony et al., 2001). BCAAs stimulate protein synthesis *in vitro* preparations of skeletal muscle (Nie et al., 2018). Circulating BCAAs have been shown to act as potent nutrient signals in muscle where they induce protein synthesis (Shimomura et al., 2006). The provision of EAAs such as valine to fish in the diet is paramount, as these animals are unable to synthesize them, but the synthesis of valine takes place in microorganisms and plants (Neinast et al., 2019). This synthesis is initiated by glycolytic pyruvate (Wang et al., 2023).

The valine biosynthesis pathway comprises four stages (Figure 6.2). Pyruvate is the starting point of the valine biosynthetic pathway, which originates from the glycolytic pathway (Chandel, 2021). The pyruvate is converted to α-acetolactate by acetohydroxy acid synthase. In the second step, acetohydroxy acid isomeroreductase catalyzes the conversion of α-acetolactate to α, β-dihydroxyisovalerate, via auto-displacement of methyl groups. Dihydroxy-acid dehydratase is the enzyme for the third step in which α-ketoisovalerate is formed from α, β-dihydroxyisovalerate (Wang et al., 2023). It is worth noting that this enzyme is inhibited by valine (Leyval et al., 2003). In the last step, α-ketoisovalerate is catalyzed by branched-chain amino acid transaminase (BCAT) to synthesize valine (Wang et al., 2023).

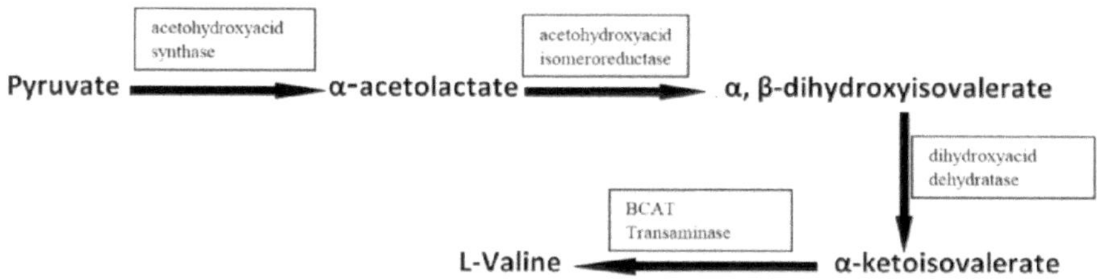

FIGURE 6.2 The main steps of valine synthesis.

Source: Adapted from Wang et al. (2023).

6.3 EMPIRICAL METHODS USED FOR VALINE REQUIREMENT

Fish nutrition plays an essential and important role in the expansion of the aquaculture sector, as it influences not only production costs, but also fish growth, health, and waste generation (Hardy and Kaushik, 2021; Prabu et al., 2017; Turchini et al., 2019). In aquaculture nutrition, protein is considered the most expensive component of fish feed (Boisen, 2003), knowing that fish do not have a dietary requirement for crude protein, but need a well-proportioned dietary source of certain AAs, which are indispensable and crucial for optimum growth (Miles and Chapman, 2007). So, to ensure optimal fish development with good fish flesh quality, it is essential to provide them with a feed containing all the necessary AAs, both in quality with the presence of all the EAAs, and in quantity (Hua et al., 2019; Lall and Anderson, 2005; Li et al., 2021), which promotes the optimization of the AA composition of the aquaculture feed according to the needs of aquatic animals, but also for a better improvement in the use of proteins and also to reduce nitrogen excretion in the environment.

The availability of information on individual EEA requirements is very important for formulating cost-effective feeds for fish nutrition (Hua et al., 2019; Trushenski et al., 2006; Yun et al., 2015), which also makes it possible to replace or supplement animal protein sources with EAAs through the adoption of these alternative protein sources (through plant, insect, and microbial meals (Macusi et al., 2023)) in order to reduce production costs and contribute to the healthy development and increased sustainability of the aquaculture industry, especially valine, which is a limiting AA (Diógenes et al., 2016; Mansano et al., 2020) and potentially limiting in conventional fish feeds (Rodrigues et al., 2020). Valine participates in the synthesis of body proteins, in addition to performing other important physiological functions, including the repair and growth of muscle tissue and the maintenance of nitrogen balance in the body (Silva et al., 2022; Suryawan et al., 2011). In this context, the determination of valine needs is important given its different roles in fish nutrition.

The need to provide essential dietary AAs for growth and maintenance is important, especially in young, fast-growing fish whose quantitative requirements are much higher (Coewy, 1994). The determination of AA requirements for each of the dietary EAA is considered the fundamental basis of AA nutrition of fish (Lall and Anderson, 2005). Valine has been identified as an EAA for fish (Nose, 1979) and has been reported to play important roles in various physiological functions in animals, including protein synthesis, energy production, immune functions, and tissue repair (Brosnan and Brosnan, 2006; Castillo and Gatlin, 2018; Zhang et al., 2017). It should be remembered that the first definitive studies on protein and AA nutrition of fish were conducted by Halver and co-workers in the late 1950s and early 1960s in Chinook salmon (*Oncorhynchus tshawytscha*), used to determine the qualitative AA requirements of the Chinook salmon (Halver et al., 1957), and that two methods have been used to determine the nutritional requirements of animals: the factorial method and the empirical method (Hauschild et al., 2010). According to Furuya et al. (2004), the factorial method determines the total requirement of a nutrient as the sum of maintenance and production requirements; this method is used to determine energy, protein, and macro-mineral requirements, while the empirical method determines the amount of a nutrient needed by evaluating a population's response to the gradual increase of this nutrient during the study.

A number of empirical methods have been applied to establish the quantitative dietary EAA requirements of fish (Bureau and Encarnação, 2006). These methods are based on studies in which all the AAs in the diet are fed to the fish at appropriate levels, with the exception of the AA under study. Two important empirical methods are operated to determine valine requirements: the dose-response method and the deletion method.

6.3.1 DOSE-RESPONSE METHOD

For the estimation of the ideal profile of AAs, some methods have been formerly tested in fish, such as the dose–response (Diógenes et al., 2016; Khan et al., 2020; NRC, 2011; Peres and Oliva–Teles, 2009; Wilson, 2002; Favero et al., 2009). In the dose-response studies carried out for the formulation of experimental diets, two methods are used in animal and fish nutrition. The first is the quality supplementation method, commonly practiced in animal nutrition, including fish (Abimorad et al., 2007; Bicudo et al., 2010; Fernandes et al., 2000, 2001; Kan et al., 2019; Klein et al., 2014; Kpogue et al., 2013; Rawles et al., 2013; Sankian et al., 2017; Wang et al., 2017), which is widely exploited for the determination of AA requirements and involves the progressive addition of an AA test to a basal diet deficient in the limiting AA (D'Mello, 1982). However, this method has been criticized for successively increasing levels of limiting AAs and excessive use of synthetic AAs that can alter the balance and relationship with other AAs in the diet, compromising the animals' productive responses (Siqueira et al., 2013). The other dose-response method is the diet dilution method, which has been well studied in animals such as pigs and poultry and is an alternative to the supplementation method (Fisher and Morris, 1970; Gous and Morris, 1985; Gous et al., 2018), but in fish nutrition, it has been reported in a few studies such as Abboudi et al. (2007) and Liebert and Benkendorff (2007). The dose-response method remains criticized because the other fixed AAs can be underestimated compared to what is tested, which leads to performance limitations (Khan et al., 2019).

For the estimation of valine requirements in animals and fish in several studies, the dose-response method was used, considered in principle as a method for determining requirements (Cowey, 1995). Dietary requirements for valine were used for this method in trout by Rodehutscord et al. (1997), in Indian major carp by Abidi and Khan, and for juvenile blunt bream by Ren et al. (2015). One of the disadvantages of the dose-response method is that feeding trials are costly and take several weeks or months before appreciable responses are observed.

6.3.2 DELETION METHOD

The second method of estimation of EAA requirements can be determined experimentally by the deletion technique, initially developed for pigs and recently successfully applied to several fish species, even for salmonids (Green and Hardy, 2002; Lall and Anderson, 2005; Rollin et al., 2003), juvenile gilthead bream, *Sparus aurata* (Peres and Oliva-Teles 2009), juvenile silver-black moue, *Sparidentex hasta* (Marammazi et al., 2017), and juvenile Nile tilapia (Diógenes et al., 2016), Nile tilapia (Mansano et al., 2020), Pacu (Boaratti et al., 2020; Oliveira et al., 2021). The concept of this technique is based on the concept that each EAA is equally limiting to protein accretion (Wang and Fuller, 1989), and therefore changes in N gain due to partial removal of one EAA in turn in an otherwise balanced diet will provide an estimation of the optimum EAA profile (Peres and Oliva-Teles, 2009). Among the advantages of this deletion method are that it allows the determination of the optimum EAA profile in a single trial by using the fish of the same lot, the same diet, the same rearing environment, and the same trial period, thus avoiding the variation in the estimates (Mansano et al., 2020).

So, given the advantages of the deletion method, it has been used as an alternative to determine the optimal valine ratio in fish diets. Several studies have used this method to determine the valine requirements of different fish species, for juvenile Nile tilapia (Diógenes et al., 2016; Nascimento, 2015; Xiao et al., 2017), for pacu (Abimorad et al., 2010; Oliveira et al., 2021), for silvery-black porgy (Marammazi et al., 2016), and for Tambaqui (Silva et al., 2022).

6.4 VALINE DIETARY REQUIREMENT

In fish nutrition, the determination of nutritional requirements in terms of limiting EAAs in practical fish diets is of major importance (Abimorad et al., 2010) because fish do not have a metabolic need for crude protein, but require a balanced quantity of EAAs. Understanding fish protein requirements is inadequate in the absence of data on EAA requirements (Rahimnejad and Lee, 2013) in order to have a positive effect on fish yield and a reduction in the excretion of nitrogen compounds in the aquatic environment where fish live (Bomfim et al., 2010; Ebeneezar et al., 2019;

Lin et al., 2013; Silva et al., 2018; Sousa et al., 2018; Sousa, 2021), especially as the challenge for the development of sustainable aquaculture is to meet the nutritional needs of fish while minimizing environmental impact and guaranteeing a quality product.

According to Vijayaram et al. (2023), the balance of AAs in terms of protein quantity and quality in formulated diets can maximize performance, reduce nitrogen excretion, and improve the immune response of fish to disease, with economic and environmental benefits for aquaculture. Thus, the determination of the EAA requirement for aquaculture farmed fish is extremely important for the formulation of a well-balanced, inexpensive, and environmentally friendly product (Small and Soares, 1999), which shows that for aquaculture, new knowledge about the importance of AAs in fish nutrition should be used in the search for promising foods and in the formulation of diets (Li et al., 2009; Prasanna et al., 2021). Among these EAAs, valine plays a very fundamental role in protein synthesis and the optimum growth of fish. It is required for the repair and growth of tissue, energy production, and maintaining N-balance in the body (Ahmad and Khan, 2006), knowing that dietary AA requirements can be classified as qualitative and quantitative (Wu, 2013). Thus, valine requirements for fish can be defined as the amount of this branched-chain AA necessary to achieve specific production objectives, such as maximizing weight gain and improving consumption index, a reduction in nitrogen excretion, and a commercial gain.

Given the importance of this AA for farmed fish species and for better optimization of its food use, dietary requirements for valine have been estimated for different species of fish commonly grown, with large variations from 2.5 to 4.0% of DP (Wilson 2002). AA requirements can be estimated in % of protein or in g/kg of diet, and dietary valine requirements have been worked out in a number of fish species, and Table 6.1 summarizes the AA requirements for various fish species.

TABLE 6.1 Valine Requirement of Different Fish Species

Fish	% Dietary Valine g/kg	% Dietary Protein	References
Tilapia larvae	2.80	–	Santiago and Lovell (1988)
Tilapia Fingerling	1.15 and 1.27	4.11 and 4.53%	Xiao et al. (2017)
Tilapia adult	1.18	–	Diogenes et al. (2016)
Tilapia Juvenile	3.78	–	Diogenes et al. (2016)

TABLE 6.1 *(Continued)*

Fish	% Dietary Valine g/kg	% Dietary Protein	References
Atlantic salmon	5.80	–	Rollin et al. (2003)
Juvenile			
Tilapia juvenile	7.6	–	Nascimento et al. (2020)
Tilapia adult	7.3	–	Rodrigues et al. (2021)
Indian major carp	1.63 and 1.5%	4.0 and 3.75%	Abidi and Khan (2004); Ahmad and Khan (2006)
Fingerling	1.52%	3,8%	
Blunt snout bream	1.32 and 1.26%	3.88 and 3.71%	Ren et al. (2015)
Rainbow trout	1.41 and 1.50%	3.85 and 4.10%	Bae et al. (2012)
Red sea bream	0.9%	2%	Rahimnejad and Lee (2013)
Channel catfish fingerling	0.71%	2.96%	Wilson et al. (1980)
Channel catfish	–	3%	Wilson et al. (1980)
Golden pompano	–	4.6%	Tan et al. (2016); Huang et al. (2017)
Chinook salmon fingerling	–	3.2%	–
juvenile hybrid grouper	1.56%	3.16%	Zhou et al. (2021)
Tambaqui	3.3%	4.53%	Silva et al. (2022)
Blunt snout bream Juvenile	1.26–1.32%	3.71–3.88%	Ren et al. (2015)
golden pompano Juvenile	1.98 –2%	4.6%	Huang et al. (2017)
Catla catla Fingerling	–	3.09%	Zehra and Khan (2014)
Gass carp juvenile	1.56	–	Luo et al. (2010)
Grass carp	1.4	–	Luo et al. (2014
Hybrid grouper	–	3.16%	Zhou et al. (2020)
Carp	–	2.3%	Ogino (1980)
Cirrhinus mrigala	3.9 g	–	Benakappa and Varghese (2003)
Rainbow trout	–	2.4%	Ogino (1980)
Tilapia juvenile	3.46	–	Mansano et al. (2020)
Tiapia adult	10.78 g/kg	–	Rodrigues et al. (2020)
silvery-black porgy	21.5 g/kg	–	Marammazi et al. (2017)
Gilthead seabream juveniles	3.21 (g/16 g N)	–	Peres and Oliva-Teles (2009)

The differences observed in the literature concerning valine nutritional requirements can be the result of several factors: the species evaluated, the physiological state of the animals, environmental conditions, the size and age of the fish (Wilson, 2002); differences in methodology, including growth conditions as well as the protein source (Benakappa & Varghese, 2003). Another fundamental factor that may influence the demand for valine in fish is the presence, absence, and quantity of other branched-chain AAs in the aquaculture feed tested (Ahmed & Khan, 2006), given the antagonism or synergy between this group of AAs. It should also be noted that in diets, a deficient or excess level of valine has been shown to negatively affect fish growth performance. Dietary valine deficiency has resulted in limited growth and poor non-specific immune responses in rainbow trout (*Oncorhynchus mykiss*) (Ahmad et al., 2021), while an excess of dietary valine was reported to negatively affect the growth performance and innate immunity in golden pompano (*Trachinotus ovatus*) (Huang et al., 2018). Further, due to their similar chemical structure, BCAAs share the same enzymes in their catabolic pathways, resulting in interaction between them (Harper et al., 1984). This fact makes it more controversial to define exact individual BCAA requirements, as an excess of one would increase catabolism of the others and thus alter requirements (Van Milgen et al., 2012).

6.5 SOURCES OF DIETARY VALINE (NATURAL AND SYNTHETIC)

In fish, proteins represent around 70% of dry weight (Wilson, 2002). Fish do not need protein, but rather a balanced mixture of AAs in the diet (Wilson, 1985), responsible for fish development and structural functions. The balance of the AA composition of fish feed can be managed during fish production by taking into account the specific needs of fish species, which will have an effect on fish quality, production cost, and environmental impact (Hua et al., 2019). AAs are the basic protein components necessary for fish at all their development stages, acting as energy substrates and participating in different metabolic pathways. At least 50% of the AAs consumed by fish are deposited in muscle (Kaushik and Seiliez, 2010), and since fish are unable to synthesize these AAs, including valine, they must be provided through the diet (NRC, 2011).

Valine protein sources include both natural and synthetic sources:

6.5.1 NATURAL SOURCES

6.5.1.1 ANIMAL SOURCE: FISH MEAL

Fishmeal is considered the highest quality source of protein for fish feeds because of its best profile of EAAs that covers the needs of fish (Medale and Kaushik, 2009), and because of its protein content (64 to 72%) (Cho and Kim, 2011). However, its production is limited and expensive (Anderson et al., 2016; Damusaru et al., 2018; Tomas et al., 2005), and it is used especially for carnivorous fish (Hua et al., 2019). Many studies have established the authenticity of fishmeal as the most appropriate protein source for fish in aquaculture, due to its balanced AA profile, high digestibility, and palatability (Che et al., 2017; Dam et al., 2019; Elesho et al., 2021; Hardy, 2010). According to Lall and Anderson, the valine (amino acid (g/16 g, N)) content in fishmeal is between 5.1 and 5.43, while according to Cho and Kim (Cho and Kim, 2011; Medale and Kaushik, 2009), krill fain contains 5.4 (Medale and Kaushik, 2009).

6.5.1.2 VEGETABLE SOURCE: SOYBEAN MEAL

A specific understanding of the needs of an EAA diet is essential to have a balanced AA diet, especially in the case of plant proteins, which are the most promising option due to their widespread availability and low cost (Kari et al., 2020). Plants have become the main source of protein for farmed fish (Bowyer et al., 2013; Gatlin et al., 2007; Meng et al., 2020; Oliva-Teles et al., 2015; Pervin et al., 2020; Tacon et al., 2009) and contribute to the development of sustainable aquaculture (Kalla et al., 2001; Hardy, 2010). Among the main sources of vegetable protein used in aquaculture feeds, there are oilseeds including soya soybean meal (SB), with low price and balanced AAs (Damusaru et al., 2018; Gatlin et al., 2007; Hardy, 1999; Moniruzzaman et al., 2018); and rapeseed meal, cottonseed meal, and sunflower seed meal); legumes (peas, lupins, and faba beans); cereals (wheat, rice, maize, barley, and corn) (Oliva Teles et al., 2015; Tacon and Hasan, 2011). The valine content in SB ranges from 2.21 to 3.0 (NRC, 2011; Lee et al., 2021). According to NRC (2011), the valine content (% of dry matter) for other plant-based protein sources of the AA valine is 2.1 for rapeseed meal, 1.4 for lupin meal, 3.0 for corn gluten, and 4.5 for wheat gluten.

6.5.1.3 MICROALGAE

For sustainable aquaculture production, following the global expansion of the aquaculture industry, demand for alternative sources of fish feed is increasing (Ansai et al., 2021; Annamalai et al., 2021; Chen et al., 2021; Eroldogan et al., 2022; Seong et al., 2021; Shah et al., 2018; Siddik et al., 2023). However, in recent years, microalgae are being investigated as a potential feed ingredient for fingerlings and adult fishes (Hodar et al., 2020), knowing that the use of microalgae in fish nutrition can relieve pressure on wild fish stocks (Trevi et al., 2023). Moreover, microalgae also have a well-balanced AA profile, obviating the need for high-cost AA supplements in the diet in aquaculture (Neal et al., 2018; Sarker, 2018; Sørensen, 2017). In terms of quality, microalgae have a rich and varied AA composition, containing all the EEAs (Barkia et al., 2019). The microalgae are used as supplement feed in aquaculture in fish farming (Haoujar et al., 2020), and the most frequently utilized species in aquaculture feed are *Chlorella, Tetraselmis, Isochrysis, Spirulina, Pavlova, Chlamydomonas, Phaeodactylum, Dunaliella, Chaetoceros, Nannochloropsis, Arthrospira, Botryococcus, Porphyridium, Haematococcus, Schizochytrium, Porphyridium, Scenedesmus* (Haoujar et al., 2020; Madeira et al., 2017; Nagappan et al., 2021; Yaakob et al., 2014). Based on studies conducted in several studies (Molino et al., 2018; Patil et al., 2007; Tredici et al., 2009; Sarker et al., 2016; Shi et al., 2018), valine content in some microalgae was identified in g/Kg of dry weight: Spirulina (8.69), *Chlorella* (9.40), *Dunaliella* (9.73), *Chlamydomonas* (16.00), *Scenedesmus* (10.86), *Haematococcus* (5.11), *Porphyridium* (5.32), *Isochrysis* (6.00), *Nannochloropsis* (20.08), *Schizochytrium* (5.43).

6.5.2 SYNTHETIC SOURCE

In fish farming, to ensure good growth with metabolic responses in fish, a well-balanced AA profile in the diet is one of the most influential factors, and it is for this purpose that fishmeal is used, which provides the ideal AA profile as a source of proteins in aquaculture (Cho and Kin, 2010; Kaushik and Seiliez, 2010; Ween et al., 2017). Currently, fish nutrition research on nutritional requirements has focused on the importance of determining AA requirements, as fish have no metabolic protein requirements, but rather an adequate balance between essential and non-EAAs (Bomfim et al., 2010; Costa Sousa et al., 2021; Ebeneezar et al., 2019; Lin et al., 2013; NRC, 2011; Silva et al., 2018). In addition to fishmeal, most of the alternative ingredients commonly used in fish diets may be deficient in one or more AAs (Nunes, 2014; Wang et al., 2022). To compensate for any deficiencies in the diet of AAs, the latter are supplemented in fish feeds in their synthetic form, which improves the profile of AAs in the diet and compensates for any deficiencies, optimizing protein utilization and fish growth performance (Wang et al., 2022). Valine cannot be synthesized *de novo* by fish and other animals and must be obtained by protein degradation of the diet, particularly fishmeal (Greiner et al., 2019; Karau and Grayson, 2014). To overcome the problem of AA profile, synthetic L-valine dietary supplementation is widely used to synthesize proteins in the body as an alternative to reduce the dependence on fishmeal in fish feed, unlike D-valine, which forms the bacterial cell walls. To ensure the production of L-valine, several methods have been adopted.

6.5.2.1 EXTRACTION METHOD

Unlike animals and humans, plants and microorganisms are capable of biosynthesis of valine (Oldiges, 2014; Neinast et al., 2019). L-valine is obtained from extraction from natural protein sources, including extraction from protein-hydrolysates (fish silage). The proteins present in the fish silage can also be hydrolyzed to free AAs, of which valine is one (Ghaly et al., 2013; Santana et al., 2023). This method exploits the differences in physicochemical properties (such as chemical affinity and pH) of the AAs to separate them (Zhang et al., 2014). Soy is considered an important source of vegetable protein in animal feed (Preece et al., 2017). It is a good nutrient source in animal diets, containing approximately 50% protein, and it contains EAAs (Dale et al., 2009), as valine extraction can be carried out on wheat and corn (Hermann, 2003; Leuchtenberger et al., 2005; Pasupuleti et al., 2010).

However, the quantity of AAs (including valine) in the original protein limits the quantity of AAs produced. As a result, the extraction method is not profitable for producing large amounts of specific AAs (Wang et al., 2020), and with the high demand in the animal nutrition market for L-valine, the development of cost-effective methods for producing this L-valine will become the focus of several studies and stimulate the production of this AA by other more cost-effective methods (Hao et al., 2022). Recent research has proposed several new strategies to improve the production of L-valine, while its production is obtained by chemical synthesis and microbial culture (Leuchtenberger et al., 2005; Oldiges et al., 2014) such as culture conditions in fermentation and systematic metabolic engineering. The production of L-valine is done by two processes; microbial fermentation is the main way to produce L-valine (Gao et al., 2021; Hao et al., 2022), while mutagenesis and metabolic engineering strategies are used to improve product formation (Schwentner et al., 2018). Microbial production of L-valine was achieved by exploiting various industrial microorganisms, including *Corynebacterium glutamicum* (Ikeda and Ngakawa, 2003; Bartek et al., 2011; Chen et al., 2015; Zhang et al., 2018; Liu et al., 2021), and *Escherichia coli* (Geraskina et al., 2018; Hao et al., 2022; Liu et al., 2023).

6.5.2.2 MICROBIAL FERMENTATION

The development of the aquaculture sector has stimulated the growth of the global AA market. As a result, there is a growing interest in the development and optimization of efficient processes for the production of valine. Extraction methods have been largely replaced by the production of L-valine by microbial fermentation (Blombach et al., 2007; Yamamoto et al., 2017). For this purpose, the bacterium *Corynebacterium glutamicum* is considered an attractive microorganism for the production by microbial fermentation of L-valine (Han et al., 2020; Zhang et al., 201). This genus of Gram-positive bacteria has been used for more than 50 years for the industrial production of AAs due to its remarkable capacity (Leuchtenberger et al., 2002). Industrial fermentation to produce L-valine by *C. glutamicum* is mainly due to mutation selection (Liu et al., 2021; Park and Lee 2010; Wang et al., 2018). Fermentation carried out by these microorganisms requires ideal conditions for their development and growth. These conditions include a culture medium containing the necessary nutrients and optimal environmental conditions (e.g., temperature, pH, agitation, oxygen) (Tokuyama et al., 2020). The advantage of fermentation is that large amounts of AAs can be produced at low cost with relatively small facilities, and the use of fermentation to produce AAs has increased the AA market (Şanlier et al., 2019).

During the fermentation process, sugars (as a source of carbon) are added to a culture medium, helping these microorganisms to multiply and produce AAs through the intervention of their enzymes, which speed up reactions to break down and synthesize new substances (D'Este et al., 2018; Duan et al., 2023; Michet et al., 2015), and the production of AAs, for example, L-glutamate, L-lysine, and L-valine (Leuchtenberger et al., 2005; Liebl, 1991). One of the advantages of fermentation production over other methods is that it only produces L-shaped AAs, which avoids further purification steps, in addition to being used in mild conditions preventing product degradation (Ugimoto, 2010). According to Wendich (2019), the production of L-valine by fermentation is estimated at 500 tons.

6.5.2.3 L-VALISSSTION BASED ON SYSTEMATIC METABOLIC ENGINEERING

With significant demand in the animal market in recent years and the rapid development of synthetic biology and genome analysis technology, and the development of microbial strains capable of efficiently producing L-valine, the production of AAs has gone from random mutation to metabolic engineering (Liu et al., 2021). Systems metabolic engineering is an emerging discipline that combines the concepts of systems biology, synthetic biology, and evolutionary engineering (Lee and Kim, 2015), as defined by Lee (2011), and metabolic systems engineering strategies have been successfully applied, mainly in two major producer micro-organisms: *Corynebacterium glutamicum* and *Escherichia coli* (Ma and al., 2017).

A better understanding of these microbes responsible for the production of valine was achieved through the development of innovative techniques that allowed the analysis of the genome of bacterial strains (Ikeda and Nakagawa, 2003). Thus, metabolic engineering strategies directly involve point mutations in genes relevant to the target AA (Volker, 2006), which have emerged as the most perfect solution for improving the production of L-valine and other AAs and their precursors (Hao et al., 2010; Wendisch 2014). In addition, metabolic engineering strategies aim to improve AA product yields by inhibiting competitive pathways leading to the production of undesirable by-products that are often difficult to separate from the target AA (Kondo et al., 2013; Wang et al., 2018; Wieschalka et al., 2013). *Corynebacterium glutamicum* producing strains of L-valine are characterized by increased biosynthesis controlled by acetohydroxyacid synthase (AHAS, ilvnge), disrupted pathways to undesirable byproducts (lactate, succinate), an amplified glycolytic pathway, and optimal performance (Hasegawa et al., 2013). For information, modified strains of *C. glutamicum* accumulate l-valine at a rate of 227 g L^{-1} in the absence of oxygen in just 48 hours from the use of metabolic technology (Hasegawa et al., 2012).

6.6 INTERACTION OF VALINE WITH OTHER AMINO ACIDS

Valine is a branched-chain essential amino acid (BCAA), which cannot be synthesized by aquatic animals, so it must be provided by fish feed proteins (Karau and Grayson, 2014; Soumeh et al., 2015). Much research on fish nutrition has focused on studying valine requirements individually, neglecting interactions with other AAs, particularly leucine and isoleucine from the same BCAA family (Xu et al., 2022). In addition, an imbalance in dietary BCAAs (deficiency or excess) can also reduce nutrient utilization and thus growth performance in fish, partly due to a negative effect of BCAA imbalance and antagonism on the activities of intestinal digestive and absorption enzymes (Xu et al., 2022).

Valine is a BCAA, which cannot be synthesized by aquatic animals, so it must be provided by fish feed proteins (Karau and Grayson, 2014; Soumeh et al., 2015). Many research studies on fish nutrition have focused on studying valine requirements individually, neglecting interactions with other AAs, especially leucine and isoleucine from the same BCAA family, as the antagonism between the Val and Leu diets that has been reported in the lake trout *Salvelinus namaycush*, whose growth and feed use of fish was reduced when Val took high-Leu dietary supplements (5% of feed) (Hughes et al., 1984). In the study of Han et al. (2014) on the interactive effects of dietary valine and leucine on the Japanese flounder *Paralichthys olivaceu*, it was shown that for fish fed with a high level of feeding Val in a low Leu content (1.6% of the diet) can contribute to protecting cells from oxidative stress of this fish. Wang et al. (2018) discovered that a significant interaction between Ile and Val was found on digestive enzyme and immune-related enzyme activities of Japanese flounder.

Another recent study conducted by Shi et al. (2021) on the same fish, but with the interaction between dietary valine (Val) and isoleucine

(Ile) on growth, blood parameters, and tolerance to low salinity stress in Japanese plaice *Paralichthys olivaceus*, showed that plaice fed the diet supplemented with 2% isoleucine and 2.27% valine were superior to other groups on growth and blood parameters, and tolerance against low salinity stress. The AAs valine, leucine, and isoleucine (BCAA) share enzymes to catabolize the first two steps, BCAT and branched-chain keto-acid dehydrogenase (Cole, 2015). Diets rich in corn as a protein source have relatively high leucine concentrations (Holen et al., 2022; Kwon et al., 2019); this high leucine content could stimulate the activity of metabolic enzymes and promote valine catabolism, thus reducing serum valine concentration (Wiltafsky, 2010). In fish nutrition, it is important to design diets with a balanced branched-chain AA profile to avoid interactions between leucine and valine, since in case of imbalances there is a negative effect on performance, as reported by a study on Nile tilapia by researchers Xu et al. (2022), which showed that an antagonistic effect between high leucine and high valine in the diet leads to inhibition of tilapia growth, whereas the results of research by Shi et al. (2021) on Japanese flounder prove otherwise, because a high level of dietary valine adjusted to a high content of dietary isoleucine significantly improves the growth performance of fish.

During studies conducted on juveniles of red drum *Sciaenops ocellatus* by Castillo and Gatlin (2017), whose objective was the determination of the minimum needs in Leu, Ile, and Val combined every 3 for the red drum, with the aim of producing aquaculture feed composed of alternative protein ingredients while maintaining an ideal AA profile for optimal performance of the red drum, the authors were able to evaluate minimum needs in Leu, Ile, and Val combined every 3 for maximum growth of red drum juveniles at 15.7 g/kg for leucine, 11.1 g/kg for isoleucine, and 12.4 g/kg for valine, for dry feeding. They highlighted that this combination of BCAAs could give better performance for the drum juveniles.

6.7 VALINE AND IDEAL PROTEIN CONCEPT

In aquaculture, fishmeal has been used as the main source of protein in the diet of fish and other animals for decades due to its high protein content and balanced AA profile. Protein inclusion rates generally vary from 30% to 50% in aquaculture feed and represent the highest share of feed costs for fish (Glencross et al., 2020; Oliva-Teles et al., 2015; Tacon and Metian, 2015). These proteins, being one of the main nutrients in the feeding of fishes, play a major role in the cost of formulating feed and directly influence the performance of aquatic animals, such as the consumption index, weight gain, and quality of flesh. However, the amount of protein in food is an important but not ideal condition for effective and balanced nutrition in aquaculture (Henry et al., 2015), since it is recommended to pay particular attention to the essential requirements for AAs, that is, those that are not synthesized by fish, including arginine, histidine, isoleucine, leucine, lysine, methionine, phenylalanine, threonine, tryptophan, and valine (Hou et al., 2015; Hua and Bureau, 2019; NRC, 2011; Zheng and Wu, 2021). These AAs greatly influence the growth patterns, reproductive performance, and development of a fish (Wilson, 2003). In this context, in fish nutrition, the ideal protein concept was adopted, whereby the ideal protein is considered to have a balanced AA composition that best meets the needs of the fed fish. This concept of an ideal protein was suggested by Wilson (1991), founded on the direct correlation between the profile of EAAs in fish tissue and the profile of dietary requirements in EAAs (Miles and Chapman, 2008), because fish do not have specific requirements for protein, but rather need the EAAs that make up protein (Teles et al., 2019), since in fish, the difference lies in the strong need for protein in aquaculture feeds compared to those in terrestrial feed (Nunes et al., 2014).

The concept of ideal protein was applied by Wilson (1991) to estimate the dietary requirements of EAAs of *I. punctatus* and reported close agreement with the values of EAA requirements determined by empirical methods. While the AA profile of an ideal protein may be expressed in relation to lysine, which is the first limiting AA in typical animal feed, this does not take into account the possible interactions between AAs (e.g., between branched chain AAs; Kim et al., 2001; van Milgen and Dourmad, 2015). In recent years, following the boom in the production of synthetic AAs including L-valine, great opportunities have been offered to put into practice the concept of ideal protein in fish nutrition (Furuya, 2023; NRC, 2011; Sakomura and Rostagno, 2016) during the formulation of fish feeds.

The formulation of a fish feed based on an ideal protein is an effective means to use less protein in aquaculture feed to meet particularly EAA requirements (Boisen et al., 2000; Bureau and Encarnação, 2006; Emmert and Baker, 1997). The application of the concept in this case makes it possible to maximize AA utilization and lower feed costs by promoting the deposition of these AAs. In addition, it has been accepted that the AA composition of the aquaculture feed should correspond to the AA composition of the fish fed in the feed (Mambrini and Kaushik, 1995), and the formulation of feed with optimal ratios of AAE is a fairly effective method for saving DP supplementation (Peres and Oliva-Teles, 2009). According to Miles and Chapman (2008), if fish feed contains the exact amount of EAA required by a fish species, then the ideal protein for that species is met, and so there is no AA deficiency or excess. Growing fish fed such a diet would effectively use AAs instead of turning them into energy. AAs would then be used for maintenance, health, and synthesis of new structural proteins, which would lead to maximum dietary efficiency and growth.

In an ideal protein that is consistent with an ideal AA balance, without excess or deficiency, the AA lysine is used as a reference for the protein profile (Baker and Han, 1994; Bomfim et al., 2008; Furuya et al., 2005; Silva et al., 2018) because lysine is considered the first limiting EAA in many protein sources used for feed, and lysine-rich ingredients are often expensive (Ovie and Eze, 2013). It is for this reason that lysine is used as the reference AA used in the determination of the ideal protein standard; it is one of the most limiting AAs in fish feed (Small and Soares, 2000), knowing also that lysine is primarily oriented towards the deposition of body proteins (Miles and Chapman, 2008; Ovie and Eze, 2013), in addition to being a limiting AA in certain sources of vegetable protein alternatives to fishmeal, mainly in cereals, such as gluten (Michelato et al., 2016; Mukhtar et al., 2017; Silva et al., 2018). Using lysine as a reference, the target concentrations for each of the other EAAs are expressed as a percentage of lysine, which is set at 100% (Miles and Chapman, 2008).

Valine has been identified as an EAA for fish (Nose, 1979), a potentially limiting EAA in fish nutrition, as reported in several studies in several fish species, namely studies by Ahmed and Khan (2006) on Indian main carp (*Cirrhinus mrigala*), Bae et al. (2012) on rainbow trout (*Oncorhynchus mykiss*), Dong et al. (2012) Jian carp (*Cyprinus carpio* var. Jian), Ren et al. (2015) on blunt bream (*Megalobrama amblycephala* Yih), Diógenes et al. (2016), Castillo and Gatlin (2018) on the red drum (*Sciaenops ocellatus*), Xiao et al. (2018) on the Nile tilapia (*Oreochromis niloticus*), Rodrigues (2020) on tilapia, Oliveira et al. (2021) in their study on pacu fish (*Piaractus mesopotamicus*), in hybrid grouper (*Epinephelus fuscoguttatus, Epinephelus lanceolatus*) by Zhou et al. (2021), and tambaqui (*Colossoma macropomum*) by Silva et al. (2022).

The ideal protein concept used by Wilson (1991) to estimate *I. punctatus* dietary requirement for EAAs was in close agreement with the EAA requirement values determined by empirical methods. This ideal protein concept is based on the principle that EAAs required by fish are supplied in specific proportions to each other. Each of these AAs present in the diet is related to the lysine requirement that is used as a reference, while the target concentrations for each of the other EAAs are expressed as a percentage of lysine, which is set at 100%. The results of this ratio were 0.70, obtained from work on pigs, by Veira et al. (2021). During the work of Gao et al. (2021) on the effect of optimizing the ratio of dietary valine-lysine at the end of gestation or lactation on biochemical indices and performance of lactating primiparous sows, they showed that optimizing the valine/lysine ratio to 0.87:1 (instead of 0.70:1) by supplementing gilts with valine at the end of gestation and during lactation was beneficial. Siebert et al. (2021) conducted studies on weaned piglets to determine the optimal ratio for these animals and estimated this ratio to be 0.68 (68% for valine and 100% for lysine).

6.8 PHYSIOLOGICAL AND NUTRITIONAL FUNCTIONS OF VALINE

To achieve optimal growth, but also to ensure fish health in aquaculture, adequate nutrition is essential. For many years, research on fish nutrition has focused primarily on determining optimal nutritional requirements for better growth of different fish species (NRC, 2011). Optimizing the nutritional needs of cultivable fish species is crucial to reduce feed costs by allowing the stipulation of optimized amounts of several nutrients for better growth and well-being of fish (Ahmed et al., 2004). The AAs from the aquaculture feed proteins are essential for the development of fish, since they are the basic components of proteins and allow the synthesis of proteins, thus promoting the growth of fish by anabolic processes (Li et al., 2021). These AAs are also important signaling molecules (Ahmad et al., 2021; Habte-Tsion, 2020).

In aquaculture, for purely economic reasons, fish producers give more importance to the content or digestibility of raw food proteins without taking into account the intake of AAs in the diet, because many fish species have high DP and fishmeal requirements (Wu, 2021).

Valine is one of the EAAs (Nones, 1979), necessary to maintain normal growth and regulate various physiological processes in fish (NRC, 2011). With the expansion of aquaculture, and to ensure food security in the world, in recent years, research has highlighted the functional roles of specific AAs and the advancement of nutritional skills and costs. Effective feeding of farmed fish species is of extreme concern for the commercial success of aquaculture (Ahmad and Khan 2006; Peres and Oliva-Teles 2009). Due to the involvement of branched AAs in various metabolic pathways, valine nutrition is beginning to garner more interest in its dietary optimization given its role and importance in animal and nutritional physiology (Wang et al., 2023).

Valine is one of the EAAs in fish (Nones, 1979), whose feed content depends on the type of ingredients used. This AA, belonging to the branched-chain amino acid (BCAA) family, provides important physiological functions in animals, including protein synthesis, energy production, immune functions, and tissue repair (Brosnan and Brosnan, 2006; Castillo and Gatlin, 2018; Zhang et al., 2017). Its most important physiological function is the synthesis of participating proteins, so it is closely related to fish growth (Ahmad et al., 2021; Wu, 2021) and plays an important role in growth regulation (Ahmed et al., 2019). Skeletal muscle, a key metabolic site in fish, performs a great function in adaptive regulation in response to dietary nutrient deficiencies (Hidalgo et al., 2017). According to Suryawan et al. (2011), valine has important physiological functions in muscle tissue growth and protein synthesis in the body. The work of Luo et al. (2010) showed that valine imbalance affects juvenile growth in grass carp, and the same results were reported in Indian carp (*Labeo rohita*) (Abidi and Khan, 2004), juvenile Nile tilapia (Diógenes et al., 2016), and Atlantic salmon (*Salmo salar*) (Rollin et al., 2003).

Fish rely on the mobilization of BCAA muscle catabolism to regulate starvation during nutrient deprivation (Hidalgo et al., 2017; Li et al., 2018). Valine, in the form of leucine and isoleucine, can participate in several cascades of cellular signals affecting anabolism and catabolism and maintain the physiological dynamic balance of fish under food shortage conditions (Ahmad et al., 2018). According to Cao et al. (2023), the results of their work indicated that fasting causes catabolism of proteins and AAs in common carp and a negative balance of protein turnover, and valine participates in the regulation of ubiquitin-the protease system and the autophagy-lysosomal system, which reduces the breakdown of muscle proteins under starvation conditions. Their study revealed that supplementation with valine (in addition to the other two branched-chain AAs) can increase muscle mass by replenishing new muscle fibers and can also increase the size of existing muscle fibers to promote skeletal muscle. Chung and Baker (1992) describe valine deficiency as reducing the use of other limiting AAs for protein deposition and is, therefore, essential. Zhou et al. (2020) also found that optimal dietary valine in fish contributes to and promotes protein synthesis by TOR/S6K1 signaling (Weichhart 2009).

In addition, valine is increasingly recognized as an anabolic nutrient signal, communicating the presence of an ingested protein-containing meal to peripheral tissues, and promoting protein synthesis in fish muscle (Castillo and Gatlin, 2017; Kawanag et al., 2014, 2015; Lansard et al., 2011). Immunity is the ability of an organism to resist attacks of pathogens. In this context, valine can also improve the ability of the organism to resist some impact on the immune system of fish, and the effect of Val on fish immunity has been studied extensively; for example, Chen et al. (2017) found that Val engages in macrophage phagocytosis and takes a major part in inhibiting arginase activity to stimulate nitric oxide production. Valine also provides different immune functions in fish and contributes to the regulation of several key metabolic pathways in response to infectious pathogens (Petro and Bhatacharjee, 1981), and can regulate the expression of immune genes and participate in immune responses to promote the development of immune organs, develop immunity of the body, and maintain homeostasis in aquatic animals (Allameh and Toghyani, 2019; Xiao et al., 2018; Zhou et al., 2021). In addition, valine has anti-inflammatory properties, which are used as a dietary supplement in the aquaculture diet (Zhao et al., 2023). Thus, adequate supplementation of valine in the diet could stimulate host defense mechanisms, that is, natural killer cell activity and neutrophil phagocytic function.

There is a close link between valine and antioxidant status (Zhou et al., 2021). Studies have shown that valine can improve the antioxidant status of fish through the Nrf2/Keap1 signaling pathway (Feng et al., 2015). It has been found that appropriate dietary valine levels could enhance the antioxidant capacity in grass carp (*Ctenopharyngodon idella*) (Feng et al., 2015) and juvenile hybrid grouper (Zhou et al., 2021). Valine also contributes to improving the digestive tract of fish, especially the gut, to improve fish growth by increasing digestion and absorption capacity, thereby influencing the balance of the intestinal microflora (Dong et al., 2013), which will allow better digestion of food by increasing both digestive capacity and absorption capacity and also influencing the balance of the intestinal microflora (Ahmad et al., 2021). Valine is involved in the synthesis of the amino neurotransmitters serotonin and the catecholamines dopamine and norepinephrine, in addition to protein synthesis; these neurotransmitters are derived from the aromatic AAs tryptophan, phenylalanine, and tyrosine, energy production, and glutamate compartmentalization (Fernstrom 2005).

6.9 NUTRITIONAL DISORDERS OF VALINE DEFICIENCY

In aquaculture, dietary AAs are essential for fish because of their various roles as energy substrates, for the synthesis of endogenous proteins, and for the regulation of metabolic pathways (Wu et al., 2014). The success of production in this sector depends on the nutritional balance of food received by fish (Hixson, 2014), because a lack of nutrients, especially EAAs, may affect growth, development, and fish health (Trosvik et al., 2013). Fish can suffer from altered protein synthesis, reduced weight gain, stunting, and a host of other health problems, which can lead to death (Kaushik and Seiliez 2010). In fish, dietary problems, primarily related to nutritional imbalances, may be reported (Hardy 2012; Oliva-Teles 2012), but valine has been identified as an AAE for fish (Nose 1979). Knowing that an excess of Val alimentaire would have a negative effect on the growth yield and innate immunity of Pompano Doré (*Trachinotus ovatus*) as indicated by Huang et al. (2018), although a dietary deficiency in valine may result in limited dietary use, therefore poor growth in several fish species, such as *Pagrus major*, mrigal carp (Ahmed and Khan 2006), red sea bream (*Pagrus major*) by Rahimnejad and Lee (2013), and Jian carp, (*Cyprinus carpio* var. Jian) reported by Dong et al. (2012), and poor non-specific immune responses in rainbow trout (*Oncorhynchus mykiss*) in the work of Ahmad et al. (2021), hybrid grouper (*Epinephelus fuscoguttatus* x *E. lanceolatus*) by Zhou et al. (2020), and Nile tilapia (*Oreochromis niloticus*) by Xiao et al. (2018), while authors Rahimnejad and Lee (2013, 2014) observed an increase in plasma total protein levels with increasing levels of valine in the diet in red sea bream (*Pagrus major*).

Valine can also affect the antioxidant status and immune system of animals. Luo et al. (2014) reported that Val's imbalances may have adverse effects on the immune state of the intestinal mucosa, while Feng et al. (2015) reported that valine deficiency in the diet impaired the structural integrity of fish gill, and that valine can have an effect on the regulation of the expression of antioxidant enzymes by the TOR and NF-E2-signaling pathways related to factor 2 (Nrf2). The gut is a site of critical importance for digestion and nutrient uptake (Deng et al., 2010), which is closely related to food use efficiency and thus growth performance (Dimitroglo et al., 2009; Zhou et al., 2010), but is considered one of the most important organs of immunity (Lokesh, 2012). Allameh et al. (2019) reported that in broilers, valine supplementation on low protein diets improved the height of small intestine villi. In the study of juvenile hybrid grouper by Zhou et al. (2021), morphological results of the foregut and midgut indicated that optimal dietary valine content increased the height and width of folds and the height of microvilli in the gut of hybrid grouper, which may also explain the role of valine in the rapid growth of fish. Other studies have indicated that the effect of dietary imbalance of valine on fish development and growth can be caused by loss of appetite and low dietary use (Ahmed and Khan 2006; Dong et al., 2013; Rahimnejad and Lee 2013).

6.10 FACTORS AFFECTING VALINE REQUIREMENT

In aquaculture, fish have no specific protein requirements, but rather well-balanced AA requirements to ensure optimal growth (Webster and Lim, 2002; Wilson and Poe, 1985; Wilson, 2002), because fish do not have the ability to synthesize valine; this AA must be provided by proteins as mixtures of AAs in the diet of fish (NRC, 2011). Optimal dietary requirements for EFA are dependent on many factors such as genetic strains, growth potential, environmental conditions, and fish welfare (Mansanob et al., 2020; Monentcham et al., 2010). Dietary energy density is one of the key factors affecting AA requirements, but there are uncertainties regarding metabolizable energy levels of the main food components, which tend to exclude the expression of AA requirements in terms of metabolizable energy (Cowey, 1994). According to Khan et al. (2019), although the optimal profile of dietary AAs is often assumed to be constant, recent evidence has shown that the relative contribution of each AAE may change over life stages. Several studies have shown that valine requirements differ with fish growth (Santiago and Lovell, 1988;

Xiao et al., 2017) with estimated levels of 0.78 and 1.21%, while work on juvenile tilapia by Diógenes et al. (2016), the authors obtained increasing estimates of valine requirements based on fish weight. Abidi and Khan (2004) reported that Val requirement varies with experimental conditions, differences in species, formulated feed composition, and AA source. A previous study showed that the Val requirement level was 15.6 g/kg at 32% protein in grass carp diet (Luo et al., 2010) but 14.0 g/kg at 29% protein (Luo et al., 2014). The Val requirement for 1.57–5.68 g and 498 g rainbow trout were 18.2 g/kg and 14.1–15.0 g/kg, respectively (Ahmad et al., 2021; Bae et al., 2012).

The different literature results for nutritional requirements for valine relate primarily to size and age requirements (Diogenes et al., 2016), and species with a formulated dietary composition and source of AA (Abidi and Khan, 2004) fish, while other factors (other than age and species) may influence needs, especially fish physiological status, environmental conditions, size, and age of aquatic animals (Diogenes et al., 2016; Mansanob et al., 2020; Monentcham et al., 2010; Wilson, 2002); differences in methodology, including growth conditions and the nature of the protein source (Benakappa and Varghese, 2003); and the statistical model applied. Another important factor that may affect the demand for valine in fish is the composition of the diet tested in relation to the presence, absence, and quantity of other branched-chain AAs (Ahmed and Khan, 2006), taking into account the antagonism or synergy between this group of AAs.

Valine requirements can also be affected by the composition of the feed consumed by the fish, as indicated by the study carried out by Xiao et al. (2017) on tilapia, whose requirements vary depending on the formulation of the feed tested. The same results were observed in studies on Grass carp (*Ctenopharyngodon idella*), carried out by Luo et al. in 2010 and 2014, where valine requirements were dependent on the protein content of the feed. These requirements were 1.56 for a protein content of 32% and 1 for 29% protein, respectively. Research into the nutritional requirements of fish has focused on determining EAA requirements, as fish do not have a metabolic need for protein, but rather for an adequate balance of EAAs, including valine. The majority of studies conducted on valine have been more oriented towards the determination of the AA requirements according to the species of fish and also according to the different ages of the fish. There are not enough studies that cover all the factors that affect the EAA valine requirements for fish. In this context, it is essential to conduct studies to determine the effects of other factors affecting valine requirements for fish species and at different stages of rearing.

6.11 FUTURES PERSPECTVES AND CONCLUSION

With the expansion of aquaculture to ensure global food security in recent years (Dewali et al., 2023), research has highlighted the functional roles of specific AAs as well as their nutritional costs. Effective nutrition of farmed fish species is of utmost importance for the commercial success of aquaculture (Ahmad and Khan 2006; Peres and Oliva-Teles 2009). As a result, metabolism and AA function studies have become essential for the sustainable development of low fishmeal diets for fish, due to the depletion of fish stocks in the oceans (Glencross, 2020). In aquaculture, dietary AAs, normally supplied by fish meal, are essential for fish for their various roles as energy substrates, for endogenous protein synthesis, and for regulating metabolic pathways (Wu et al., 2014). The success of production in this sector depends on the nutritional balance of the feed received by the fish (Hixson, 2014). In fish nutrition, the determination of nutritional requirements in terms of limiting EAAs in the diets provided to fish has been favored (Abimorad et al., 2010), because fish have no metabolic need for crude protein, but require a balanced amount of EAAs, as understanding of fish protein requirements is inadequate in the absence of knowledge about EAA requirements (Rahimnejad and Lee, 2013). According to authors Vijayaram et al. (2023), the quantity and quality of protein in formulated diets, particularly the balance of AAs, can maximize performance, reduce nitrogen excretion, and improve the immune response of the fish against disease, with economic and environmental benefits for aquaculture, showing that for aquaculture, carrying out future studies on the role of AAs in fish nutrition should be used in food research promisingly and in the formulation of diets (Li et al., 2009; Prasanna et al., 2021).

Since valine is an EAA of fish (Nones, 1979), its content in aquaculture animal feed depends on the type of ingredients used. Its most important physiological role is the synthesis of participating proteins, so it is closely related to the growth of fish (Ahmad et al., 2021; Wu,

2021) and plays an important function in growth regulation (Ahmed et al., 2019). It should also be noted that diets should have an optimal content of this AA, since an insufficient or excessive level of valine has been shown to have a negative effect on the growth performance of fish (Ahmad et al., 2021).

To ensure proper management in fish feeding, the nutritional requirements of valine can be defined as the quantity of this AA necessary to achieve specific production objectives, such as maximizing weight gain and improving feed index, consumption, reduced nitrogen excretion, and commercial gain. Thus, in aquaculture, it is important to determine the valine requirements of farmed fish species for normal growth and feed utilization. The differences observed in the literature concerning nutritional requirements for valine can be explained by the effect of several factors: the species evaluated, the physiological state of the animals, environmental conditions, size and age of the fish (Wilson, 2002), and the nature of the protein source (Benakappa and Varghese, 2003). In this context, and to ensure fish production in line with high consumer demand, valine is synthesized as an alternative means of ensuring AA balance, especially valine in aquaculture feeds. With limited fisheries, rising fishmeal prices, and the need to ensure fish production in line with strong consumer demand, other sources of valine-rich ingredients are being used, or valine is being synthesized as an alternative means of ensuring AA balance, especially valine, in aquaculture feeds.

Research carried out on fish nutrition has focused on the study of the requirements for valine, a BCAA, without taking into consideration the interdependencies with other AAs, especially those of the BCAA family (Xu et al., 2022). Furthermore, an imbalance of dietary BCAAs (deficiency or excess) can also reduce nutrient utilization and therefore growth performance in fish, in part due to a negative effect of BCAA imbalance and antagonism on the activities of intestinal digestive and absorption enzymes (Xu et al., 2022). In fish, dietary problems, mainly related to nutritional imbalances in valine, may be reported (Hardy, 2012; Oliva-Teles, 2012). Excess dietary Val has been reported to negatively affect growth performance and innate immunity in *Pompano aureus* (*Trachinotus ovatus*) (Huang et al., 2018), while a dietary deficiency of valine resulted in poor growth performance and partial dietary use in several fish species.

KEYWORDS

- **aquafeed formulation**
- **branched-chain amino acids**
- **essential amino acids**
- **fish growth performance**
- **sustainable aquaculture**
- **valine in fish nutrition**

REFERENCES

Abidi, S. F., & Khan, M. A. (2004). Dietary valine requirement of Indian major carp, *Labeo rohita* (Hamilton) fry. *Journal of Applied Ichthyology, 20*, 118–122. https://doi.org/10.1046/j.1439-0426.2003.00526.x.

Abimorad, E. G., Favero, G. C., Squassoni, G. H., & Carneiro, D. (2010). Dietary digestible lysine requirement and essential amino acid to lysine ratio for pacu (*Piaractus mesopotamicus*). *Aquaculture Nutrition, 16*, 370–377. https://doi.org/10.1111/j.1365-2095.2009.00674.x.

Adeva-Andany, M. M., López-Maside, L., Donapetry-García, C., Fernández-Fernández, C., & Sixto-Leal, C. (2017). Enzymes involved in branched-chain amino acid metabolism in humans. *Amino Acids, 49*(6), 1005–1028. https://doi.org/10.1007/s00726-017-2412-7.

Ahmad, I., Ahmed, I., & Dar, N. A. (2021). Dietary valine improved growth, immunity, enzymatic activities and expression of TOR signaling cascade genes in rainbow trout, *Oncorhynchus mykiss* fingerlings. *Scientific Reports, 11*, 22089. https://doi.org/10.1038/s41598-021-01142-4.

Ahmad, I., Ahmed, I., Fatma, S., & Peres, H. (2021). Role of branched-chain amino acids on growth, physiology and metabolism of different fish species: A review. *Aquaculture Nutrition*. https://doi.org/10.1111/anu.13267.

Ahmad, I., Ahmed, I., Fatma, S., & Peres, H. (2021). Role of branched-chain amino acids on growth, physiology and metabolism of different fish species: A review. *Aquaculture Nutrition, 27*(5), 1270–1289.

Ahmed, I., & Khan, M. A. (2006). Dietary branched-chain amino acid valine, isoleucine and leucine requirements of fingerling Indian major carp, *Cirrhinus mrigala* (Hamilton). *British Journal of Nutrition, 96*, 450–460.

Ahmed, I., Khan, M. A., & Jafri, A. K. (2004). Dietary threonine requirement of fingerling Indian major carp, *Cirrhinus mrigala* (Hamilton). *Aquaculture Research, 35*, 162–170. https://doi.org/10.1111/j.1365-2109.2004.00997.x.

Allameh, S., & Toghyani, M. (2019). Effect of dietary valine supplementation to low protein diets on performance, intestinal morphology and immune responses in broiler chickens. *Livestock Science, 229*, 137–144.

Anthony, J. C., Anthony, T. G., Kimball, S. R., & Jefferson, L. S. (2001). Signaling pathways involved in translational control of protein synthesis in skeletal muscle by leucine. *The Journal of Nutrition, 131*(3), 856S–860S. https://doi.org/10.1093/jn/131.3.856S.

Bae, J. Y., Park, G., Yun, H., Hung, S. S. O., & Bai, S. C. (2012). The dietary valine requirement for rainbow trout, *Oncorhynchus mykiss*, can be estimated by plasma free valine and ammonia concentrations after dorsal aorta cannulation. *Journal of Applied Animal Research, 40*(1), 73–79. https://doi.org/10.1080/09712119.2011.628395.

Baker, D. H., & Han, Y. (1994). Ideal amino acid profile for chicks during the first three weeks posthatching. *Poultry Science, 73*(1), 1441–1447.

Baki, B., Ozturk, D. K., & Kerim, M. (2017). Determination of essential amino acid changes related to the growth of sea bass (*Dicentrarchus labrax*). *Turkish Journal of Fisheries and Aquatic Sciences, 17*, 1425–1429. https://doi.org/10.4194/1303-2712-v17_6_36.

Bartek, T., Blombach, B., Lang, S., Eikmanns, B. J., Wiechert, W., Oldiges, M., Nöh, K., & Noack, S. (2011). Comparative 13C metabolic flux analysis of pyruvate dehydrogenase complex-deficient, L-valine-producing *Corynebacterium glutamicum*. *Applied and Environmental Microbiology, 77*, 6644–6652.

Biswas, D., Duffley, L., & Pulinilkunnil, T. (2019). Role of branched-chain amino acid–catabolizing enzymes in intertissue signaling, metabolic remodeling, and energy homeostasis. *The FASEB Journal*. https://doi.org/10.1096/fj.201802842RR.

Blombach, B., Arndt, A., Auchter, M., & Eikmanns, B. J. (2009). L-valine production during growth of pyruvate dehydrogenase complex-deficient *Corynebacterium glutamicum* in the presence of ethanol or by inactivation of the transcriptional regulator SugR. *Applied and Environmental Microbiology, 75*(4), 1197–1200. https://doi.org/10.1128/AEM.02351-08.

Blombach, B., Schreiner, M. E., Holátko, J., Bartek, T., Oldiges, M., & Eikmanns, B. J. (2007). L-valine production with pyruvate dehydrogenase complex-deficient *Corynebacterium glutamicum*. *Applied and Environmental Microbiology, 73*(7), 2079–2084. https://doi.org/10.1128/AEM.02826-06.

Boaratti, A. Z., Nascimento, T. M. T., Khan, K. U., Mansano, C. F. M., Oliveira, T. S., Queiroz, D. M. A., Romaneli, R. S., Rodrigues, A. T., Sakomura, N. K., & Fernandes, J. B. K. (2020). Assessment of the ideal ratios of digestible essential amino acids for pacu (*Piaractus mesopotamicus*) juveniles by the amino acid deletion method. *Journal of the World Aquaculture Society*. https://doi.org/10.1111/jwas.12740.

Boisen, S. (2003). A new concept for feed evaluation based on standardized digestible amino acids and potential physiological energy of nutrient fractions in pig feeds. In W. B. Souffrant (Ed.), *Progress in Research on Energy and Protein Metabolism and Nutrition* (pp. 129–132).

Boisen, S., Hvelplund, T., & Weisbjerg, M. (2000). Ideal amino acid profiles as a basis for feed protein evaluation. *Livestock Production Science, 64*(2, 3), 239–251. https://doi.org/10.1016/s0301-6226(99)00146-3.

Bomfim, M. A. D., Lanna, E. A. T., Donzele, J. L., Quadros, M., Ribeiro, F. B., & Sousa, M. P. (2010). Níveis de lisina, com base no conceito de proteína ideal, em rações para alevinos de tilápias-do-Nilo. *Revita Brasileira de Zootecnia, 39*, 1–8. http://dx.doi.org/10.1590/S1516-35982010000100001.

Bomfim, M. A. D., Lanna, E. A. T., Donzelle, J. L., Abreu, M. L. T., Ribeiro, F. B., & Quadros, M. (2008). Redução de proteína bruta com suplementação de aminoácidos, com base no conceito de proteína ideal, em rações para alevinos de tilápia-do-Nilo. *Revista Brasileira de Zootecnia, 37*(10), 1713–1720.

Brosnan, J. T., & Brosnan, M. E. (2006). Branched-chain amino acids: Enzyme and substrate regulation. *The Journal of Nutrition, 136*(1), 207S–211S. https://doi.org/10.1093/jn/136.1.207s.

Bureau, D. P., & Encarnação, P. M. (2006). Adequately defining the amino acid requirements of fish: The case example of lysine. In L. E. Cruz Suárez et al. (Eds.), *Avances en Nutrición Acuícola VIII*. Universidad Autónoma de Nuevo León.

Cao, X., Cui, H., Ji, X., Li, B., Lu, R., Zhang, Y., & Chen, J. (2023). Determining the potential roles of branched-chain amino acids in the regulation of muscle growth in common carp (*Cyprinus carpio*) based on transcriptome and microRNA sequencing. *Aquaculture Nutrition*. https://doi.org/10.1155/2023/7965735

Castillo, S., & Gatlin, D. M. (2018). Dietary requirements for leucine, isoleucine and valine (branched-chain amino acids) by juvenile red drum *Sciaenops ocellatus*. *Aquaculture Nutrition, 24*(4), 1056–1065.

Chandel, N. S. (2021). Glycolysis. *Cold Spring Harbor Perspectives in Biology, 13*(5), a040535. https://doi.org/10.1101/cshperspect.a040535.

Chen, C., Li, Y., Hu, J., Dong, X., & Wang, X. (2015). Metabolic engineering of *Corynebacterium glutamicum* ATCC13869 for L-valine production. *Metabolic Engineering, 29*, 66–75. https://doi.org/10.1016/j.ymben.2015.03.004.

Cho, J., & Kim, I. (2011). Fish meal—nutritive value. *Journal of Animal Physiology and Animal Nutrition, 95*(6), 685–692.

Chung, T. K., & Baker, D. H. (1992). Ideal amino acid pattern for 10-kilogram pigs. *Journal of Animal Science, 70*, 3102–3111.

Costa Sousa, M., Bomfim, M. A. D., Ribeiro, F. B., Siqueira, J. C., Marchão, R. S., Sousa, T. J. R., & Takishita, S. S. (2021). Lysine requirements of tambatinga (♀ *Colossoma macropomum* × ♂ *Piaractus brachypomus*) fingerlings using different diet formulation techniques. *Aquaculture Nutrition*. https://doi.org/10.1111/anu.13320.

Cowey, C. B. (1994). Amino acid requirements of fish: A critical appraisal of present values. *Aquaculture, 124*(1–4), 1–11. https://doi.org/10.1016/0044-8486(94)90349-2.

Craig, S. (2017). Understanding fish nutrition, feeds, and feeding. Virginia Cooperative Extension, Virginia Tech. www.ext.vt.edu (accessed on 3 June 2025).

Dale, B. E., Allen, M. S., Laser, M., & Lynd, L. R. (2009). Protein feeds coproduction in biomass conversion to fuels and chemicals. *Biofuels, Bioproducts and Biorefining, 3*(3), 219–230.

Deng, J., Mai, K., Ai, Q., Zhang, W., Tan, B., Xu, W., et al. (2010). Alternative protein sources in diets for Japanese flounder *Paralichthys olivaceus* (Temminck and Schlegel): II. Effects on nutrient digestibility and digestive enzyme activity. *Aquaculture Research, 41*(7), 861–870.

D'Este, M., Alvarado-Morales, M., & Angelidaki, I. (2018). Amino acids production focusing on fermentation technologies—A review. *Biotechnology Advances, 36*(1), 14–25. https://doi.org/10.1016/j.biotechadv.2017.09.001.

Dimitroglou, A., Merrifield, D. L., Moate, R., Davies, S. J., Spring, P., Sweetman, J., & Bradley, G. (2009). Dietary mannan oligosaccharide supplementation modulates intestinal microbial ecology and improves gut morphology of rainbow trout, *Oncorhynchus mykiss* (Walbaum). *Journal of Animal Science, 87*(10), 3226–3234. https://doi.org/10.2527/jas.2008-1428.

Dimou, A., Tsimihodimos, V., & Bairaktari, E. (2022). The critical role of the branched chain amino acids (BCAAs) catabolism-regulating enzymes, branched-chain aminotransferase (BCAT) and branched-chain α-keto acid dehydrogenase (BCKD), in human pathophysiology. *International Journal of Molecular Sciences, 23*(7), 4022. https://doi.org/10.3390/ijms23074022.

DiógeneS, A. F., et al. (2016). Establishing the optimal essential amino acid ratios in juveniles of Nile tilapia (*Oreochromis niloticus*) by the deletion method. *Aquaculture Nutrition, 22*(2), 435–443.

Dong, M., Feng, L., Kuang, S. Y., Liu, Y., Jiang, J., Hu, K., Jiang, W. D., Li, S. H., Tang, L., & Zhou, X. Q. (2012). Growth, body composition, intestinal enzyme activities and microflora of juvenile Jian carp (*Cyprinus carpio* var. Jian) fed graded levels of dietary valine. *Aquaculture Nutrition, 19*(1), 1–14. https://doi.org/10.1111/j.1365-2095.2011.00926.x.

Duan, M., Chen, S., Liu, X., Liu, J., & Zhu, D. (2023). The application of *Corynebacterium glutamicum* in L-threonine biosynthesis. *Fermentation, 9*(9), 822. https://doi.org/10.3390/fermentation9090822.

Ebeneezar, S., Vijayagopal, P., Srivastava, P. P., Gupta, S., Sikendra, K., Varghese, T., Prabu, D. L., Chandrasekar, S., Varghese, E., Sayooj, P., Tejpal, C., & Wilson, L. (2019). Dietary lysine requirement of juvenile silver pompano, *Trachinotus blochii* (Lacepede, 1801). *Aquaculture, 511*, 1–8. https://doi.org/10.1016/j.aquaculture.2019.734234.

Emmert, J. L., & Baker, D. H. (1997). Use of the ideal protein concept for precision formulation of amino acid levels in broiler diets. *Journal of Applied Poultry Research, 6*(4), 462–470. https://doi.org/10.1093/japr/6.4.462.

Falco, F., Stincone, P., Cammarata, M., & Brandelli, A. (2020). Amino acids as the main energy source in fish tissues. *Aquaculture and Fisheries Studies, 3*(1), 1–11.

Feng, L., Luo, J. B., Jiang, W. D., et al. (2015). Changes in barrier health status of the gill for grass carp (*Ctenopharyngodon idella*) during valine deficiency: Regulation of tight junction protein transcript, antioxidant status and apoptosis-related gene expression. *Fish & Shellfish Immunology, 45*, 239–249.

Fernstrom, J. D. (2005). Branched-chain amino acids and brain function. *The Journal of Nutrition, 135*(6), 1539S–1546S. https://doi.org/10.1093/jn/135.6.1539S

Furuya, W. M., Cruz, T. P. D., & Gatlin, D. M. (2023). Amino acid requirements for Nile tilapia: An update. *Animals, 13*(5), 900. https://doi.org/10.3390/ani13050900

Furuya, W. M., Pezzato, L. E., Barros, M. M., Pezzato, A. C., Furuya, V. R., & Miranda, E. C. (2004). Use of ideal protein concept for precision formulation of amino acid levels in fish-meal-free diets for juvenile Nile tilapia (*Oreochromis niloticus* L.). *Aquaculture Research, 35*(12), 1110–1116.

Gaber, M., Salem, M., Zaki, M., & Nour, A. (2016). Amino acid requirements of gilthead bream (*Sparus aurata*) juveniles. *World Journal of Engineering and Technology, 4*, 18–24. https://doi.org/10.4236/wjet.2016.43B004.

Gao, H., Tuyishime, P., Zhang, X., et al. (2021). Engineering of microbial cells for L-valine production: Challenges and opportunities. *Microbial Cell Factories, 20*, 172. https://doi.org/10.1186/s12934-021-01665-5.

Gao, K., Li, G., Zhu, C., Wang, L., Yang, X., Wen, X., & Jiang, Z. (2019). Effect of optimizing dietary valine-to-lysine ratio in late gestation or lactation on biochemical indices and performance of lactating primiparous sows. *Animal Feed Science and Technology*. https://doi.org/10.1016/j.anifeedsci.2019.04.015.

Geraskina, N. V., Sycheva, E. V., Samsonov, V. V., Eremina, N. S., Hook, C. D., Serebrianyi, V. A., et al. (2019). Engineering *Escherichia coli* for auto inducible production of L-valine: An example of an artificial positive feedback loop in amino acid biosynthesis. *PLoS ONE, 14*(4), e0215777. https://doi.org/10.1371/journal.pone.0215777.

Ghaly, A. E., Ramakrishnan, V. V., Brooks, M. S., Budge, S. M., & Dave, D. (2013). Fish processing wastes as a potential source of proteins, amino acids and oils: A critical review. *Journal of Microbial & Biochemical Technology, 5*(4), 107–129.

Gorissen, S. H. M., & Phillips, S. M. (2019). Branched-chain amino acids (leucine, isoleucine, and valine) and skeletal muscle. In *Nutrition and Skeletal Muscle* (pp. 263–278). https://doi.org/10.1016/b978-0-12-810422-4.00016-6.

Greiner, L., Graham, A., Goncalves, M., Orlando, U., & Touchette, K. J. (2019). Evaluation of the optimal standardized ileal digestible valine:lysine ratio in lactating sow diets. *Journal of Animal Science, 97*(7), 2965–2971.

Habte-Tsion, H. M. (2020). A review on fish immuno-nutritional response to indispensable amino acids in relation to TOR, NF-κB and Nrf2 signaling pathways: Trends and prospects. *Comparative Biochemistry and Physiology Part B: Biochemistry and Molecular Biology, 241*, 110389. https://doi.org/10.1016/j.cbpb.2019.110389.

Han, G., Xu, N., Sun, X., Chen, J., Chen, C., & Wang, Q. (2020). Improvement of L-valine production by atmospheric and room temperature plasma mutagenesis and high-throughput screening in *Corynebacterium glutamicum*. *ACS Omega, 5*(10), 4751–4758. https://doi.org/10.1021/acsomega.9b02747.

Hao, N., Yan, M., Zhou, H., Liu, H. M., Cai, P., & Ouyang, P. K. (2010). The effect of AmtR on growth and amino acids production in *Corynebacterium glutamicum*. *Applied Biochemistry and Microbiology, 46*(6), 561–566.

Hao, Y., Pan, X., Xing, R., You, J., Hu, M., Liu, Z., Li, X., Xu, M., & Rao, Z. (2022). High-level production of L-valine in *Escherichia coli* using multi-modular engineering. *Bioresource Technology, 359*, 127461. https://doi.org/10.1016/j.biortech.2022.127461.

Hardy, R. W. (2012). The nutritional pathology of teleosts. In R. J. Roberts (Ed.), *Fish Pathology* (4th ed., pp. 402–424). Wiley-Blackwell.

Hasegawa, S., Suda, M., Uematsu, K., Natsuma, Y., Hiraga, K., Jojima, T., et al. (2013). Engineering of *Corynebacterium glutamicum* for high-yield L-valine production under oxygen deprivation conditions. *Applied and Environmental Microbiology, 79*, 1250–1257.

Hasegawa, S., Uematsu, K., Natsuma, Y., Suda, M., Hiraga, K., Jojima, T., et al. (2012). Improvement of the redox balance increases L-valine production by *Corynebacterium glutamicum* under oxygen deprivation conditions. *Applied and Environmental Microbiology, 78*, 865–875.

Hauschild, L., Pomar, C., & Lovatto, P. A. (2010). Systematic comparison of the empirical and factorial methods used to estimate the nutrient requirements of growing pigs. *Animal, 4*(5), 714–723. https://doi.org/10.1017/S1751731109991546.

Hermann, T. (2003). Industrial production of amino acids by coryneform bacteria. *Journal of Biotechnology, 104*(1). https://doi.org/10.1016/S0168-1656(03)00149-4.

Hidalgo, M. C., Morales, A. E., Arizcun, M., Abellán, E., & Cardenete, G. (2017). Regional asymmetry of metabolic and antioxidant profile in the sciaenid fish shi drum (*Umbrina cirrosa*) white muscle. Response to starvation and refeeding. *Redox Biology, 11*, 682–687.

Hixson, S. (2014). Fish nutrition and current issues in aquaculture: The balance in providing safe and nutritious seafood, in an environmentally sustainable manner. *Journal of Aquaculture Research & Development, 5*, 234. https://doi.org/10.4172/2155-9546.1000234.

Hou, Y., Yin, Y., & Wu, G. (2015). Dietary essentiality of "nutritionally non-essential amino acids" for animals and humans. *Experimental Biology and Medicine (Maywood), 240*(8), 997–1007. https://doi.org/10.1177/1535370215587913.

Hua, K., & Bureau, D. P. (2019). Estimating changes in essential amino acid requirements of rainbow trout and Atlantic salmon as a function of body weight or diet composition using a novel factorial requirement model. *Aquaculture, 513*, 734440.

Huang, Z., Tan, X. H., Zhou, C. P., Yang, Y. K., Qi, C. I., Zhao, S. Y., & Lin, H. Z. (2018). Effect of dietary valine levels on the growth performance, feed utilization and immune function of juvenile golden pompano, *Trachinotus ovatus*. *Aquaculture Nutrition, 24*(1), 74–82. https://doi.org/10.1111/anu.12535.

Ikeda, M., & Nakagawa, S. (2003). The *Corynebacterium glutamicum* genome: Features and impacts on biotechnological processes. *Applied Microbiology and Biotechnology, 62*(2, 3), 99–109. https://doi.org/10.1007/s00253-003-1328-1.

Islam, M. M., Nautiyal, M., Wynn, R. M., Mobley, J. A., Chuang, D. T., & Hutson, S. M. (2010). Branched-chain amino acid metabolon: Interaction of glutamate dehydrogenase with the mitochondrial branched-chain aminotransferase (BCATm). *Journal of Biological Chemistry, 285*(1), 265–276. https://doi.org/10.1074/jbc.M109.048777.

Jakubowski, H., & Flatt, P. (2023). *Fundamentals of Biochemistry II: Bioenergetics and Metabolism*. LibreTexts. https://libretexts.org (accessed on 3 June 2025).

Karau, A., & Grayson, I. (2014). Amino acids in human and animal nutrition. In *Advances in Biochemical Engineering/Biotechnology* (Vol. 143, pp. 189–228). https://doi.org/10.1007/10_2014_269.

Kaushik, S. J., & Seiliez, I. (2010). Protein and amino acid nutrition and metabolism in fish: Current knowledge and future needs. *Aquaculture Research, 41*(3), 322–332.

Kawanago, M., Takemura, S., Ishizuka, R., Kousaka, T., & Shioya, I. (2014). Leucine affects growth and hepatic growth-related factor gene expression in Japanese amberjack *Seriola quinqueradiata*. *North American Journal of Aquaculture, 76*, 415–422. https://doi.org/10.1080/15222055.2014.933752.

Khan, K. U., Mansano, C. F., Nascimento, T. M. T., Boaratti, A. Z., Rodrigues, A. T., Queiroz, D. M. A., & Fernandes, J. B. K. (2020). Whole-body amino acid pattern of juvenile, preadult, and adult pacu, *Piaractus mesopotamicus*, with an estimation of its dietary essential amino acid requirements. *Journal of the World Aquaculture Society, 51*, 224–234.

Khan, M. A., & Abidi, S. F. (2007). Dietary isoleucine requirement of fingerling Indian major carp, *Labeo rohita* (Hamilton). *Aquaculture Nutrition, 13*, 424–430.

Kim, S., Baker, D., & Easter, R. (2001). Dynamic ideal protein and limiting amino acids for lactating sows: The impact of amino acid mobilization. *Journal of Animal Science, 79*(9), 2356–2366. https://doi.org/10.2527/2001.7992356x.

Kondo, A., Ishii, J., Hara, K. Y., Hasunuma, T., & Matsuda, F. (2013). Development of microbial cell factories for bio-refinery through synthetic bioengineering. *Journal of Biotechnology, 163*, 204–216.

Lall, S. P., & Anderson, S. Amino acid nutrition of salmonids: Dietary requirements and bioavailability. In D. Montero, B. Basurco, I. Nengas, M. Alexis, & M. Izquierdo (Eds.), *Mediterranean Fish Nutrition* (pp. 73–90). Zaragoza: CIHEAM. (Cahiers Options Méditerranéennes; No. 63).

Lansard, M., Panserat, S., Plagnes-Juan, E., Dias, K., Seiliez, I., & Skiba-Cassy, S. (2011). L-leucine, L-methionine, and L-lysine are involved in the regulation of intermediary metabolism-related gene expression in rainbow trout hepatocytes. *The Journal of Nutrition, 141*, 75–80. https://doi.org/10.3945/jn.110.124511.

Le Gouvello, R., & Simard, F. (2017). *Durabilité Des Aliments Pour Le Poisson En Aquaculture: Relaxions Et Recommendations Sur Les Aspects Technologies, Economies, Sochaux Et Environment.* Gland, Suisse: UICN; Paris, France: Comité français de l'UICN.

Lee, J. K., Cho, S. H., Park, S. U., Kim, K., & Lee, S. (2003). Dietary protein requirement for young turbot (*Scophthalmus maximus* L.). *Aquaculture Nutrition, 9*, 283–286.

Lee, J. W., Kim, T. Y., Jang, Y. S., Choi, S., & Lee, S. Y. (2011). Systems metabolic engineering for chemicals and materials. *Trends in Biotechnology, 29*(8), 370–378.

Lee, S. Y., & Kim, H. U. (2015). Systems strategies for developing industrial microbial strains. *Nature Biotechnology, 33*(10), 1061–1072.

Lee, S., Kim, K., Park, H. G., Kim, C. H., & Hong, K. E. (2001). Protein requirement of juvenile Manchurian trout *Brachymystax lenok. Fish Science, 67*, 46–51.

Leuchtenberger, W., Huthmacher, K., & Drauz, K. (2005). Biotechnological production of amino acids and derivatives: Current status and prospects. *Applied Microbiology and Biotechnology, 69*, 1–8. https://doi.org/10.1007/s00253-005-0155-y.

Leyval, D., Uy, D., Delaunay, S., Goergen, J. L., & Engasser, J. M. (2003). Characterization of the enzyme activities involved in the valine biosynthetic pathway in a valine-producing strain of *Corynebacterium glutamicum. Journal of Biotechnology, 104*(1–3), 241–252. https://doi.org/10.1016/s0168-1656(03)00162-7.

Li, H., Xu, W., Jin, J., et al. (2018). Effects of starvation on glucose and lipid metabolism in gibel carp (*Carassius auratus gibelio* var. CAS III). *Aquaculture, 496*, 166–175.

Li, P., & Wang, X. (2004). Nutritional value of fisheries by-catch and byproduct meals in the diet of red drum (*Sciaenops ocellatus*). *Aquaculture, 236*, 485–496.

Li, P., Mai, K., Trushenski, J., & Wu, G. (2009). New developments in fish amino acid nutrition: Towards functional and environmentally oriented aqua feeds. *Amino Acids, 37*, 43–53. https://doi.org/10.1007/s00726-008-0171-1.

Li, X., Zheng, S., & Wu, G. (2021). Nutrition and functions of amino acids in fish. In G. Wu (Ed.), *Amino Acids in Nutrition and Health: Amino Acids in The Nutrition of Companion, Zoo and Farm Animals* (pp. 133–168). Springer International Publishing. https://doi.org/10.1007/978-3-030-54462-1_8.

Li, X., Zheng, S., & Wu, G. (2021). Nutrition and functions of amino acids in fish. In G. Wu (Ed.), *Amino Acids in Nutrition and Health* (Vol. 1285, pp. 133–168). Springer, Cham.

Liebl, W. (1991). The genus *Corynebacterium*—nonmedical. In A. Balows, H. G. Trüper, M. Dworkin, W. Harder, & K. H. Schleifer (Eds.), *The Prokaryotes* (Vol. 2, pp. 1157–1171). Springer, New York, NY.

Lin, Y., Gong, Y., Yuan, Y., Gong, S., Yu, D., Li, Q., & Luo, Z. (2013). Dietary L-lysine requirement of juvenile Chinese sucker, *Myxocyprinus asiaticus*. *Aquaculture Research, 44*, 1539–1549. https://doi.org/10.1111/j.1365-2109.2012.03161.x.

Liu, J., Xu, J. Z., Wang, B., Rao, Z. M., & Zhang, W. G. (2021). L-valine production in *Corynebacterium glutamicum* based on systematic metabolic engineering: Progress and prospects. *Amino Acids, 53*(9), 1301–1312. https://doi.org/10.1007/s00726-021-03066-9.

Liu, S., Wang, B. B., Xu, J. Z., & Zhang, W. G. (2023). Engineering of shikimate pathway and terminal branch for efficient production of L-tryptophan in *Escherichia coli*. *International Journal of Molecular Sciences, 24*, 11866. https://doi.org/10.3390/ijms241411866.

Lokesh, J., Fernandes, J. M., Korsnes, K., BerghØ, Brinchmann, M. F., & Kiron, V. (2012). Transcriptional regulation of cytokines in the intestine of Atlantic cod fed yeast-derived mannan oligosaccharide or β-glucan and challenged with *Vibrio anguillarum*. *Fish & Shellfish Immunology, 33*, 626–631.

Luo, L., Wang, Y., Li, Q., Wang, F., Chen, S., Liu, B., & Wen, H. (2010). Research on dietary valine requirement of juvenile grass carp (*Ctenopharyngodon idella*). *Chinese Journal of Animal Nutrition, 22*(3), 616–624.

Lynch, C. J., & Adams, S. H. (2014). Branched-chain amino acids in metabolic signaling and insulin resistance. *Nature Reviews Endocrinology, 10*, 723–736.

Ma, Q., Zhang, Q., Xu, Q., Zhang, C., Li, Y., Fan, X., Xie, X., & Chen, N. (2017). Systems metabolic engineering strategies for the production of amino acids. *Synthetic and Systems Biotechnology, 2*(2), 87–96. https://doi.org/10.1016/j.synbio.2017.07.003.

Mambrini, M., & Kaushik, S. J. (1995). Indispensable amino acid requirements of fish: Correspondence between quantitative data and amino acid profiles of tissue proteins. *Journal of Applied Ichthyology, 11*(3, 4), 240–247.

Mann, G., Mora, S., Madu, G., & Adegoke, O. A. J. (2021). Branched-chain amino acids: Catabolism in skeletal muscle and implications for muscle and whole-body metabolism. *Frontiers in Physiology, 12*, 702826. https://doi.org/10.3389/fphys.2021.702826.

Manoli, I., & Venditti, C. P. (2016). Disorders of branched chain amino acid metabolism. *Transl Sci Rare Dis. 1*(2), 91–110. doi: 10.3233/TRD-160009.

Mansano, C. F. M., do Nascimento, T. M. T., Peres, H., Rodrigues, F. H. F., Khan, K. U., Romaneli, R. S., & Fernandes, J. B. K. (2020). Determination of the optimum dietary essential amino acid profile for growing phase of Nile tilapia by deletion method. *Aquaculture*, *735*, 204. https://doi.org/10.1016/j.aquaculture.2020.735204.

Marammazi, J. G., Yaghoubi, M., Safari, O., Peres, H., & Mozanzadeh, M. T. (2017). Establishing the optimum dietary essential amino acid pattern for silvery-black porgy (*Sparidentex hasta*) juveniles by deletion method. *Aquaculture Nutrition, 23*(6), 1483–1491. https://doi.org/10.1111/anu.12524.

Michel, A., Koch-Koerfges, A., Krumbach, K., Brocker, M., & Bott, M. (2015). Anaerobic growth of *Corynebacterium glutamicum* via mixed-acid fermentation.

Applied and Environmental Microbiology, 81(21), 7496–7508. https://doi.org/10.1128/AEM.02413-15.

Michelato, M., Xavier, T. O., de Moura, L. B., de Almeida, F. L. A., Pedrosa, V. B., Furuya, V. R. B., & Furuya, W. M. (2016). Dietary lysine requirement to enhance muscle development and fillet yield of finishing Nile tilapia. *Aquaculture, 457*, 124–130.

Miles, R. D., & Chapman, F. A. (2007). The concept of ideal protein in formulation of aquaculture feeds. Gainesville: Department of Fisheries and Aquatic Sciences; UF/IFAS Extension, University of Florida. http://edis.ifas.ufl.edu (accessed on 3 June 2025).

Mukhtar, B., Malik, M. F., Shah, S. H., Azzam, A., Slahuddin, & Liaqat, I. (2017). Lysine supplementation in fish feed. *International Journal of Applied Biology and Forensics, 1*(2), 26–31.

Nascimento, T. M. T. (2015). *Níveis de aminoácidos digestíveis para juvenis de tambaqui, Colossoma macropomum e turnover isotópico do carbono (δ13C) e do nitrogênio (δ15N) no tecido muscular* (Doctoral dissertation, São Paulo State University).

Nascimento, T. M., Mansano, C. F., Peres, H., Rodrigues, F. H., Khan, K. U., Romaneli, R. S., Sakomura, N. K., & Fernandes, J. B. (2020). Determination of the optimum dietary essential amino acid profile for growing phase of Nile tilapia by deletion method. *Aquaculture, 523*, 735204. https://doi.org/10.1016/j.aquaculture.2020.735204.

National Research Council (NRC). (2011). *Nutrient Requirements of Fish and Shrimp.* National Academy Press.

Neinast, M. D., Jang, C., Hui, S., Murashige, D. S., Chu, Q., Morscher, R. J., & Anthony, T. G. (2019). Quantitative analysis of the whole-body metabolic fate of branched-chain amino acids. *Cell Metabolism, 29*(2), 417–429.e14. https://doi.org/10.1016/j.cmet.2018.10.013.

Neinast, M., Murashige, D., & Arany, Z. (2019). Branched chain amino acids. *Annual Review of Physiology, 81*, 139–164. https://doi.org/10.1146/annurev-physiol-020518-114455.

Nie, C., He, T., Zhang, W., Zhang, G., & Ma, X. (2018). Branched chain amino acids: Beyond nutrition metabolism. *International Journal of Molecular Sciences, 19*(4), 954. https://doi.org/10.3390/ijms19040954.

Nose, T. (1979). Summary report on the requirements of essential amino acids for carp. In J. E. Halver & K. Tiews (Eds.), *Finfish Nutrition and Fish Feed Technology* (pp. 145–156). Heinemann.

NRC (National Research Council). (2011). *Nutrient Requirements of Fish and Shrimp.* National Academy Press.

Nunes, A. J. P., Sá, M. V. C., Browdy, C. L., & Vazquez-Anon, M. (2014). Practical supplementation of shrimp and fish feeds with crystalline amino acids. *Aquaculture, 431*, 20–27. https://doi.org/10.1016/j.aquaculture.2014.04.003.

Nunes, J., Sá, M. V., Browdy, C. L., & Vazquez-Anon, M. (2014). Practical supplementation of shrimp and fish feeds with crystalline amino acids. *Aquaculture, 431*, 20–27.

Oldiges, M., Eikmanns, B. J., & Blombach, B. (2014). Application of metabolic engineering for the biotechnological production of l-valine. *Applied Microbiology and Biotechnology, 98*(13), 5859–5870. https://doi.org/10.1007/s00253-014-5782-8

Oliva-Teles, A. (2012). Nutrition and health of aquaculture fish. *Journal of Fish Diseases, 35*(2), 83–108.

Oliveira, T. S., Khan, K. U., Boaratti, A. Z., Rodrigues, A. T., Reis, M. P., Sakomura, N. K., & Fernandes, J. B. K. (2021). Evaluation of the optimum dietary essential amino acid pattern for adult pacu (*Piaractus mesopotamicus*). *Aquaculture*, 540, 736686. https://doi.org/10.1016/j.aquaculture.2021.736686.

Ovie, S. O., & Eze, S. S. (2013). Lysine requirement and its effect on the body composition of *Oreochromis niloticus* fingerlings. *Journal of Fisheries and Aquatic Science, 8*(2), 94–100.

Pasupuleti, V., Holmes, C., & Demain, A. (2010). Applications of protein hydrolysates in biotechnology. In V. K. Pasupuleti & A. L. Demain (Eds.), *Protein Hydrolysates in Biotechnology* (pp. 1–9). Springer. https://doi.org/10.1007/978-1-4020-6674-0_1.

Pereira, A. G., Fraga-Corral, M., Garcia-Oliveira, P., Otero, P., Soria-Lopez, A., Cassani, L., & Simal-Gandara, J. (2022). Single-cell proteins obtained by circular economy intended as a feed ingredient in aquaculture. *Foods, 11*(18), 2831. https://doi.org/10.3390/foods11182831.

Peres, H., & Oliva-Teles, A. (2009). The optimum dietary essential amino acid profile for gilthead seabream (Sparus aurata) juveniles. *Aquaculture, 296*(1, 2), 81–86. https://doi.org/10.1016/j.aquaculture.2009.04.046.

Petro, T. M., & Bhatacharjee, J. K. (1981). Effect of dietary essential amino acid limitations upon the susceptibility to *Salmonella typhimurium* and the effect upon humoral and cellular immune responses in mice. *Infection and Immunity, 32*(1), 251–259.

Prabhu, K., & Shetty, J. K. (2014). *Quick Review of Biochemistry for Undergraduates: Question and Answers* (1st ed.). https://doi.org/10.13140/RG.2.1.4966.6961.

Prasanna, J., Sreeramana, A., & Praveen, B. M. (2021). Introduction to fish nutrition, feed formulation, and feeding conversion. *Advances in Animal and Veterinary Sciences, 12*, 208–216.

Preece, K., Hooshyar, N., & Zuidam, N. (2017). Whole soybean protein extraction processes: A review. *Innovative Food Science & Emerging Technologies, 43*, 163–172.

Pundir, C. S., Lata, S., & Narwal, V. (2018). Biosensors for determination of d and l-amino acids: A review. *Biosensors and Bioelectronics, 117*, 373–384. https://doi.org/10.1016/j.bios.2018.06.033.

Rahimnejad, S., & Lee, K. J. (2013). Dietary valine requirement of juvenile red sea bream *Pagrus major*. *Aquaculture, 416, 417*, 212–218.

Rawles, S. D., Green, B. W., McEntire, M. E., Gaylord, T. G., & Barrows, F. T. (2018). Reducing dietary protein in pond production of hybrid striped bass (*Morone chrysops* × *M. saxatilis*): Effects on fish performance and water quality dynamics. *Aquaculture, 490*, 217–227.

Ren, M. C., Habte-Tsion, H. M., Liu, B., Zhou, Q. L., Xie, J., Ge, X. P., Liang, H. L., & Zhao, Z. X. (2015). Dietary valine requirement of juvenile blunt snout bream (*Megalobrama amblycephala* Yih, 1955). *Journal of Applied Ichthyology, 31*(6), 1086–1092.

Rodrigues, A. T., et al. (2020). Ideal profile of essential amino acids for Nile tilapia (*Oreochromis niloticus*) in the finishing growth phase. *Aquaculture Research, 51*(11), 4724–4735.

Rodrigues, R. B., Lui, T. A., Neu, D. H., Boscolo, W. R., & Bittencourt, F. (2018). Valine in diets for Nile tilapia. *Revita Ciencia Agronomical, 49*(3), 467–474. https://doi.org/10.5935/1806-6690.20180053.

Rollin, X., Mambrini, M., Abboudi, T., Larondelle, Y., & Kaushik, S. J. (2003). The optimum dietary indispensable amino acid pattern for growing Atlantic salmon (*Salmo salar* L.) fry. *British Journal of Nutrition, 90*(5), 865–876.

Şanlier, N., Gökcen, B. B., & Sezgin, A. C. (2019). Health benefits of fermented foods. *Critical Reviews in Food Science and Nutrition, 59*(3), 506–527.

Santana, T. M., Dantas, F. d. M., Monteiro Dos Santos, D. K., Kojima, J. T., Pastrana, Y. M., De Jesus, R. S., & Gonçalves, L. U. (2023). Fish viscera silage: Production, characterization, and digestibility of nutrients and energy for tambaqui juveniles. *Fishes, 8*(2), 111. https://doi.org/10.3390/fishes8020111.

Schwentner, A., Feith, A., Münch, E., Busche, T., Rückert, C., Kalinowski, J., & Blombach, B. (2018). Metabolic engineering to guide evolution–Creating a novel mode for L-valine production with *Corynebacterium glutamicum*. *Metabolic Engineering, 47*, 31–41. https://doi.org/10.1016/j.ymben.2018.02.015.

Shimomura, Y., Yamamoto, Y., & Bajotto, G. S. (2006). Nutraceutical effects of branched-chain amino acids on skeletal muscle. *Journal of Nutrition, 136*(2), 529S–533S.

Siebert, D., Khan, D. R., & Torrallardona, D. (2021). The optimal valine to lysine ratio for performance parameters in weaned piglets. *Animals, 11*(5), 1255. https://doi.org/10.3390/ani11051255

Silva, J. C., Bomfim, M. A. D., Lanna, E. A. T., Ribeiro, F. B., Siqueira, J. C., Sousa, T. J. R., Marchão, R. S., & Nascimento, D. C. N. (2018). Lysine requirement for tambaqui juveniles. *Semina: Ciências Agrárias, 39*(5), 2157–2168. http://dx.doi.org/10.5433/1679-0359.2018v39n5p2157

Silva, J. V. D., Ribeiro, F. B., Siqueira, J. C. D., Bomfim, M. A. D., Nascimento, D. C. N. D., & Marchão, R. S. (2022). Dietary valine requirement of tambaqui (*Colossoma macropomum*) with different body weights. *Ciência e Agrotecnologia, 46*.

Simmer, H. H. (1981). Medizin und Chemie um die Mitte des 19. Jahrhunderts in Erlangen: Eugen Franz Freiherr von Gorup-Besanez (1817–1878) [Medicine and chemistry around the middle of the 19th century in Erlangen. Eugen Franz Freiherr von Gorup-Besanez (1817–1878) (author's transl.)]. *Journal of Clinical Chemistry and Clinical Biochemistry, 19*(7), 497–509. (German)

Small, B. C., & Soares, J. H. (2000). Quantitative dietary requirement of juvenile striped bass. *Aquaculture Nutrition, 6*(4), 207–212.

Suryawan, A., Orellana, R. A., Fiorotto, M. L., & Davis, T. A. (2011). Leucine acts as a nutrient signal to stimulate protein synthesis in neonatal pigs. *Journal of Animal Science, 89*(7), 2004–2016. https://doi.org/10.2527/jas.2010-3400

Teles, A. O., Couto, A., Enes, P., & Peres, H. (2019). Dietary protein requirements of fish–A meta-analysis. *Reviews in Aquaculture*. https://doi.org/10.1111/raq.12391.

Teodósio, R., Aragão, C., Conceição, L. E. C., Dias, J., & Engrola, S. (2022). Metabolic fate is defined by amino acid nature in gilthead seabream fed different diet formulations. *Animals, 12*(13), 1713. https://doi.org/10.3390/ani12131713.

Tokuyama, K., Shimodaira, Y., Terawaki, T., Kusunose, Y., Nakai, H., Tsuji, Y., & Shimizu, H. (2020). Data science-based modeling of the lysine fermentation process. *Journal of Bioscience and Bioengineering*. https://doi.org/10.1016/j.jbiosc.2020.06.011.

Trosvik, K., Webster, C., Thompson, K., Metts, L., Gannam, A., & Twibell, R. (2013). Effects on growth performance and body composition in Nile tilapia, *Oreochromis niloticus*, fry fed organic diets containing yeast extract and soybean meal as a total replacement of fish meal without amino acid supplementation. *Biological Agriculture & Horticulture, 29*(3), 173–185. https://doi.org/10.1080/01448765.2013.810123.

Ugimoto, M. S. (2010). Amino acids, production processes. In M. C. Flickinger (Ed.), *Encyclopedia of Bioprocess Technology* (pp. 1–11). John Wiley & Sons.

van Milgen, J., & Dourmad, J. Y. (2015). Concept and application of ideal protein for pigs. *Journal of Animal Science and Biotechnology, 6*(1), 15. https://doi.org/10.1186/s40104-015-0016-1.

Van Milgen, J., Gloaguen, M., Le Floc'H, N., Brossard, L., Primot, Y., & Corrent, E. (2012). Meta-analysis of the response of growing pigs to the isoleucine concentration in the diet. *Animal, 6*(10), 1601–1608.

Veira, A. M., Santos, L. S., Gobi, J. P., Silva, W. C., Fraga, A. Z., & Hauschild, L. (2021). Individual responses of growing pigs to diets with valine and isoleucine to lysine ratios. *Animal Feed Science and Technology, 279*, 115037. https://doi.org/10.1016/j.anifeedsci.2021.115037.

Vijayaram, S., Ringø, E., Zuorro, A., van Doan, H., & Sun, Y. (2023). Beneficial roles of nutrients as immunostimulants in aquaculture: A review. *Aquaculture and Fisheries*.

Volker, F. (2006). Genetic regulation of *Corynebacterium glutamicum* metabolism. *Journal of Microbiology and Biotechnology, 16*(7), 999–1009.

Wang, C., Peng, Y., Zhang, Y., Xu, J., Jiang, S., Wang, L., & Yin, Y. (2023). The biological functions and metabolic pathways of valine in swine. *Journal of Animal Science and Biotechnology, 14*(1), 135. https://doi.org/10.1186/s40104-023-00927-z.

Wang, S. D., Li, X., Zhang, M. Z., Jiang, H. B., & Li, M. (2022). Dietary supplementation of crystalline amino acid improves growth performance and health of yellow catfish that reduced by plant proteins replacement of fishmeal. *Aquaculture Nutrition, 2022*, Article ID 7145090. https://doi.org/10.1155/2022/7145090.

Wecke, C., Pastor, A., & Liebert, F. (2013). Ideal isoleucine and valine to lysine ratios in low protein diets for growing pigs. In *Energy and Protein Metabolism and Nutrition in Sustainable Animal Production* (pp. 193–194). https://doi.org/10.3920/978-90-8686-781-3_59.

Ween, O., Stangeland, J. K., Fylling, T. S., & Aas, G. H. (2017). Nutritional and functional properties of fishmeal produced from fresh by-products of cod (*Gadus morhua* L.) and saithe (*Pollachius virens*). *Heliyon, 3*(7), e00343. https://doi.org/10.1016/j.heliyon.2017.e00343.

Weichhart, T., & Saemann, M. D. (2009). The multiple facets of mTOR in immunity. *Trends in Immunology, 30*(5), 218–226.

Wendisch, V. F. (2014). Microbial production of amino acids and derived chemicals: Synthetic biology approaches to strain development. *Current Opinion in Biotechnology, 30*, 51–58.

Wendisch, V. F. (2019). Avancées et perspectives de l'ingénierie métabolique pour la production d'acides aminés. *Metabolic Engineering*. https://doi.org/10.1016/j.ymben.2019.03.008.

Wieschalka, S., Blombach, B., Bott, M., & Eikmanns, B. J. (2013). Bio-based production of organic acids with *Corynebacterium glutamicum*. *Microbial Biotechnology, 6*(1), 87–102.

Wilson, R. P. (1991). Amino acid nutrition of fish: A new method of estimating requirement values. In *Proceedings of the 12th US-Japan Symposium on Aquaculture Nutrition* (pp. 49–54). Newport, Oregon.

Wilson, R. P. (2002). Amino acids and proteins. In J. E. Halver & R. W. Hardy (Eds.), *Fish Nutrition* (3rd ed., pp. 143–179). Academic Press.

Wilson, R. P., Poe, W. E., & Robinson, E. H. (1980). Leucine, isoleucine, valine, and histidine requirements of fingerling channel catfish. *The Journal of Nutrition, 110*(4), 627–633. https://doi.org/10.1093/jn/110.4.627.

Wongkittichote, P., Ah Mew, N., & Chapman, K. A. (2017). Propionyl-CoA carboxylase – A review. *Molecular Genetics and Metabolism, 122*(4), 145–152. https://doi.org/10.1016/j.ymgme.2017.10.002.

Wu, G. (2009). Amino acids: Metabolism, functions, and nutrition. *Amino Acids, 37*(1), 1–17.

Wu, G. (2013). *Amino Acids: Biochemistry and Nutrition.* CRC Press.

Wu, G. (2021). *Amino Acids in Nutrition and Health.* Advances in Experimental Medicine and Biology. https://doi.org/10.1007/978-3-030-54462-1.

Wu, G., Bazer, F. W., Dai, Z., Li, D. F., Wang, J. J., & Wu, Z. L. (2014). Amino acid nutrition in animals: Protein synthesis and beyond. *Annual Review of Animal Biosciences, 2*(1), 387–417. https://doi.org/10.1146/annurev-animal-022513-114113.

Xiao, W., Li, D. Y., Zhu, J. L., et al. (2018). Dietary valine requirement of juvenile Nile tilapia, *Oreochromis niloticus. Aquaculture Nutrition, 24*(1), 315–323.

Xiao, W., Li, D. Y., Zhu, J. L., Zou, Z. Y., Yue, Y. R., & Yang, H. (2017). Dietary valine requirement of juvenile Nile tilapia, *Oreochromis niloticus. Aquaculture Nutrition, 24*(1), 315–323. https://doi.org/10.1111/anu.12562.

Yamamoto, K., Tsuchisaka, A., & Yukawa, H. (2017). Branched-chain amino acids. In *Advances in Biochemical Engineering/Biotechnology* (Vol. 159, pp. 103–128). https://doi.org/10.1007/10_2016_28.

Yang, L., Chu, Z., Liu, M., et al. (2023). Amino acid metabolism in immune cells: Essential regulators of effector functions, and promising opportunities to enhance cancer immunotherapy. *Journal of Hematology & Oncology, 16*, 59. https://doi.org/10.1186/s13045-023-01453-1.

Zehra, S., & Khan, M. A. (2014). Dietary valine requirement of fingerling *Catla catla. Journal of Applied Aquaculture, 26*(3), 232–251. https://doi.org/10.1080/10454438.2014.937659.

Zhang, H., Li, Y., Wang, C., & Wang, X. (2018). Understanding the high L-valine production in *Corynebacterium glutamicum* VWB-1 using transcriptomics and proteomics. *Scientific Reports, 8*(1), 3632. https://doi.org/10.1038/s41598-018-21926-5.

Zhang, S., Zeng, X., Ren, M., et al. (2017). Novel metabolic and physiological functions of branched chain amino acids: A review. *Journal of Animal Science and Biotechnology, 8*, 10. https://doi.org/10.1186/s40104-016-0139-z.

Zhao, G., Gong, G., Wang, P., & Wang, L. (2014). Enzymatic synthesis of L-aspartic acid by *Escherichia coli* cultured with a cost-effective corn plasm medium. *Annals of Microbiology, 64*(4), 1615–1621.

Zhao, L., Li, Y., Li, Z., Wu, S., Huang, K., Chen, J., & Li, C. (2019). Effect of the valine-to-lysine ratio on the performance of sows and piglets in a hot, humid environment. *Journal of Thermal Biology, 81*, 89–97. https://doi.org/10.1016/j.jtherbio.2019.02.021.

Zhao, Y., Li, J. Y., Jiang, Q., et al. (2020). Leucine improved growth performance, muscle growth, and muscle protein deposition through AKT/TOR and AKT/FOXO3a signaling pathways in hybrid catfish *Pelteobagrus vachelli* × *Leiocassis longirostris*. *Cell, 9*(2), 327.

Zhou, Z., Wu, X., Gatlin, D., Wang, X., Mu, W., Ye, B., & Ma, L. (2021). Dietary valine levels affect growth, protein utilization, immunity and antioxidant status in juvenile hybrid grouper (*Epinephelus fuscoguttatus* ♀ × *Epinephelus lanceolatus* ♂). *British Journal of Nutrition, 125*(4), 408–419. https://doi.org/10.1017/S0007114520002858.s

CHAPTER 7

Dietary Phenylalanine Requirement of Different Fish Species

KOUSAR JAN,[1] IMTIAZ AHMED,[1,*] and ISHTIYAQ AHMAD[2]

[1]Department of Zoology, Fish Nutrition Research Laboratory, University of Kashmir, Hazratbal, Srinagar, Jammu and Kashmir, India

[2]Department of Zoology, Fish Nutrition Research Laboratory, University of Kashmir, Srinagar, Jammu and Kashmir, India; Division of Fish Genetics and Biotechnology, Faculty of Fisheries, Sher-e-Kashmir University of Agricultural Sciences and Technology of Kashmir (SKUAST-Kashmir), Rangil, Ganderbal, Srinagar, Jammu and Kashmir, India

*Corresponding author

ABSTRACT

Appropriate nutrition plays a vital role in increasing fish production through aquaculture. However, this system cannot succeed and sustain without practical feeds that are species-specific, nutritionally balanced, easily affordable, and environmentally friendly. Of all the nutrients needed by fish, protein is one of the most expensive components of fish feed since it is used for tissue growth and repair, as well as providing both essential and non-essential amino acids (EAAs) required to build protein and energy for growth and maintenance. Among all the nutrients, protein is one of the main essential nutrients required by fish for maintaining their life and growth. It is not only essential for growth and maintenance but also important for enzymatic activities and hormone production, which are vital for normal metabolism. Therefore, it is imperative to determine the precise protein requirement of different fish species. Since the quality of protein is actually determined on the basis of amino acid (AA) mixture, it

Fish Nutrition: Protein and Amino Acid Requirements. Imtiaz Ahmed, Ishtiyaq Ahmad, and Shabihul Fatma Sayed (Eds.)

DOI: 10.1201/9781998511570-7

is significant to understand the role of individual AAs in the growth of fish. AAs play a vital role in almost all the bodily functions of fish. Among 22 AAs, 10 are considered essential and 12 are considered non-essential in fish. Among EAAs, phenylalanine, in the form of an aromatic AA, is required for protein synthesis and growth promotion in fishes. Fish have different specific phenylalanine requirements based on their species, life stage, size, and habitat. Fish nutritionists and academics studying aquaculture frequently use feeding trials to ascertain the dietary needs for phenylalanine and other important AAs. In these experiments, diets containing different amounts of the tested AAs are prepared, and the fish's growth performance, feed efficiency, and other pertinent characteristics are monitored. Usually, the ideal dietary phenylalanine needed for a given fish species is stated as a ratio to other necessary AAs or as a percentage of the overall diet. Furthermore, phenylalanine needs in fish can be influenced by water temperature, water quality, and the presence of other nutrients in the diet. It is advised to confer with specialists in fish nutrition and aquaculture, examine scientific literature, or read publications on aquaculture research for precise and species-specific information on the phenylalanine requirements of fish. This chapter reviews the research on the phenylalanine requirement of diverse fish species, which would be helpful in boosting their production.

7.1 INTRODUCTION TO AQUACULTURE

In the last 10 years, there has been a notable growth in fish consumption worldwide, reaching a record of 20.20 kg of fish per person annually (FAO, 2022). The share of global aquaculture in fish production increased from 25.7% to 46% between 2000 and 2018, with projections indicating that this share might reach 50% by 2030 (FAO, 2020; Zhang et al., 2023). The most crucial component of a successful aquaculture operation is feed, primarily fishmeal, which is usually used as the main ingredient in formulated feeds owing to its well-known nutritional profile, high protein content, balanced amino acids (AAs) composition, extreme digestibility, excellent palatability, and low levels of anti-nutritional factors (Gatlin et al., 2007). Nevertheless, as fish meal is being used more often in many cultural contexts, the demand for it will soon exceed supply globally, increasing its price (Ahmed et al., 2023). Since it is responsible for the majority of ongoing expenses that influence yield and determine the price of farmed fish, it is therefore a topic of enormous concern (Mukherjee et al., 2010). New technology, new production techniques, and new raw materials for fish feeding must be adopted in order to increase fish production in a sustainable manner and avoid the exploitation of natural resources (Fiordelmondo et al., 2022). Nowadays, it is becoming more and more important to find fresh and innovative raw materials for fish feeding. It is also critical to comprehend how much of these materials may be replaced without negatively affecting the growth and nutritional quality of fish. Out of all the nutrients needed by fish, protein is considered the most expensive component of fish feed because it is necessary for tissue growth and repair, as well as providing EAAs needed to build body protein and energy for maintenance (Kim et al., 2017). It is wise to calculate the demand in terms of the AA contents in order to accurately determine the protein requirement of a species (Ahmed, 2022). However, the majority of plant proteins are incompatible with the EAA profile of fish, and the introduction of new protein sources may result in an unbalanced dietary EAA profile (excessive or deficiency), which lowers the fish's resistance to disease, their ability to grow, and utilize feed (Kaushik and Seiliez, 2010; Trosvik et al., 2013). According to Harper and Yoshimura (1993), the more imbalance there is between the amount of AAs required by the species and the AA composition of the protein in the diet, the less efficiently feed is used.

The biomolecules known as AAs are the building blocks of tissue protein and are essential for growth, immunity, behavior, development of the larva, reproduction, metabolism, and physiological status of the fish (Gao et al., 2016; Yaghoubi et al., 2018). They also act as precursors for the synthesis of numerous essential and physiologically active compounds, including neurotransmitters, peptide hormones, and nucleotides. In addition, they are involved in cell signaling, appetite stimulation, gene regulation, phosphorylation energy utilization, immunity, osmoregulation, detoxification from ammonia, anti-oxidative defense, metamorphosis, pigmentation, gut and neuronal development, stress responses, reproduction, normal function of the pancreas and liver, and the inhibition of aggressive behavior in aquatic animals (Wu, 2010, 2013; Wu et al., 2014; Yaghoubi et al., 2018). The AAs play a vital role in maintaining the overall development, muscle deposition, feed costs, and nitrogen loading of fish species (Small and Soares, 1999). It is thus important to determine the EAA requirements for different cultivable fish species because the majority of alternative protein sources are deficient in some or all EAAs. This way, alternative protein feedstuffs can be appropriately supplemented with the particular deficient AA as per the requirement of candidate species (Yun et al., 2015). As a result, figuring out the fundamental AA requirements of the main cultivated fish is crucial for creating fish feed formulations.

7.2 PHENYLALANINE INVOLVED IN FISH GROWTH

During protein synthesis, metabolism, and growth, aromatic AAs play an essential function. The distribution of AAs influences fish meat's aromatic qualities in part (Hall and Ahmad, 1992). Phenylalanine is an essential aromatic AA, necessary for fish physiology, metabolism, and protein synthesis (Gao et al., 2016; NRC, 2011). Being aromatic in nature, phenylalanine plays a vital role in the metamorphosis of fish larvae (Pinto et al., 2010). Besides, inclusion of the right quantity of phenylalanine also enhances the fish's ability to digest, absorb, and use antioxidants (Li et al., 2015). Both phenylalanine and tyrosine (aromatic amino acids) are involved in performing various functions in protein synthesis, and the former can easily transform into the latter in the liver and kidney of fish (Gao et al., 2016). Tyrosine is sometimes referred to as a semi-EAA, since in situations where tyrosine levels are low, additional phenylalanine supplementation is required to meet the growth and metabolic requirements of fish production. Being a precursor to tyrosine, phenylalanine can also be converted into thyroid hormone, epinephrine, norepinephrine, and tri-iodothyronine in the liver and kidney (Ravikumar et al., 2000; Zhang et al., 2023), which supports fish growth, tissue differentiation, nutrient metabolism, and ability to withstand stress (Li et al., 2009). Recently, it has been reported that dietary tyrosine supplementation in fish can lower the amount of phenylalanine required in the diet (Zehra and Khan, 2014). Tyrosine may therefore become necessary when dietary phenylalanine levels fall below the minimum threshold amount (Ahmed, 2022; Kim, 1993).

7.3 PHENYLALANINE INVOLVED IN METABOLISM

Phenylalanine is required for metabolism in many different types of organisms, including fish. It is noteworthy that distinct organisms may possess different enzymes involved in the metabolism of phenylalanine, such as phenylalanine hydroxylase. It is involved in protein synthesis and acts as a precursor for the synthesis of other significant molecules, including neurotransmitters and melanin in fish as well as other animals (Ahmed, 2022).

Depending on the evolutionary background, ecological niche, and physiological adaptations, different fish species may have different metabolic pathways. Fish phenylalanine metabolism is carried out through a sequence of metabolic processes. Being an important AA, fish cannot produce phenylalanine on their own and must get it through feeding/diet, which includes meals high in protein (Zehra and Yousif, 2021). During the digestive process, AAs, including phenylalanine, are produced from the breakdown of proteins in the meal. Following absorption in the digestive system, these AAs reach the bloodstream (Saha et al., 2021). The bloodstream then carries it to all of the fish's bodily tissues and cells (Li et al., 2020). Another AA that is synthesized from phenylalanine is tyrosine. The enzyme that transforms phenylalanine into tyrosine is called phenylalanine hydroxylase. Tyrosine is synthesized when this reaction gives phenylalanine a hydroxyl group. For the enzyme to function, tetrahydrobiopterin (BH4), a cofactor, and molecular oxygen are needed. Undoubtedly useful for protein synthesis, tyrosine can also be further broken down into a variety of physiologically active substances (Harper, 2020). For example, according to Li et al. (2009), it is a precursor to the manufacture of melanin and neurotransmitters such as dopamine, adrenaline, and norepinephrine. For the production of proteins, phenylalanine and tyrosine are essential. In the course of translation, these AAs are integrated into polypeptide chains (Li et al., 2021). Excretion or metabolism may occur from any excess phenylalanine not needed for protein synthesis or not converted to tyrosine. All fish species can therefore differ in their distinct excretion systems.

7.4 EMPIRICAL METHODS USED FOR PHENYLALANINE REQUIREMENT

Fish EAA requirements are often estimated using empirical approaches (Fagbenro, 2000). These techniques entail feeding trials with different amounts of EAA grades in the meal and monitoring the fish's physiological characteristics and growth performance (Glencross, 2006; Shearer, 2000). Under carefully controlled experimental settings, these techniques assist researchers in determining the phenylalanine need for particular fish species. The fish species, life cycle, and environmental circumstances must all be taken into account while planning and analyzing these studies. Additionally, to better understand the molecular reactions to dietary phenylalanine, advances in bioinformatics and molecular techniques may be added to conventional empirical methodologies. The following are a few popular empirical techniques for figuring out the fish phenylalanine requirement:

1. **Dose-Response Studies:** In dose-response studies, fish diets containing different concentrations of phenylalanine are fed and the fish's reactions such as growth rate, feed efficiency or protein deposition are monitored. The point at which further increases do not yield additional growth advantages is used to generate a dose-response curve and identify the ideal level of phenylalanine (Castillo et al., 2015; Do Nascimento et al., 2020).
2. **Dietary Dilution Studies:** Different protein levels, including phenylalanine concentrations are incorporated into diet plans. The objective is to dilute the protein content in order to produce a gradient of phenylalanine levels. The phenylalanine requirement is then determined by monitoring physiological reactions and growth performance (Yang et al., 2023).
3. **Broken-Line Regression Analysis:** In regression analysis, the association between phenylalanine consumption and the observed responses is statistically modelled. Using this technique, fish are fed diets that gradually raise the amount of phenylalanine until further increases no longer yield growth or performance gains. The break-point, which represents the phenylalanine need, is then found by using broken-line regression to analyze the data (Guo et al., 2022).
4. **Slope-ratio assay:** Diets containing different proportions of phenylalanine to a reference AA typically one that is not limiting in the diet is provided to fish. For every ratio, the growth response is quantified. The ideal phenylalanine-to-reference AA ratio is found by analyzing the data (El-Haroun and Bureau, 2007; Poppi et al., 2011).
5. **Isocaloric and Isoproteic Diets:** In order to maintain a consistent calorie and protein composition across various diets, the later are designed to be isocaloric and isoproteic. These diets contain varying amounts of phenylalanine. The goal of observing fish growth and performance is to determine the optimal dietary phenylalanine amount for growth under controlled conditions (Ahmed, 2022; Sharf and Khan, 2023).

6. **Tissue Amino Acid Analysis:** Fish tissue samples, such as liver or muscle are taken at the conclusion of feeding trials. To determine whether the phenylalanine requirements have been satisfied and how it relates to growth performance, the AA composition of tissues is examined (do Nascimento et al., 2020; Kasozi et al., 2019).
7. **Response Surface Methodology (RSM):** RSM is a statistical method for modelling and examining the association between a number of variables and answers. RSM can be used to investigate the interactions between phenylalanine and other dietary components in the setting of phenylalanine needs. It aids in determining the ideal set of elements for the best outcome (Cankiriligil et al., 2020; Korkmaz and Tokur, 2022).
8. **Blood Plasma Analysis:** Another technique used by researchers to evaluate the amount of phenylalanine in fish blood is blood plasma analysis. Variations in plasma phenylalanine levels can reveal if dietary phenylalanine intake is sufficient or inadequate (Sayed and Ahmed, 2022).
9. **Enzyme Activity and Gene Expression Studies:** The activity of enzymes involved in the metabolism of phenylalanine and the expression of genes linked to the use of phenylalanine may also be studied by researchers. Fish's metabolic reactions to different phenylalanine levels can be understood by evaluating the changes in gene expression patterns or enzyme activity (Zhang et al., 2023).

7.5 DIETARY PHENYLALANINE REQUIREMENT IN FISHES

In aquaculture, creating balanced feed that promotes the best possible growth, reproduction, and general health of farmed fish requires a thorough understanding of the particular needs of EAAs, including phenylalanine. The distinct AA requirements of various fish species are still being researched by scientists and aquaculture nutritionists in an effort to enhance feed formulations and raise the sustainability and efficiency of aquaculture methods. Phenylalanine is essential for several metabolic activities, as well as the formation of new proteins and tissue repair. Fish species, life stage, size, and environmental variables can all affect how much phenylalanine they need in their diet (Ahmed, 2022). Aquaculture nutritionists formulate commercial fish feeds considering the ideal AA content, which includes phenylalanine, for fish's development, health, and reproduction. While sufficient supply of EAAs has garnered a lot of attention from planners and nutritionists, efforts are being made to develop nutritionally balanced, cost-effective feed that could be easily available at an affordable price to farmers. Varying fish species have erratic dietary requirements for phenylalanine that have been researched thus far; however, a lot of work has indicated the role of tyrosine in establishing the phenylalanine requirement of different fish species. Hence, the evaluation for the replacement of tyrosine value for phenylalanine is a pre-requisite before establishing the phenylalanine requirement of any fish species. The tyrosine replacement value for phenylalanine has been reported in the range from 40% to 60% (Ahmed, 2022; NRC, 2011; Zehra and Khan, 2014). For several cultivated fish species, dietary phenylalanine requirements have been determined, such as *Oncorhynchus tshawytscha* (Chance et al., 1964), *Cyprinus carpio* (Nose, 1979), *Oreochromis niloticus* (Santiago and Lovell, 1988), *Chanos chanos* (Borlongan, 1992), *O. mykiss* (Kim, 1993), *Labeo rohita* (Khan and Abidi, 2007), *Cirrhinus mrigala* (Ahmed, 2009), *Diplodus sargus* (Saavedra et al., 2010), *Catla catla* (Zehra and Khan, 2014), *Megalobrama amblycephala* (Ren et al., 2015), *Ctenopharyngodon idella* (Gao et al., 2016; Li et al., 2015), *O. niloticus* (Xiao et al., 2020), *O. niloticus* (Zehra and Yousif, 2021), *Heteropneustes fossilis* (Ahmed, 2022).

All these findings indicated that dietary supplementation of phenylalanine has a direct impact on fish functions (Yang et al., 2023). Research has shown that proper phenylalanine supplementation in the diet could definitely stimulate the growth performance of numerous farmed fish species including *O. mykiss* (Kim, 1993), *Bidyanus bidyanus* (Ngamsnae and Gunasekera, 1999), *C. mrigala* (Ahmed, 2009), *C. catla* (Zehra and Khan, 2014), *Sciaenops ocellatus* (Castillo et al., 2015), *M. amblycephala* (Ren et al., 2015), *O. niloticus* (Zahran et al., 2021), and *H. fossilis* (Ahmed, 2022). Phenylalanine, one of the ketogenic AAs, can increase the retention of both protein and fat content in a variety of aquatic species such as *B. bidyanus* (Ngamsnae and Gunasekera, 1999), *S. ocellatus* (Castillo et al., 2015), *C. idella* (Gao et al., 2016), and *O. niloticus* × *O. aureus* (Xiao et al., 2020). Phenylalanine supplementation may also enhance fish's immunological condition (Feng et al., 2015), antioxidant ability (Feng et al., 2017), and absorptive capacity (Xiao et al., 2020). But excessive and inadequate phenylalanine can both impact negatively in fish development and well-being (Ren et al., 2015; Zehra and Khan, 2014). For every species of farmed fish, it is therefore necessary to determine the ideal dietary phenylalanine requirement. The best phenylalanine concentrations for different farmed fish species diets have been shown to range from 4.7 to 36.6 g/kg in the past such as *Ictalurus punctatus* (4.7–4.9 g/kg) (Robinson et al., 1980), *B. bidyanus* (22.0 g/kg) (Ngamsnae and Gunasekera, 1999), *C. catla* (11.2 g/kg) (Zehra and Khan, 2014), *S. ocellatus* (16.9 g/kg) (Castilo et al., 2015), *M. amblycephala* (9.3–10.1 g/kg) (Ren et al., 2015), *C. idella* (17.6 g/kg) (Gao et al., 2016), *O. niloticus* (11.7–12.1 g/kg) (Jiang et al., 2016), and *Salmo salar* (9.0–16.0 g/kg) (Hua and Bureau, 2019). In general, fish's optimal glucose metabolism also depends on the maintenance of dietary AA levels at an acceptable level (Huang et al., 2021; Liang et al., 2017) (Table 7.1).

TABLE 7.1 Phenylalanine Requirements in Various Fish Species

Fish Species	Phenylalanine Requirement g kg^{-1}	Protein Requirement g kg^{-1}	References
Oncorhynchus tshawytscha	17.0	44.0	Chance et al. (1964)
Cyprinus carpio	13.0	33.0	Nose (1979)
Oreochromis niloticus	11.0	38.0	Santiago and Lovell (1988)
Chanos chanos	19.0	42.2	Borlongan (1992)
Oncorhynchus mykiss	15.0	43.0	Kim (1993)
Labeo rohita fries	11.6	54.0	Khan and Abidi (2007)
Cirrhinus mrigala fingerlings	21.5	53.8	Ahmed (2009)
Carassius gibelio juveniles	10.9	30.2	Ma et al. (2010)
Catla catla fingerlings	10.1	–	Zehra and Khan (2014)
Ctenopharyngodon idella	8.31	27.5	Feng et al. (2015)
Ctenopharyngodon idella	10.4	–	Li et al. (2015)
Megalobrama amblycephala juveniles	18.6	54.7	Ren et al. (2015)
Ctenopharyngodon idella juveniles	17.6	65.9	Gao et al. (2016)
Oreochromis niloticus × *O. aureus* juveniles	8.78	–	Xiao et al. (2020)
Oreochromis niloticus fingerlings	20.6	58.8	Zehra and Yousif (2021)
Heteropneustes fossilis fingerlings	19.2	48.0	Ahmed (2022)
Heteropneustes fossilis	15.83	–	Sayed and Ahmed (2022)
Epinephelus fuscoguttatus × *Epinephelus lanceolatus* juveniles	12.7	27.3	Yang et al. (2023)
Channa punctatus fingerlings	19.8	–	Sharf and Khan (2023)

7.6 SOURCES OF DIETARY PHENYLALANINE (NATURAL AND SYNTHETIC)

In aquaculture, the focus is typically on formulating feeds with premium natural components to satisfy the nutritional needs of fish. It has been suggested that the objective in formulating fish feeds is to offer a comprehensive, balanced feed to the fish that promotes the best possible development, health, and reproduction (Prabu et al., 2017). Therefore, the primary method to improve fish production in aquaculture is to supply a diet containing all the EAAs which fish are not in a position to synthesize. Fish mostly get phenylalanine from natural sources in their diets, and it is an important AA (Yi et al., 2023). Fish feeds are often designed to contain protein sources that naturally offer a balanced profile of AAs, including phenylalanine; hence, it is uncommon to add synthetic phenylalanine supplements directly into fish diets. The principal sources of phenylalanine in fish are as follows:

7.6.1 NATURAL SOURCES

1. **Fish Meal:** In commercial fish feeds, fish meal derived from tiny, oily fish such as anchovies, sardines and herring is a key source of phenylalanine. It contains a lot of phenylalanine that can fulfil the necessary requirement of phenylalanine (Cho and Kim, 2011).
2. **Crustaceans and Insects:** Apart from phenylalanine, the natural sources of phenylalanine include shrimp, krill and insects that live in the water. Numerous fish species frequently consume these kinds of organisms as their foods (Nogales-Merida et al., 2019; Tilami et al., 2020).
3. **Plankton and Algae:** Plankton and algae could be the other sources of phenylalanine for fish because they also represent a good quality of protein enriched with EAAs (Kolmakova and Kolmakov, 2019).
4. **Natural Prey Items:** The phenylalanine intake of wild fish is influenced by their consumption of a range of naturally occurring prey items, such as smaller fish and invertebrates (Molinero et al., 1994).

7.6.2 COMMERCIAL FISH FEEDS

1. **Formulated Feeds:** Commercial fish feeds are designed to meet the specific nutritional requirements of different species of fish. The utilization of fish meal, soybean meal (SB), cereals, and other protein sources in these meals helps to provide a well-balanced AA profile, which includes phenylalanine (Bhosale et al., 2010; Lall and Dumas, 2022).
2. **Plant Proteins:** Certain plant-based protein sources, such as aquatic macrophytes and SB are included in some fish diets (Ahmed et al., 2023). Plant proteins contribute to the overall AA composition even though they could include less phenylalanine than animal proteins (Day et al., 2022; Krajcovicova-Kudlackova et al., 2005).

7.6.3 SYNTHETIC SOURCES (RARELY USED)

1. **Amino Acid Supplements:** Supplements containing synthetic AAs can be added to fish meals to correct certain imbalances or inadequacies (Falco et al., 2020). However, depending on natural and protein-rich foods is more prevalent than directly supplementing with phenylalanine. Despite being widely utilized in lab settings to ascertain the AA requirements of various fish species, synthetic AAs are not advised for use in conventional feed due to their higher cost. However, by adding substances containing these AAs, the knowledge gained from these experiments can be readily applied to commercial feed formulations.

7.7 INTERACTION OF PHENYLALANINE WITH OTHER AMINO ACIDS

An important AA, phenylalanine interacts in different ways with other AAs in fish. These interactions are essential to both the general physiology of the organism and the structure and function of proteins (Li et al., 2009). Knowing how phenylalanine interacts with other AAs in fishes specifically can help us in better understanding the molecular mechanisms behind protein structure and other physiological functions of fishes (Feng et al., 2017). The distinct ecological and environmental characteristics of fishes may have an impact on the particular functions that these interactions play in their biology. The following are some significant features of fish phenylalanine interactions with other AAs:

1. **Protein Synthesis:** During the synthesis of proteins, phenylalanine is integrated into polypeptide chains. Under the guidance of mRNA and ribosomes, it interacts with other AAs in the expanding peptide chain (Sharf and Khan, 2023; Yi et al., 2022).
2. **Hydrophobic Interactions:** Phenylalanine's non-polar side chain makes it a hydrophobic AA. It frequently interacts with other hydrophobic AAs, aiding in the process by which proteins fold into three-dimensional structures (Man et al., 2019). Protein interior stabilization, appropriate folding and structural integrity are all dependent on hydrophobic interactions (Hu et al., 2020).
3. **Aromatic Interactions:** The side chain of phenylalanine has an aromatic ring. Together with other aromatic AAs like tryptophan and tyrosine, this aromatic ring can engage in pi-pi stacking interactions (Al Mughram et al., 2021; Ricker et al., 2023). Protein structures are stabilized by aromatic interactions, which also affect ligand binding or protein recognition in particular cases.
4. **Ion-Pair Interactions:** Ion-pair interactions between phenylalanine and charged AAs are also possible. For instance, it might interact through electrostatic attractions with negatively charged AAs (like glutamic acid and aspartic acid). These interactions can impact protein's stability and function by affecting the total charge distribution inside the protein (Barlow and Thornton, 1983; Nick Pace et al., 2000).
5. **Biological Signaling:** Phenylalanine is one of the AAs that is involved in a number of signaling pathways (Guo et al., 2022; Yi et al., 2023). They may interact with non-AA signaling molecules in addition to other AAs as part of signal transduction cascades.
6. **Enzymatic Activities:** Phenylalanine participates in catalytic events by interacting with other AAs at the active sites of enzymes

(Fazial and Tan, 2021). Enzymes participating in different metabolic pathways depend on these interactions to function properly.

7. **Metabolic Pathways:** Phenylalanine is one of the AA that is involved in metabolic processes. Tyrosine contain a crucial chemical that further contributes to the creation of neurotransmitters and other physiologically active substances which is actually been made possible by phenylalanine itself (Ahmed, 2022).

7.8 PHENYLALANINE AND IDEAL PROTEIN CONCEPT

As a fundamental component of many different proteins in the body, phenylalanine is essential for the production of proteins. Regarding fish nutrition, phenylalanine has a crucial role in fish growth and development, as well as the production of particular proteins (NRC, 2011; Xiao et al., 2020). The AA composition of an optimal diet to fulfill an organism's unique requirements for protein synthesis is the goal of the ideal protein concept. For fish to thrive, reproduce, and function at their best in aquaculture, including fish farming, diets with the optimal protein profile must be formulated (Ahmed and Ahmad, 2020). It is crucial to take into account the individual requirement of EAAs, such as phenylalanine, in order to attain the optimal protein concept in fish diets. To address the nutritional demands of fish for maximum growth and health, the perfect protein concept entails creating a diet that contains all important AAs in the proper amounts (Prabu et al., 2017). In order to get the optimal protein concept in fish feed, the following factors are to be considered:

1. **Amino Acid Profile:** Understanding the specific AA requirements of the target fish species is crucial. This includes assessing the ideal balance of EAAs such as phenylalanine (Aragao et al., 2004; Kaushik and Seiliez, 2010).

2. **Digestibility:** Proteins in the diet and their ability to be digested are equally crucial. A fish's ability to effectively absorb and use AAs for growth is dependent on its protein digestibility (Ahmed et al., 2023; Koch et al., 2016).

3. **Protein Source:** AA compositions vary throughout different protein sources. To produce a balanced AA composition, it is

critical to choose protein sources that must fulfil the AA needs of the fish (Dabrowski et al., 2010).

4. **Protein-to-Energy Ratio:** DP-to-energy balance is also crucial (Konnert et al., 2022; Nguyen, 2017). The correct balance of diets can promote healthy growth without causing excessive energy deposition. Besides growth and maintenance, the balanced proportion of AAs in the diet also serves as an energy source to the fish.
5. **Fish Life Stage:** Depending on the stage of life, fish have different dietary requirements (e.g., fry, fingerlings, adults). For optimum growth and development, diets should be designed with the individual life stage in mind (Lall and Dumas, 2022).

Research on fish nutrition is still developing, and ongoing investigations contribute to our understanding of the precise AA requirements, including phenylalanine, for many fish species. To improve the health and production of farmed fish, nutritionists who specialize in aquaculture and fish farming collaborate to formulate diets that are in line with the concept of optimum protein AA requirements of the fish.

7.9 PHYSIOLOGICAL AND NUTRITIONAL FUNCTIONS OF PHENYLALANINE

Phenylalanine is one of the crucial components of protein synthesis in fish (Wu et al., 2014). The synthesis of polypeptide chains, the building blocks of proteins, also depends on this necessary AA. Proteins are necessary for several physiological processes, such as immune system function, growth, tissue repair, and enzyme synthesis (Li et al., 2007). Fish growth performance may be enhanced by phenylalanine in relation to its capacity to enhance fish feed utilization. According to earlier research, an increase in phenylalanine levels was associated with a higher feed consumption rate in fish species such as *C. mrigala* (Ahmed, 2009) and *H. fossilis* (Sayed and Ahmed, 2022). Besides this, it was studied that in response to outside stimuli, the target of rapamycin (TOR) pathway controls both cell growth and metabolism (Wullschleger et al., 2006). Phenylalanine has been shown in the past to control the expression of TOR (Feng et al., 2015), and investigations in human skeletal muscle have also demonstrated a strong correlation between phenylalanine and TOR (Vendelbo et al., 2014). To better understand how phenylalanine influences these processes, it is necessary to investigate its effects on the metabolism of fish species. Nevertheless, there are reports available on the involvement of phenylalanine in the TOR signaling pathways and glucose and lipid metabolism in carnivorous fishes (Yi et al., 2022). Furthermore, studies have demonstrated that phenylalanine in fish can affect plasma glucose concentration through diet (Ren et al., 2015). The growth and metabolism of aquatic animals, in particular, depend heavily on lipid metabolism in aquaculture (Brownsey et al., 2006; Semenkovich, 1997). Past studies have revealed that dietary phenylalanine levels have an impact on lipid content (Gao et al., 2016; Li et al., 2015; Zehra and Yousif, 2021). A study conducted by Liang et al. (2017) and Huang et al. (2021) also discovered that AAs impact lipid metabolism. On the other hand, Xiao et al. (2016) reported that mice may also experience a decrease in lipid levels when phenylalanine is deficient in the diet. Enhancement in absorption capacity and digestion has also been reported with the involvement of phenylalanine in the diet of fishes (Gao et al., 2016; Li et al., 2015; Zhang et al., 2023).

Apart from phenylalanine, another aromatic AA, tyrosine, is conditionally considered an EAA. It has a crucial role in the production of neurotransmitters like adrenaline, norepinephrine, and dopamine (Ahmed, 2022). Neurotransmitters are essential for the nervous system's signal transmission. Another compound, dopamine, enhanced the production of extracellular superoxide dismutase (SOD) protein and cell surface SOD activity in cultured astrocytes (Takano et al., 2013). The synthesis of thyroid hormones, which regulate metabolic processes and impact various aspects of body growth, metabolism, including feed efficiency, oxygen consumption, thermogenesis, environmental adaptation, protein, carbohydrate, and lipid synthesis, are mainly dependent on phenomenology (Gropper and Smith, 2013; Sayed and Ahmed, 2022). The synthesis of several hormones, including thyroid hormones, involves phenylalanine. Fish metabolism and energy balance are influenced by these hormones (Peter, 2011). In erythrocytes, thyroxine elevated Glutathione (GSH) levels and Glutathione peroxidase (GPx) activity (Virgili et al., 1999). In addition, phenylalanine is used in the pigment production process in certain fish species. For instance, melanin, which affects the color of the skin and scales in some ornamental fish, is derived from phenylalanine (Leclercq et al., 2010). Fish release phenylalanine in order to synthesize melanin (Yao et al., 2022). Interleukin 1-beta (IL-1b) and interleukin-6 (IL-6) are two examples of the cytokines that have been shown to inhibit the body's synthesis of melanin. Therefore, more research is needed to determine whether phenylalanine and these cytokines are related (Zhang et al., 2023).

Complementary to phenylalanine are substances like flavonoids that possess antioxidant qualities. According to Ozougwu (2016), antioxidants neutralize damaging free radicals, hence protecting cells from oxidative stress. Antioxidant enzymes, such as catalase (CAT) and SOD, are also present in fish to defend against oxidative damage (Lin et al., 2018). Li et al. (2015) discovered that in juvenile carp, phenylalanine deficiency or excess negatively impacted CAT activity, whereas excess phenylalanine negatively impacted SOD gene expression. The expression of CAT and SOD genes could only be elevated when supplied with phenylalanine in appropriate quantity. A study conducted by Li et al. (2015) showed that feeding grass carp, *C. idella*, 9.57 g/kg of phenylalanine could lower the amount of malondialdehyde (MDA) in their gills. Phenylalanine also takes part in improving the immune system of fish. It plays an important role in aiding the production of antibodies and other immune-related chemicals that help in the fish's defense against infections (de Oliveira Coelho and Pinto, 2020). Fish intestinal lymphoid tissues, including monocytes, lymphocytes, macrophages, and granulocytes, are primarily responsible for regulating the intestinal immunological barrier (Gomez et al., 2013). Conversely, nutrients have the ability to alter the intestinal immune system through changes in the intestinal wall's structural integrity. Phenylalanine is especially crucial when growth is rapid, as indicated in the juvenile and fingerling stages, where it has been observed to aid in the synthesis of proteins required for general growth and tissue development (Ahmed, 2009). Phenylalanine is one of the AAs that may be involved in fish stress response. The metabolism of AAs can be impacted by stress; however, stress can be lessened by including enough phenylalanine in the diet. In gilthead seabream, *Sparus aurata*, and meagre, *Argyrosomus regius*, the food supplementation with a high phenylalanine content was found to reduce plasma stress markers (Salamanca et al., 2020).

Fish growth hormone (GH), insulin-like growth factor-I (IGF-1), growth hormone receptor 1 (GHR1), TOR and ribosomal protein S6 kinase 1 (S6K1) can all have their transcription levels regulated by phenylalanine (Feng et al., 2017). Because GH and IGF-1 are the primary genes in the somatotropic axis (Pandey et al., 2013) and have been linked to both the growth regulation and nutritional status in teleosts (Pedroso et al., 2006), it is possible to use the transcriptional level of IGF-1 as a marker of fish growth (Inoki et al., 2002). In the meantime, S6K1 has the ability to promote protein synthesis in several cell types (Li et al., 2015; Marabita et al., 2016). Phenylalanine may also have an impact on serine/threonine kinase (AKT) expression (Guo et al., 2022). Moreover, it has been demonstrated that AAs may control the gluconeogenic process by means of enzymes like phosphoenolpyruvate carboxykinase (PEPCK) and glucose-6-phosphatase (G6Pase) (Dai et al., 2015; Goswami et al., 2004). Additionally, research has demonstrated that AAs can influence fish [glucokinase (GK) and pyruvate kinase (PK) expression (Huang et al., 2021; Liang et al., 2017), with some concentrations inhibiting both expressions (Suzuki et al., 2021).

7.10 NUTRITIONAL DISORDERS OF PHENYLALANINE DEFICIENCY

Fish that are lacking in certain nutrients may experience stunted growth and development, which is especially problematic for species that are actively procreating. Since phenylalanine is an EAA required for multiple physiological processes, a dietary shortage in fish can result in a variety of nutritional issues. Fish development, immunity, feed intake, and survival in their natural environments are all significantly impacted by the intake of dietary phenylalanine (Li et al., 2008; Pinto et al., 2009; Sayed and Ahmed, 2022). Several fish species fed phenylalanine-deficient diets have shown reduced feed utilization, antioxidant and growth performance. These findings have been followed by a loss of weight gain and a negative nitrogen balance (Ahmed, 2009; Saavedra et al., 2010; Syed and Ahmed, 2023; Xiao et al., 2020; Zehra and Khan, 2014; Zhang et al., 2023). A diet low in phenylalanine may also cause fish to have less protein and change in lipid composition (Ahmed, 2009; Khan and Abidi, 2007; Kim, 1993; Yi et al., 2022; Zehra and Yousif, 2021). Neurotransmitters like dopamine, epinephrine, and norepinephrine are synthesized using phenylalanine as a precursor. For healthy neuron activity and signal transmission, neurotransmitters are necessary. A lack of phenylalanine can affect the neural system and cause neurological conditions that lead to aberrant behavior (Akram et al., 2020). Changes in feeding habits, swimming styles, or reactivity to environmental cues are a few examples of this (Lall and Tibbetts, 2009). A lack of phenylalanine may weaken the immune system of fishes, leaving them more vulnerable to illnesses and infections (Yi et al., 2023). Aside from this, fish reproduction depends on a sufficient quantity of phenylalanine in the diet. A deficiency of phenylalanine can cause aberrant egg and larval developments, decreased fertility, or declined egg production. Phenylalanine also has a role in the creation of other enzymes. Fish metabolic processes may be impacted by a deficiency-induced drop-in enzyme activity (Li et al., 2015; Sayed and Ahmed, 2022). Fish need a well-balanced diet that satisfies their unique AA requirements in order to reduce the danger of phenylalanine shortage and related illnesses. Comprehending the dietary requirements of the particular fish species under cultivation is also crucial for formulating efficient feeding strategies. Frequent observation of fish health, development rates, and behavior can aid in spotting indicators of nutritional deficits, enabling diet modifications as necessary. To maximize the nutritional management of fish populations, speaking with an expert in fish nutrition or an aquatic veterinarian can be very beneficial.

7.11 FACTORS AFFECTING PHENYLALANINE REQUIREMENT IN FISHES

A multitude of factors influence the phenylalanine need of fishes. In various fish species, specific physiological, environmental, and dietary factors can have varying effects on these factors. Fish phenylalanine requirements are influenced by a number of important factors, including:

1. **Species and its Developmental Stage:** The physiological needs of various fish species for phenylalanine may vary. Furthermore, the requirement for phenylalanine can be influenced by the fish's developmental stage, including its larval, adult or juvenile stages (Pinto et al., 2009).
2. **Dietary Protein Content and Composition:** Since phenylalanine is an EAA present in proteins; therefore, the total amount of protein in the diet is very important. Phenylalanine use can be impacted by the makeup of DP, including the inclusion of other important AAs (Ren et al., 2015; Xiao et al., 2020).
3. **Protein Quality and Source:** What counts in a diet is the quality of the protein sources. It has been reported that the AA profiles

of different protein sources might differ and the availability of phenylalanine can be affected by the digestibility of various proteins (Webster and Thompson, 2015).

4. **Energy-to-Protein Ratio:** In the diet, it is crucial to maintain a balance between protein and energy ratio. A mismatched energy-to-protein ratio can impact how AAs including phenylalanine are used for different metabolic functions (Wang et al., 2018; Wilson and Halver, 1986).
5. **Environmental Factors:** Fish metabolic rates can be impacted by environmental factors such as water temperature, dissolved oxygen content and water quality. The dietary needs for EAAs, such as phenylalanine can also be impacted by changes in these variables (Ahmed, 2022; Cowey, 1994; Feng et al., 2017).
6. **Genetic Factors:** Within a species, fish strains or breeds may differ genetically, affecting how they use phenylalanine and how their metabolic processes are impacted (Zehra and Khan, 2014).
7. **Stress and Disease:** The dietary status of fish can be impacted by diseases and other stresses. According to, Fish that are stressed or ill may have changed metabolic pathways and higher demand for specific nutrients, such as phenylalanine (Salamanca et al., 2020; Sharf and Khan, 2023).
8. **Feed formulation and processing:** An essential component of commercial fish feed is their formulation. The AA profile, including phenylalanine needs to be carefully balanced by nutritionists when formulating fish feed. It has been suggested that the ways in which feeds are processed can also affect the availability of nutrients (Ren et al., 2015; Xiao et al., 2020).
9. **Feeding Practices:** The total amount of nutrients that fish consume can be affected by feeding frequency, timing and techniques. It takes suitable and consistent feeding techniques to achieve phenylalanine requirements (Ahmed, 2022). Therefore, it is suggested that feeding schedule and feeding regime is also considered when determining the nutrient requirement including phenylalanine in fish

10. **Metabolic Rate:** Fish metabolism is impacted by a number of variables, including activity level, growth rate and reproductive health. Phenylalanine needs may be higher in fish with higher metabolic rates (Gingerich et al., 2010; Metcalfe et al., 1995). Thus, the requirement may also be varied during the different phases of fish growth.
11. **Interaction with Other Nutrients:** Phenylalanine can be utilized in a variety of metabolic pathways depending on how it interacts with other nutrients, including vitamins and minerals. To guarantee ideal growth, health and general performance, it is crucial to take all of these variables into account when formulating balanced fish feed for a particular fish species. To ascertain the specific phenylalanine requirements for various fish species in real-world aquaculture settings, regulated nutritional studies and tests can be conducted.

7.12 FUTURE PERSPECTIVES

Advances in technology, continued research, and a better comprehension of the physiological and metabolic requirements of different fish species are all contributing factors to the future prospects of dietary phenylalanine requirements in fish. To fully comprehend the unique phenylalanine needs of various fish species, additional research is required. Customizing food formulations to meet the distinct requirements of individual species will become crucial for optimum development and well-being, as aquaculture grows and diversifies. Researchers' comprehension of the effects of fish genetics on phenylalanine metabolism will be facilitated by developments in nutrigenomics science. By using this knowledge, precise nutrition plans that are adapted to the unique genetic makeup of various fish species may be developed. A thorough understanding of the metabolic pathways including phenylalanine can be obtained by the application of systems biology and metabolomics techniques. Through examining the metabolome, one can gain a comprehensive picture of fish nutrition needs by learning how phenylalanine interacts with other metabolites and pathways. One interesting approach is the creation of specialized fish feed based on the requirements of individuals. This entails taking into consideration variances among populations and individuals in addition to species-specific requirements, accounting for variables such as genetics, growth rates, and environmental conditions. Future studies will examine the effects of innovative protein constituents on phenylalanine requirements, as the aquaculture sector looks for sustainable substitutes for conventional protein sources. This includes supplying fish diets with plant-based proteins, insect meal, and other substitute protein sources.

Technological developments in biotechnology, such as genetic engineering, could present prospects for improving fish's capacity to produce or use phenylalanine more effectively. Fish strains that have undergone genetic engineering may be created with optimal AA metabolism, which would enhance growth and nutritional efficiency. Future studies will probably concentrate on creating aquaculture methods that are environmentally sustainable. This involves looking into ways to lessen the environmental effect of fish farming and meet phenylalanine requirements with eco-friendly diets. Precision aquaculture can benefit from the integration of digital technology, including sensors, real-time monitoring, and data analytics. These tools can offer insightful information about fish nutrition in real time, enabling the development of feeding plans that are flexible and dependent on phenylalanine requirements. Future studies will examine how phenylalanine metabolism in fish is impacted by changing environmental conditions, with an eye towards the possible effects of climate change on aquatic ecosystems. It will be essential to comprehend these dynamics in order to modify aquaculture methods in response to climate change. Future research in aquaculture settings may examine the effects of phenylalanine supplementation on stress response and disease resistance.

7.13 CONCLUSION

In conclusion, phenylalanine requirements in fishes must be understood in order to optimize aquaculture procedures and guarantee the well-being and growth of fish. As an EAA, phenylalanine is necessary for the synthesis of proteins, the creation of neurotransmitters, and several metabolic processes. The amount of phenylalanine that different fish species require varies. Because each species is different in terms of its physiological traits, metabolic rates, and nutritional requirements, hence study on species-specific diet formulation is necessary. Fish undergo phenylalanine requirements that vary depending on their developmental stage. It is imperative to formulate feed that take into account the unique nutritional needs of larvae, juveniles, and adults in order to ensure optimal growth and development. Phenylalanine has intricate interactions with other elements in the diet, including vitamins, minerals, and other AAs. Comprehending these interplays is crucial in devising well-rounded and all-encompassing diets that fulfil the whole nutritional requirements of fishes. Stressors, water temperature, and water quality are examples of environmental variables that can affect fish phenylalanine metabolism. In order to create methods that take into account changes in aquaculture settings, future research should take these environmental factors into consideration. Novel approaches to optimizing fish diets are presented by developments in digital technology, metabolomics, and nutrigenomics. Real-time monitoring and data analytics made possible by precision aquaculture can help to develop more efficient feeding plans that are tailored to the unique dietary needs of fish populations. Investigating alternate protein sources and lessening the environmental impact of fish farming are essential as the aquaculture sector looks for sustainable techniques. In order to practice sustainable aquaculture, fish nutrition requirements must be met while simultaneously taking the industry's environmental impact into account. Standardized fish feeding guidelines can be developed with the help of shared information and best practices. Future advancements in our comprehension of fish phenylalanine requirements will mostly depend on the continuous integration of scientific study, technological innovation, and sustainable practices. This knowledge will not only promote the health and development of farmed fish, but overall, it will also make the aquaculture sector more sustainable and productive.

ACKNOWLEDGEMENTS

The author expresses gratitude to the Head, Department of Zoology for providing the required laboratory services and to the Department of Science and Technology (DST), Government of India, New Delhi for awarding Kousar Jan a DST-INSPIRE fellowship (Grant Number: IF170251).

KEYWORDS

- **aquafeed formulation**
- **essential amino acids**
- **phenylalanine in fish nutrition**
- **protein requirements in fish**
- **species-specific dietary needs**
- **sustainable aquaculture**

REFERENCES

Ahmed, I. (2009). Dietary total aromatic amino acid requirement and tyrosine replacement value for phenylalanine in Indian major carp: *Cirrhinus mrigala* (Hamilton) fingerlings. *Journal of Applied Ichthyology, 25*(6), 719–727. https://doi.org/10.1111/j.1439-0426.2009.01284.x.

Ahmed, I. (2022). Effect of dietary phenylalanine levels on growth, hemato-biochemical composition and tyrosine replacement value for phenylalanine in stinging catfish *Heteropneustes fossilis*–Bloch 1974 fingerling. *Animal Feed Science and Technology, 288*, 115294. https://doi.org/10.1016/j.anifeedsci.2022.115294.

Ahmed, I., & Ahmad, I. (2020). Effect of dietary protein levels on growth performance, hematological profile and biochemical composition of fingerlings rainbow trout, *Oncorhynchus mykiss* reared in Indian Himalayan region. *Aquaculture Reports, 16*, 100268. https://doi.org/10.1016/j.aqrep.2019.100268.

Ahmed, I., Khan, Y. M., Lateef, A., Jan, K., Majeed, A., & Shah, M. A. (2023). Effect of fish meal replacement by Azolla meal on growth performance, hemato-biochemical and serum parameters in the diet of scale carp, *Cyprinus carpio* var. *communis*. *Journal of the World Aquaculture Society, 54*(5), 1301–1316. https://doi.org/10.1111/jwas.12995.

Akram, M., Daniyal, M., Ali, A., Zainab, R., Shah, S. M. A., Munir, N., & Tahir, I. M. (2020). Role of phenylalanine and its metabolites in health and neurological disorders. In *Synucleins—Biochemistry and Role in Diseases*. IntechOpen. https://doi.org/10.5772/intechopen.83648.

Al Mughram, M. H., Catalano, C., Bowry, J. P., Safo, M. K., Scarsdale, J. N., & Kellogg, G. E. (2021). 3D interaction homology: Hydropathic analyses of the "π–cation" and "π–π" interaction motifs in phenylalanine, tyrosine, and tryptophan residues. *Journal of Chemical Information and Modeling, 61*(6), 2937–2956. https://doi.org/10.1021/acs.jcim.1c00235.

Aragao, C., Conceicao, L. E., Fyhn, H. J., & Dinis, M. T. (2004). Estimated amino acid requirements during early ontogeny in fish with different life styles: Gilthead seabream

(*Sparus aurata*) and Senegalese sole (*Solea senegalensis*). *Aquaculture, 242*(1–4), 589–605. https://doi.org/10.1016/j.aquaculture.2004.09.015.

Barlow, D. J., & Thornton, J. M. (1983). Ion-pairs in proteins. *Journal of Molecular Biology, 168*(4), 867–885. https://doi.org/10.1016/S0022-2836(83)80079-5.

Bhosale, S. V., Bhilave, M. P., & Nadaf, S. B. (2010). Formulation of fish feed using ingredients from plant sources. *Research Journal of Agricultural Sciences, 1*(3), 284–287.

Borlongan, I. G. (1992). Dietary requirement of milkfish (*Chanos chanos* Forsskal) juveniles for total aromatic amino acids. *Aquaculture, 102*(4), 309–317. https://doi.org/10.1016/0044-8486(92)90184-M.

Brownsey, R. W., Boone, A. N., Elliott, J. E., Kulpa, J. E., & Lee, W. M. (2006). Regulation of acetyl-CoA carboxylase. *Biochemical Society Transactions, 34*(2), 223–227. https://doi.org/10.1042/BST0340223.

Cankiriligil, E. C., Berik, N., & Erbay, E. A. (2020). Optimization of hydrolyzation procedure for amino acid analysis in fish meat with HPLC-DAD by Response Surface Methodology (RSM). *Ege Journal of Fisheries and Aquatic Sciences, 37*(2), 113–123. http://doi.org/10.12714/egejfas.37.2.01.

Castillo, S., Halligan, S., & Gatlin, D. M. III. (2015). Growth responses of juvenile red drum *Sciaenops ocellatus* to dietary phenylalanine and tyrosine can be used to calculate the total aromatic amino acid requirement. *The Journal of Nutrition, 145*(10), 2341–2346. https://doi.org/10.3945/jn.115.215848.

Chance, R. E., Mertz, E. T., & Halver, J. E. (1964). Nutrition of salmonoid fishes: XII. Isoleucine, leucine, valine and phenylalanine requirements of Chinook salmon and interrelations between isoleucine and leucine for growth. *The Journal of Nutrition, 83*(3), 177–185. https://doi.org/10.1093/jn/83.3.177.

Cho, J. H., & Kim, I. H. (2011). Fish meal–nutritive value. *Journal of Animal Physiology and Animal Nutrition, 95*(6), 685–692. https://doi.org/10.1111/j.1439-0396.2010.01109.x.

Cowey, C. B. (1994). Amino acid requirements of fish: A critical appraisal of present values. *Aquaculture, 124*(1–4), 1–11. https://doi.org/10.1016/0044-8486(94)90349-2.

Dabrowski, K., Zhang, Y., Kwasek, K., Hliwa, P., & Ostaszewska, T. (2010). Effects of protein-, peptide- and free amino acid-based diets in fish nutrition. *Aquaculture Research, 41*(5), 668–683. https://doi.org/10.1111/j.1365-2109.2010.02490.x.

Dai, W., Panserat, S., Plagnes-Juan, E., Seiliez, I., & Skiba-Cassy, S. (2015). Amino acids attenuate insulin action on gluconeogenesis and promote fatty acid biosynthesis via mTORC1 signalling pathway in trout hepatocytes. *Cellular Physiology and Biochemistry, 36*(3), 1084–1100. https://doi.org/10.1159/000430281.

Day, L., Cakebread, J. A., & Loveday, S. M. (2022). Food proteins from animals and plants: Differences in the nutritional and functional properties. *Trends in Food Science & Technology, 119*, 428–442. https://doi.org/10.1016/j.tifs.2021.12.020.

de Oliveira Coelho, L., & Pinto, R. (2020). *Amino Acids as Key Mediators of Immune Status and Nutritional Condition in Fish* (Ph.D. thesis). Institute of Biomedical Sciences Abel Salazar of the University of Porto.

do Nascimento, T. M., Mansano, C. F., Peres, H., Rodrigues, F. H., Khan, K. U., Romaneli, R. S., Sakomura, N. K., & Fernandes, J. B. (2020). Determination of the optimum dietary essential amino acid profile for growing phase of Nile tilapia by deletion method. *Aquaculture, 523*, 735204. https://doi.org/10.1016/j.aquaculture.2020.735204.

El-Haroun, E. R., & Bureau, D. P. (2007). Comparison of the bioavailability of lysine in blood meals of various origins to that of L-lysine HCl for rainbow trout (*Oncorhynchus mykiss*). *Aquaculture, 262*(2–4), 402–409. https://doi.org/10.1016/j.aquaculture.2006.10.032.

Fagbenro, O. A. (2000). Validation of the essential amino acid requirements of Nile tilapia, *Oreochromis niloticus* (Linnaeus 1758), assessed by the ideal protein concept. In *Proceedings of the Fifth International Symposium on Tilapia in Aquaculture*, Rio de Janeiro, Brazil.

Falco, F., Barra, M., Wu, G., Dioguardi, M., Stincone, P., Cuttitta, A., Torri, M., Bonanno, A., & Cammarata, M. (2020). *Engraulis encrasicolus* larvae from two different environmental spawning areas of the Central Mediterranean Sea: First data on amino acid profiles and biochemical evaluations. *The European Zoological Journal, 87*(1), 580–590. https://doi.org/10.1080/24750263.2020.1823493.

FAO. (2020). *The State of World Fisheries and Aquaculture 2020: Sustainability in Action.* FAO.

FAO. (2022). *The State of World Fisheries and Aquaculture 2022: Towards Blue Transformation.* https://doi.org/10.4060/cc0461en.

Fazial, F. F., & Tan, L. L. (2021). Phenylalanine-responsive fluorescent biosensor based on graphene oxide–chitosan nanocomposites catalytic film for non-destructive fish freshness grading. *Food Control, 125*, 107995. https://doi.org/10.1016/j.foodcont.2021.107995.

Feng, L., Li, W., Liu, Y., Jiang, W. D., Kuang, S. Y., Jiang, J., Tang, L., Wu, P., Tang, W. N., Zhang, Y. A., & Zhou, X. Q. (2015). Dietary phenylalanine-improved intestinal barrier health in young grass carp (*Ctenopharyngodon idella*) is associated with increased immune status and regulated gene expression of cytokines, tight junction proteins, antioxidant enzymes and related signaling molecules. *Fish & Shellfish Immunology, 45*(2), 495–509. https://doi.org/10.1016/j.fsi.2015.05.001.

Feng, L., Li, W., Liu, Y., Jiang, W. D., Kuang, S. Y., Wu, P., Jiang, J., Tang, L., Tang, W. N., Zhang, Y. A., & Zhou, X. Q. (2017). Protective role of phenylalanine on the ROS-induced oxidative damage, apoptosis and tight junction damage via Nrf2, TOR and NF-κB signaling molecules in the gill of fish. *Fish & Shellfish Immunology, 60*, 185–196. https://doi.org/10.1016/j.fsi.2016.11.048.

Fiordelmondo, E., Ceschin, S., Magi, G. E., Mariotti, F., Iaffaldano, N., Galosi, L., & Roncarati, A. (2022). Effects of partial substitution of conventional protein sources with duckweed (*Lemna minor*) meal in the feeding of rainbow trout (*Oncorhynchus mykiss*) on growth performances and the quality product. *Plants, 11*(9), 1220. https://doi.org/10.3390/plants11091220.

Gao, Y. J., Liu, Y. J., Chen, X. Q., Yang, H. J., & Tian, L. X. (2016). Total aromatic amino acid requirement of juvenile grass carp (*Ctenopharyngodon idella*). *Aquaculture Nutrition, 22*(4), 865–872. https://doi.org/10.1111/anu.12306.

Gatlin, D. M. III, Barrows, F. T., Brown, P., Dabrowski, K., Gaylord, T. G., Hardy, R. W., Herman, E., Hu, G., Krogdahl, Å., Nelson, R., Overturf, K., Rust, M., Sealey, W., Skonberg, D., Souza, E. J., Stone, D., Wilson, R., & Wurtele, E. (2007). Expanding the utilization of sustainable plant products in aquafeeds: A review. *Aquaculture Research, 38*(6), 551–579. https://doi.org/10.1111/j.1365-2109.2007.01704.x.

Gingerich, A. J., Philipp, D. P., & Suski, C. D. (2010). Effects of nutritional status on metabolic rate, exercise and recovery in a freshwater fish. *Journal of Comparative Physiology B, 180*, 371–384. https://doi.org/10.1007/s00360-009-0419-4.

Glencross, B. (2006). The nutritional management of barramundi, *Lates calcarifer*–A review. *Aquaculture Nutrition, 12*(4), 291–309. https://doi.org/10.1111/j.1365-2095.2006.00410.x.

Gomez, D., Sunyer, J. O., & Salinas, I. (2013). The mucosal immune system of fish: The evolution of tolerating commensals while fighting pathogens. *Fish & Shellfish Immunology, 35*(6), 1729–1739. https://doi.org/10.1016/j.fsi.2013.09.032.

Goswami, C., Datta, S., Biswas, K., & Saha, N. (2004). Cell volume changes affect gluconeogenesis in the perfused liver of the catfish *Clarias batrachus*. *Journal of Biosciences, 29*, 337–347. https://doi.org/10.1007/BF02702616.

Gropper, S. S., & Smith, J. L. (2013). *Advanced nutrition and human metabolism* (6th ed.). Cengage Learning.

Guo, C., Zhang, X., Xie, S., Luo, J., Zhu, T., Yang, Y., Li, X., Huang, C., Dang, Y., Zhou, Q., & Jin, M. (2022). Dietary phenylalanine level could improve growth performance, glucose metabolism and insulin and mTOR signaling pathways of juvenile swimming crabs, *Portunus trituberculatus*. *Aquaculture Reports, 27*, 101395. https://doi.org/10.1016/j.aqrep.2022.101395.

Hall, G. M., & Ahmad, N. H. (1992). Functional properties of fish protein hydrolysates. In G. M. Hall (Ed.), *Fish processing technology* (pp. 249–265). Blackie Academic and Professional. https://cir.nii.ac.jp/crid/1571417125107110400.

Harper, A. E. (2020). Phenylalanine metabolism. In *Aspartame* (pp. 77–109). CRC Press.

Harper, A. E., & Yoshimura, N. N. (1993). Protein quality, amino acid balance, utilization, and evaluation of diets containing amino acids as therapeutic agents. *Nutrition (Burbank, Los Angeles County, Calif.), 9*(5), 460–469. https://europepmc.org/article/med/8286887 (accessed on 3 June 2025).

Hu, X., Liao, M., Gong, H., Zhang, L., Cox, H., Waigh, T. A., & Lu, J. R. (2020). Recent advances in short peptide self-assembly: From rational design to novel applications. *Current Opinion in Colloid & Interface Science, 45*, 1–13. https://doi.org/10.1016/j.cocis.2019.08.003.

Hua, K., & Bureau, D. P. (2019). Estimating changes in essential amino acid requirements of rainbow trout and Atlantic salmon as a function of body weight or diet composition using a novel factorial requirement model. *Aquaculture, 513*, 734440. https://doi.org/10.1016/j.aquaculture.2019.734440.

Huang, D., Liang, H., Ren, M., Ge, X., Ji, K., Yu, H., & Maulu, S. (2021). Effects of dietary lysine levels on growth performance, whole body composition and gene expression related to glycometabolism and lipid metabolism in grass carp, *Ctenopharyngodon idellus* fry. *Aquaculture, 530*, 735806. https://doi.org/10.1016/j.aquaculture.2020.735806.

Inoki, K., Li, Y., Zhu, T., Wu, J., & Guan, K. L. (2002). TSC2 is phosphorylated and inhibited by Akt and suppresses mTOR signalling. *Nature Cell Biology, 4*(9), 648–657. https://www.nature.com/articles/ncb839 (accessed on 3 June 2025).

Jiang, W. D., Wu, P., Tang, R. J., Liu, Y., Kuang, S. Y., Jiang, J., Tang, L., Tang, W. N., Zhang, Y. A., Zhou, X. Q., & Feng, L. (2016). Nutritive values, flavor amino acids, healthcare fatty acids and flesh quality improved by manganese referring to up-regulating the antioxidant capacity and signaling molecules TOR and Nrf2 in the muscle of fish. *Food Research International, 89*, 670–678. https://doi.org/10.1016/j.foodres.2016.09.020.

Kasozi, N., Iwe, G., Sadik, K., Asizua, D., & Namulawa, V. T. (2019). Dietary amino acid requirements of pebbly fish, *Alestes baremoze* (Joannis, 1835) based on whole body amino acid composition. *Aquaculture Reports*, *14*, 100197. https://doi.org/10.1016/j.aqrep.2019.100197.

Kaushik, S. J., & Seiliez, I. (2010). Protein and amino acid nutrition and metabolism in fish: Current knowledge and future needs. *Aquaculture Research*, *41*(3), 322–332. https://doi.org/10.1111/j.1365-2109.2009.02174.x.

Khan, M. A., & Abidi, S. F. (2007). Total aromatic amino acid requirement of Indian major carp *Labeo rohita* (Hamilton) fry. *Aquaculture*, *267*(1–4), 111–118. https://doi.org/10.1016/j.aquaculture.2007.02.025.

Kim, K. I. (1993). Requirement for phenylalanine and replacement value of tyrosine for phenylalanine in rainbow trout (*Oncorhynchus mykiss*). *Aquaculture*, *113*(3), 243–250. https://doi.org/10.1016/0044-8486(93)90477-G.

Kim, K. W., Kim, K. D., Han, H. S., Won, S., Moniruzzaman, M., Lee, J. H., Choi, Y. H., & Bai, S. C. (2017). Evaluation of the dietary protein requirement of a selectively bred (F-5 generation) strain of olive flounder, *Paralichthys olivaceus*. *Turkish Journal of Fisheries and Aquatic Sciences*, *17*(5), 1009–1016. https://doi.org/10.4194/1303-2712-v17_5_16.

Koch, J. F., Rawles, S. D., Webster, C. D., Cummins, V., Kobayashi, Y., Thompson, K. R., Gannam, A. L., Twibell, R. G., & Hyde, N. M. (2016). Optimizing fish meal-free commercial diets for Nile tilapia, *Oreochromis niloticus*. *Aquaculture*, *452*, 357–366. https://doi.org/10.1016/j.aquaculture.2015.11.017.

Kolmakova, A. A., & Kolmakov, V. I. (2019). Amino acid composition of green microalgae and diatoms, cyanobacteria, and zooplankton. *Inland Water Biology*, *12*, 452–461. https://link.springer.com/article/10.1134/S1995082919040060 (accessed on 3 June 2025).

Konnert, G. D. P., Martin, E., Gerrits, W. J. J., Gussekloo, S. W. S., Masagounder, K., Mas-Muñoz, J., & Schrama, J. W. (2022). Interactive effects of protein and energy intake on nutrient partitioning and growth in Nile tilapia. *Animal*, *16*(4), 100494. https://doi.org/10.1016/j.animal.2022.100494.

Korkmaz, K., & Tokur, B. (2022). Optimization of hydrolysis conditions for the production of protein hydrolysates from fish wastes using response surface methodology. *Food Bioscience*, *45*, 101312. https://doi.org/10.1016/j.fbio.2021.101312.

Krajcovicova-Kudlackova, M., Babinska, K., & Valachovicova, M. (2005). Health benefits and risks of plant proteins. *Bratislavske Lekarske Listy*, *106*(6, 7), 231–234.

Lall, S. P., & Dumas, A. (2022). Nutritional requirements of cultured fish: Formulating nutritionally adequate feeds. In *Feed and Feeding Practices in Aquaculture* (pp. 65–132). Woodhead publishing.

Lall, S. P., & Tibbetts, S. M. (2009). Nutrition, feeding, and behavior of fish. *Veterinary Clinics of North America: Exotic Animal Practice*, *12*(2), 361–372. https://doi.org/10.1016/j.cvex.2009.01.005.

Leclercq, E., Taylor, J. F., & Migaud, H. (2010). Morphological skin colour changes in teleosts. *Fish and Fisheries*, *11*(2), 159–193. https://doi.org/10.1111/j.1467-2979.2009.00346.x.

Li, B., Su, L., Zhang, H., Deng, H., Chen, Q., & Shi, H. (2020). Microplastics in fishes and their living environments surrounding a plastic production area. *Science of the Total Environment*, *727*, 138662. https://doi.org/10.1016/j.scitotenv.2020.138662.

Li, C., Ni, D., Song, L., Zhao, J., Zhang, H., & Li, L. (2008). Molecular cloning and characterization of a catalase gene from Zhikong scallop *Chlamys farreri*. *Fish & Shellfish Immunology*, *24*(1), 26–34. https://doi.org/10.1016/j.fsi.2007.06.010.

Li, L., Dugan, H. L., Stamper, C. T., Lan, L. Y. L., Asby, N. W., Knight, M., Stovicek, O., Zheng, N. Y., Madariaga, M. L., Shanmugarajah, K., Jansen, M. O., Changrob, S., Utset, H. A., Henry, C., Nelson, C., Jedrzejczak, R. P., Fremont, D. H., Joachimiak, A., Krammer, F., Huang, J., Khan, A. A., & Wilson, P. C. (2021). Improved integration of single-cell transcriptome and surface protein expression by LinQ-View. *Cell Reports Methods, 1*(4), 100056. https://doi.org/10.1016/j.crmeth.2021.100056.

Li, P., Mai, K., Trushenski, J., & Wu, G. (2009). New developments in fish amino acid nutrition: Towards functional and environmentally oriented aquafeeds. *Amino Acids, 37*, 43–53. https://doi.org/10.1007/s00726-008-0171-1.

Li, P., Yin, Y. L., Li, D., Kim, S. W., & Wu, G. (2007). Amino acids and immune function. *British Journal of Nutrition, 98*(2), 237–252. https://doi.org/10.1017/S000711450769936X.

Li, W., Feng, L., Liu, Y., Jiang, W. D., Kuang, S. Y., Jiang, J., Li, S. H., Tang, L., & Zhou, X. Q. (2015). Effects of dietary phenylalanine on growth, digestive and brush border enzyme activities and antioxidant capacity in the hepatopancreas and intestine of young grass carp (*Ctenopharyngodon idella*). *Aquaculture Nutrition, 21*(6), 913–925. https://doi.org/10.1111/anu.12223.

Liang, H., Habte-Tsion, H. M., Ge, X., Ren, M., Xie, J., Miao, L., Zhou, Q., Lin, Y., & Pan, W. (2017). Dietary arginine affects the insulin signaling pathway, glucose metabolism and lipogenesis in juvenile blunt snout bream *Megalobrama amblycephala*. *Scientific Reports, 7*, 7864. https://doi.org/10.1038/s41598-017-06104-3.

Lin, Y., Miao, L. H., Pan, W. J., Huang, X., Dengu, J. M., Zhang, W. X., Ge, X. P., Liu, B., Ren, M. C., Zhou, Q. L., Xie, J., Pan, L. K., & Xi, B. W. (2018). Effect of nitrite exposure on the antioxidant enzymes and glutathione system in the liver of bighead carp, *Aristichthys nobilis*. *Fish & Shellfish Immunology, 76*, 126–132. https://doi.org/10.1016/j.fsi.2018.02.015.

Man, Y., Yin, R., Cai, K., Qin, C., Wang, J., Yan, H., & Li, M. (2019). Primary amino acids affect the distribution of methyl mercury rather than inorganic mercury among tissues of two farmed-raised fish species. *Chemosphere, 225*, 320–328. https://doi.org/10.1016/j.chemosphere.2019.03.058.

Marabita, M., Baraldo, M., Solagna, F., Ceelen, J. J. M., Sartori, R., Nolte, H., Nemazanyy, I., Pyronnet, S., Kruger, M., Pende, M., & Blaauw, B. (2016). S6K1 is required for increasing skeletal muscle force during hypertrophy. *Cell Reports, 17*(2), 501–513. https://doi.org/10.1016/j.celrep.2016.09.020.

Metcalfe, N. B., Taylor, A. C., & Thorpe, J. E. (1995). Metabolic rate, social status and life-history strategies in Atlantic salmon. *Animal Behaviour, 49*(2), 431–436. https://doi.org/10.1006/anbe.1995.0056.

Molinero, A., Garcia, E., & Flos, R. (1994). Amino acid profile in natural prey of sole—*Solea solea*. *Comparative Biochemistry and Physiology Part A: Physiology, 108*(4), 649–655. https://doi.org/10.1016/0300-9629(94)90351-4.

Mukherjee, A. K., Kalita, P., Unni, B. G., Wann, S. B., Saikia, D., & Mukhopadhyay, P. K. (2010). Fatty acid composition of four potential aquatic weeds and their possible use as fish-feed neutraceuticals. *Food Chemistry, 123*(4), 1252–1254. https://doi.org/10.1016/j.foodchem.2010.05.057.

National Research Council (NRC). (2011). *Nutrient Requirements of Fish and Shrimp* (pp. 335–339). Washington, DC: National Academy Press.

Ngamsnae, P., Silva, S. S. D., & Gunasekera, R. M. (1999). Arginine and phenylalanine requirement of juvenile silver perch *Bidyanus bidyanus* and validation of the use of body amino acid composition for estimating individual amino acid requirements. *Aquaculture Nutrition, 5*(3), 173–180. https://doi.org/10.1046/j.1365-2095.1999.00102.x.

Nguyen, L. (2017). *Ideal Protein Concept and Its Application in Practical Diets for Nile tilapia Oreochromis niloticus* (Doctoral dissertation, Auburn University).

Nick Pace, C., Alston, R. W., & Shaw, K. L. (2000). Charge-charge interactions influence the denatured state ensemble and contribute to protein stability. *Protein Science, 9*(7), 1395–1398. https://doi.org/10.1110/ps.9.7.1395.

Nogales-Merida, S., Gobbi, P., Jozefiak, D., Mazurkiewicz, J., Dudek, K., Rawski, M., & Jozefiak, A. (2019). Insect meals in fish nutrition. *Reviews in Aquaculture, 11*(4), 1080–1103. https://doi.org/10.1111/raq.12281.

Nose, T. (1979). Summary report on the requirements of essential amino acids for carp. In K. Tiews & J. E. Halver (Eds.), *Finfish Nutrition and Fish Feed Technology* (pp. 145–156). Heenemann, Berlin.

Ozougwu, J. C. (2016). The role of reactive oxygen species and antioxidants in oxidative stress. *International Journal of Research in Pharmacy and Biosciences, 3*(6), 1–8.

Pandey, N. K., Singh, R. P., Saxena, V. K., Shit, N., Singh, R., Sharma, R. K., & Sastry, K. V. H. (2013). Effect of IGF1 gene polymorphism and expression levels on growth factors in Indian colored broilers. *Livestock Science, 155*(2–3), 157–164. https://doi.org/10.1016/j.livsci.2013.05.009.

Pedroso, F. L., de Jesus-Ayson, E. G. T., Cortado, H. H., Hyodo, S., & Ayson, F. G. (2006). Changes in mRNA expression of grouper (*Epinephelus coioides*) growth hormone and insulin-like growth factor I in response to nutritional status. *General and Comparative Endocrinology, 145*(3), 237–246. https://doi.org/10.1016/j.ygcen.2005.09.001.

Peter, M. S. (2011). The role of thyroid hormones in stress response of fish. *General and Comparative Endocrinology, 172*(2), 198–210. https://doi.org/10.1016/j.ygcen.2011.02.023.

Pinto, W., Figueira, L., Dinis, M. T., & Aragao, C. (2009). How does fish metamorphosis affect aromatic amino acid metabolism? *Amino Acids, 36*, 177–183. https://doi.org/10.1007/s00726-008-0045-6.

Pinto, W., Rodrigues, V., Dinis, M. T., & Aragao, C. (2010). Can dietary aromatic amino acid supplementation be beneficial during fish metamorphosis? *Aquaculture, 310*(1–2), 200–205. https://doi.org/10.1016/j.aquaculture.2010.10.026.

Poppi, D. A., Quinton, V. M., Hua, K., & Bureau, D. P. (2011). Development of a test diet for assessing the bioavailability of arginine in feather meal fed to rainbow trout (*Oncorhynchus mykiss*). *Aquaculture, 314*(1–4), 100–109. https://doi.org/10.1016/j.aquaculture.2011.01.016.

Prabu, E., Felix, S., Felix, N., Ahilan, B., & Ruby, P. (2017). An overview on significance of fish nutrition in aquaculture industry. *International Journal of Fisheries and Aquatic Studies, 5*(6), 349–355.

Ravikumar, A., Deepadevi, K. V., Arun, P., Manojkumar, V., & Kurup, P. A. (2000). Tryptophan and tyrosine catabolic pattern in neuropsychiatric disorders. *Neurology India, 48*(3), 231.

Ren, M., Liu, B., Habte-Tsion, H. M., Ge, X., Xie, J., Zhou, Q., Liang, H., Zhao, Z., & Pan, L. (2015). Dietary phenylalanine requirement and tyrosine replacement value for phenylalanine of juvenile blunt snout bream, *Megalobrama amblycephala*. *Aquaculture, 442*, 51–57. https://doi.org/10.1016/j.aquaculture.2015.02.029.

Ricker, B., Mitra, S., Castellanos, E. A., Grady, C. J., Woldring, D., Pelled, G., & Gilad, A. A. (2023). Proposed three-phenylalanine motif involved in magneto-reception signaling of an Actinopterygii protein expressed in mammalian cells. *Open Biology, 13*(11), 230019. https://doi.org/10.1098/rsob.230019.

Robinson, E. H., Wilson, R. P., & Poe, W. E. (1980). Total aromatic amino acid requirement, phenylalanine requirement and tyrosine replacement value for fingerling channel catfish. *The Journal of Nutrition, 110*(9), 1805–1812. https://doi.org/10.1093/jn/110.9.1805.

Saavedra, M., Conceicao, L. E., Barr, Y., Helland, S., Pousao-Ferreira, P., Yufera, M., & Dinis, M. T. (2010). Tyrosine and phenylalanine supplementation on *Diplodus sargus* larvae: Effect on growth and quality. *Aquaculture Research, 41*(10), 1523–1532. https://doi.org/10.1111/j.1365-2109.2009.02446.x.

Saha, S. K., Pathak, N. N., Saha, S. K., & Pathak, N. N. (2021). Proteins and other nitrogenous substances of nutritional significance. In *Fundamentals of Animal Nutrition* (pp. 65–92). Springer. https://doi.org/10.1007/978-981-15-9125-9_7.

Salamanca, N., Giraldez, I., Morales, E., de La Rosa, I., & Herrera, M. (2020). Phenylalanine and tyrosine as feed additives for reducing stress and enhancing welfare in gilthead seabream and meagre. *Animals, 11*(1), 45. https://doi.org/10.3390/ani11010045.

Santiago, C. B., & Lovell, R. T. (1988). Amino acid requirements for growth of Nile tilapia. *The Journal of Nutrition, 118*(12), 1540–1546. https://doi.org/10.1093/jn/118.12.1540.

Sayed, S. F., & Ahmed, I. (2022). Effects of dietary phenylalanine:tyrosine ratio on growth, DNA/RNA, serum biochemistry, digestive enzyme activities and physiological responses of *Heteropneustes fossilis*. *Aquaculture Studies, 23*(1), 1–17. https://doi.org/10.4194/AQUAST859.

Semenkovich, C. F. (1997). Regulation of fatty acid synthase (FAS). *Progress in Lipid Research, 36*(1), 43–53. https://doi.org/10.1016/S0163-7827(97)00003-9.

Sharf, Y., & Khan, M. A. (2023). Dietary phenylalanine requirement and tyrosine replacement value for phenylalanine for fingerling *Channa punctatus* based on growth parameters, hematology and antioxidant status. *Aquaculture, 572*, 739503. https://doi.org/10.1016/j.aquaculture.2023.739503.

Shearer, K. D. (2000). Experimental design, statistical analysis and modelling of dietary nutrient requirement studies for fish: A critical review. *Aquaculture Nutrition, 6*(2), 91–102. https://doi.org/10.1046/j.1365-2095.2000.00134.x.

Small, B. C., & Soares, J. H. (1999). Quantitative dietary threonine requirement of juvenile striped bass *Morone saxatilis*. *Journal of the World Aquaculture Society, 30*(3), 319–323. https://doi.org/10.1111/j.1749-7345.1999.tb00682.x.

Suzuki, R., Sato, Y., Fukaya, M., Suzuki, D., Yoshizawa, F., & Sato, Y. (2021). Energy metabolism profile of the effects of amino acid treatment on hepatocytes: Phenylalanine and phenylpyruvate inhibit glycolysis of hepatocytes. *Nutrition, 82*, 111042. https://doi.org/10.1016/j.nut.2020.111042.

Takano, K., Tanaka, N., Kawabe, K., Moriyama, M., & Nakamura, Y. (2013). Extracellular superoxide dismutase induced by dopamine in cultured astrocytes. *Neurochemical Research, 38*, 32–41. https://doi.org/10.1007/s11064-012-0882-2.

Tilami, S. K., Turek, J., Cerveny, D., Lepic, P., Kozak, P., Burkina, V., Sakalli, S., Tomcala, A., Sampels, S., & Mraz, J. (2020). Insect meal as a partial replacement for fish meal in a formulated diet for perch *Perca fluviatilis*. *Turkish Journal of Fisheries and Aquatic Sciences, 20*(12), 867–878. https://doi.org/10.4194/1303-2712-v20_12_03.

Trosvik, K. A., Webster, C. D., Thompson, K. R., Metts, L. A., Gannam, A., & Twibell, R. (2013). Effects on growth performance and body composition in Nile tilapia, *Oreochromis niloticus*, fry fed organic diets containing yeast extract and soybean meal as a total replacement of fish meal without amino acid supplementation. *Biological Agriculture & Horticulture, 29*(3), 173–185. https://doi.org/10.1080/01448765.2013.810123.

Vendelbo, M. H., Moller, A. B., Christensen, B., Nellemann, B., Clasen, B. F. F., Nair, K. S., Jorgensen, J. O. L., Jessen, N., & Moller, N. (2014). Fasting increases human skeletal muscle net phenylalanine release and this is associated with decreased mTOR signaling. *PLoS ONE, 9*(7), e102031. https://doi.org/10.1371/journal.pone.0102031.

Virgili, F., Canali, R., Figus, E., Vignolini, F., Nobili, F., & Mengheri, E. (1999). Intestinal damage induced by zinc deficiency is associated with enhanced CuZn superoxide dismutase activity in rats: Effect of dexamethasone or thyroxine treatment. *Free Radical Biology and Medicine, 26*(9, 10), 1194–1201. https://doi.org/10.1016/S0891-5849(98)00307-4.

Wang, W., Wang, S., Wang, W., Wang, H., Xiang, X., & Zhou, X. (2018). Effect of dietary energy to protein ratio on growth, biological indices, proximate composition, and amino acid profiles in the muscle of *Varicorhinus macrolepis*. *North American Journal of Aquaculture, 80*(2), 168–179. https://doi.org/10.1002/naaq.10020.

Webster, C. D., & Thompson, K. R. (2015). Protein, amino acids, and ingredients. In C. S. Lee, C. Lim, D. M. Gatlin III, & C. D. Webster (Eds.), *Dietary Nutrients, Additives, and Fish Health* (pp. 25–45). Wiley-Blackwell. https://doi.org/10.1002/9781119005568.ch2.

Wilson, R. P., & Halver, J. E. (1986). Protein and amino acid requirements of fishes. *Annual Review of Nutrition, 6*(1), 225–244.

Wu, G. (2010). Functional amino acids in growth, reproduction, and health. *Advances in Nutrition, 1*(1), 31–37. https://doi.org/10.1146/annurev.nu.06.070186.001301.

Wu, G. (2013). Functional amino acids in nutrition and health. *Amino Acids, 45*, 407–411. https://doi.org/10.1007/s00726-013-1500-6.

Wu, G., Bazer, F. W., Dai, Z., Li, D., Wang, J., & Wu, Z. (2014). Amino acid nutrition in animals: Protein synthesis and beyond. *Annual Review of Animal Biosciences, 2*(1), 387–417. https://doi.org/10.1146/annurev-animal-022513-114113.

Wullschleger, S., Loewith, R., & Hall, M. N. (2006). TOR signaling in growth and metabolism. *Cell, 124*(3), 471–484. https://doi.org/10.1016/j.cell.2006.01.016.

Xiao, F., Wang, C., Yin, H., Yu, J., Chen, S., Fang, J., & Guo, F. (2016). Leucine deprivation inhibits proliferation and induces apoptosis of human breast cancer cells via fatty acid synthase. *Oncotarget, 7*(39), 63679–63690. https://doi.org/10.18632/oncotarget.11626.

Xiao, W., Zou, Z., Li, D., Zhu, J., Yue, Y., & Yang, H. (2020). Effect of dietary phenylalanine level on growth performance, body composition, and biochemical parameters in plasma of juvenile hybrid tilapia, *Oreochromis niloticus* × *Oreochromis aureus*. *Journal of the World Aquaculture Society, 51*(2), 437–451. https://doi.org/10.1111/jwas.12641.

Yaghoubi, M., Mozanzadeh, M. T., Safari, O., & Marammazi, J. G. (2018). Gastrointestinal and hepatic enzyme activities in juvenile silvery-black porgy (*Sparidentex hasta*) fed essential amino acid-deficient diets. *Fish Physiology and Biochemistry, 44*, 853–868. https://doi.org/10.1007/s10695-018-0475-3.

Yang, P., Wang, H., Ma, L., Yin, H., Zhu, Z., Liu, C., Huang, W., Zhou, Z., Wu, X., & Taj, S. (2023). The optimum dietary phenylalanine requirement of hybrid grouper (*Epinephelus fuscoguttatus*♀ × *Epinephelus lanceolatus*♂) juveniles: Effects on growth performance, gut micromorphology, and antioxidation. *Aquaculture Nutrition, 2023*, 1–10. https://doi.org/10.1155/2023/9155290.

Yao, Y., Shi, L., Xiao, W., Guo, S., Liu, S., Li, H., & Zhang, S. (2022). Phenylalanine hydroxylase (PAH) plays a positive role during WSSV and *Vibrio parahaemolyticus* infection in *Litopenaeus vannamei*. *Fish & Shellfish Immunology, 120*, 515–525. https://doi.org/10.1016/j.fsi.2021.12.028.

Yi, C., Liang, H., Huang, D., Yu, H., Xue, C., Gu, J., Chen, X., Wang, Y., Ren, M., & Zhang, L. (2023). Phenylalanine plays important roles in regulating the capacity of intestinal immunity, antioxidants and apoptosis in largemouth bass (*Micropterus salmoides*). *Animals, 13*(18), 2980. https://doi.org/10.3390/ani13182980.

Yi, C., Liang, H., Xu, G., Zhu, J., Wang, Y., Li, S., Ren, M., & Chen, X. (2022). Appropriate dietary phenylalanine improved growth, protein metabolism and lipid metabolism, and glycolysis in largemouth bass (*Micropterus salmoides*). *Fish Physiology and Biochemistry*. https://doi.org/10.1007/s10695-022-01138-5.

Yun, H., Park, G., Ok, I., Katya, K., Heung, S., & Bai, S. C. (2015). Evaluation of optimum dietary threonine requirement by plasma free threonine and ammonia concentrations in surgically modified rainbow trout, *Oncorhynchus mykiss*. *Asian-Australasian Journal of Animal Sciences, 28*(4), 551–559. https://doi.org/10.5713/ajas.14.0495.

Zahran, E., Elbahnaswy, S., Ibrahim, I., & Khaled, A. A. (2021). *Nannochloropsis oculata* enhances immune response, transcription of stress, and cytokine genes in Nile tilapia subjected to air exposure stress. *Aquaculture Reports, 21*, 100911. https://doi.org/10.1016/j.aqrep.2021.100911.

Zehra, S., & Khan, M. A. (2014). Dietary phenylalanine requirement and tyrosine replacement value for phenylalanine for fingerling *Catla catla* (Hamilton). *Aquaculture, 433*, 256–265. https://doi.org/10.1016/j.aquaculture.2014.06.027.

Zehra, S., & Yousif, R. (2021). Dietary phenylalanine requirement of fingerling *Oreochromis niloticus* (Linnaeus). In *Conference Proceedings, Dubai UAE*, Jan 28–29, 2021.

Zhang, S., Wang, C. A., Liu, S., Wang, Y., Lu, S., Han, S., Jiang, H., Liu, H., & Yang, Y. (2023). Effect of dietary phenylalanine on growth performance and intestinal health of triploid rainbow trout (*Oncorhynchus mykiss*) in low fishmeal diets. *Frontiers in Nutrition, 10*, 1008822. https://doi.org/10.3389/fnut.2023.1008822.

CHAPTER 8

Tryptophan Requirement in Different Fishes

QUSEEN MUSHTAQ RESHI,[1*] AAMINA HASSAN,[1] IMTIAZ AHMED,[1] and ISHTIYAQ AHMAD[1,2]

[1]*Department of Zoology, Fish Nutrition Research Laboratory, University of Kashmir, Hazratbal Srinagar, Jammu and Kashmir, India*

[2]*Division of Fish Genetics and Biotechnology, Faculty of Fisheries, Sher-e-Kashmir University of Agricultural Sciences and Technology of Kashmir (SKUAST-Kashmir), Rangil, Ganderbal, Srinagar, Jammu and Kashmir, India*

*Corresponding author

ABSTRACT

Tryptophan is an essential amino acid (EAA) that plays a vital role in the growth, development, and general health of fish. The function and precise requirements in fish nutrition have garnered great interest in recent years. Detailed investigations regarding the understanding of tryptophan requirements in various fish species include the physiological significance of tryptophan and the effects of tryptophan excess or deficiency on fish growth and health conditions. Understanding the specific tryptophan requirements for different fish species is paramount for conceiving tailored nutrition strategies that promote sustainable aquaculture practices. Furthermore, continued research efforts aimed at unraveling the intricate metabolic pathways and physiological functions of tryptophan will contribute to the advancement of fish nutrition and welfare, ultimately fostering the growth and development of the aquaculture industry. The aim of the present chapter is to consolidate the current knowledge related

Fish Nutrition: Protein and Amino Acid Requirements. Imtiaz Ahmed, Ishtiyaq Ahmad, and Shabihul Fatma Sayed (Eds.)

DOI: 10.1201/9781998511570-8

to various aspects of tryptophan requirements in fish, which is of great significance in aquaculture and fisheries management.

8.1 INTRODUCTION

The quality of protein is intrinsically linked to the composition of its essential amino acids (EAAs) (Moraes and de Almeida, 2020). Beyond its role in proteogenesis and energy production, EAAs wield significance in various nutritional and metabolic processes in fish, together with the regulation of gene expression, appetite stimulation, cell signaling, food consumption, nutrient metabolism, digestion, growth, development, efficient energy utilization, immune response, ammonia detoxification, as well as other physiological processes like metamorphosis, reproduction, osmoregulation, and pigmentation (Li et al., 2009; Wu, 2013; Wu et al., 2014). The nutritional well-being of fish, encompassing their behavioral, structural, pathological, reproductive, and physiological aspects, is predominantly dependent on their dietary intake (Guillame et al., 2001). Understanding the specific nutritional needs of fish is crucial for optimizing nutrient utilization. In the realm of intensive fish farming, the utilization of nutritionally complex diets is indispensable. Among all the nutrients, protein stands out as a pivotal macronutrient, serving as the foundational building block for muscle and organ development. Protein, being the most expensive constituent utilized in fish feed, supplies the EAAs necessary for tissue repair, growth, the production of functional proteins, as well as an energy source (Riegel, 2012). Typically, the ingredients used in artificial feed formulation fall short in the four major EAAs: arginine, lysine, tryptophan, and methionine. Therefore, the primary focus for nutritionists in formulating well-balanced feeds rich in EAAs is to supplement these limiting amino acids (AAs) according to the specific requirements of the fish. This not only enhances growth performance and profitability but also reduces the organic load within the culture system. Meeting the optimal AA requirements is imperative for achieving cost-effectiveness and balanced diets. AA needs may increase due to metabolic shifts associated with stress response and infections. Among all the EAAs needed by fish species, particular attention is given to tryptophan (Ahmed and Khan, 2005). The appropriate inclusion of EAAs in the correct proportions can lead to the development of well-balanced aquafeeds, ultimately improving feed quality and consequently enhancing aquaculture's profitability (Li et al., 2009). The EAAs in fish and their common roles are shown in Figure 8.1.

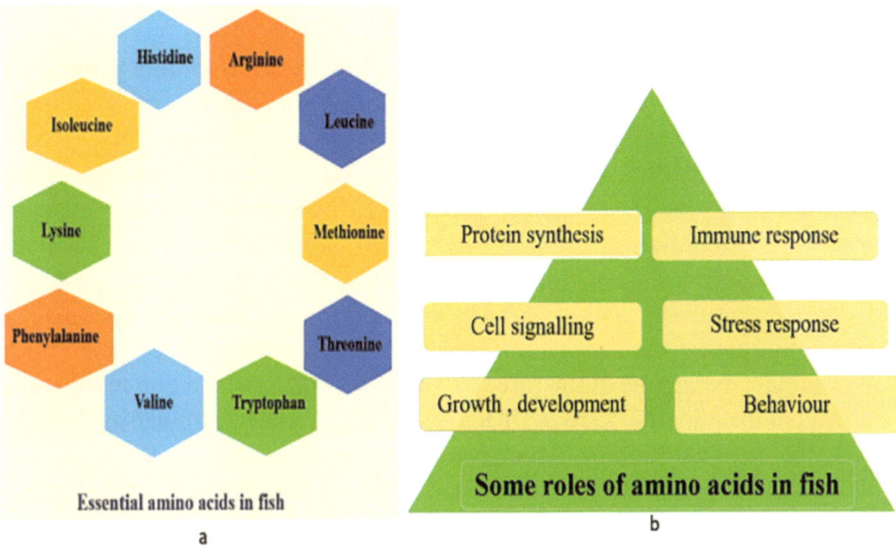

FIGURE 8.1 Showing essential amino acids in fish and their common roles.

Tryptophan exerts a versatile influence upon the physiology of teleosts (Hoseini et al., 2019). The name "tryptophan" (molecular formula as $C_{11}H_{12}N_2O_2$ and molecular mass: 204.23 Da) is coined from the time when this AA was first isolated from a tryptic digest of casein by Hopkins and Cole (1901). Tryptophan is classified within the category of hydrophobic AAs. It is formed as a result of the condensation of serine with an indole group (Nongonierma and FitzGerald, 2015). Owing to the rise in aquafeed demand, which is a result of the use of scientific knowledge, fish meal is, however, remaining mostly stagnant over the past years; therefore, the usage of fish meal is under tremendous pressure. Therefore, it is necessary to use substitute fish meal components to create more sustainable feed-stuffs that are both effective and affordable. However, formulating feed with different protein sources in place of fish meal necessitates much more thought and precise estimation of dietary needs for vital elements. Usage of low-cost and obtainable animal and plant sources of protein are usually deficient in some indispensable AAs. The findings of a limited number of researches provide values for EAAs, such as phenylalanine, tryptophan, valine, isoleucine, histidine, and threonine (Ahmad et al., 2021; Machado et al., 2019; Yaghoubi et al., 2017). To guarantee the highest quality of information on every AA, a thorough assessment is necessary to confirm that the fundamental requirements of each AA are precise.

As an EAA, tryptophan has three groups: an α-amino group, an α-carboxylic acid group, and a side chain indole (Nguyen et al., 2019). Tryptophan is special in biology for a variety of reasons. Out of the 20 AAs in the translational toolkit, it is the biggest. While the side chains of tyrosine, histidine, and phenylalanine are single-ring aromatics, its side chain is indole, an aromatic with a binuclear ring structure (Barik, 2020). To some extent because of these intricate structural characteristics, tryptophan's biosynthesis process is the most intricate and the AA that uses the greatest amount of energy. Tryptophan is the least abundant and most essential food for animals, being one of the rarest AAs in the proteome and one of the most prevalent in cells (Barik, 2020). Most eukaryotes have tryptophan as the only AA other than methionine, which has a single codon—UGG—that codes for it. Its aromatic side chain interacts with several different side chains in the protein due to its high and hydrophobic π-electron surface area, which is appropriate given its important positions in the protein structure. Also, a number of tryptophan derivatives have specific uses, including oxitriptan, tryptophylquinone, melatonin, serotonin, and tryptophol (Barik, 2020).

Tryptophan is involved in a number of biological processes and is a biochemical antecedent for multiple substances, such as serotonin and a neurotransmitter. Important processes include influencing how much feed is consumed and immunological response, as well as protein synthesis (Wen et al., 2014). The ample supply of all EAAs, including tryptophan, is crucial for the ideal fish growth and development (Hoseini et al., 2019). Tryptophan, a neutral aromatic AA, is indispensable for the well-being of animals and must be provided through their diet (Wilson, 2002). Tryptophan, being a precursor for various important metabolites serotonin, as well as melatonin, functions as a crucial constituent in different physiological processes, such as synthesis of proteins, regulation of neurotransmitters, and proper functioning of immune systems (Hoseini et al., 2019; Teixeira et al., 2023). The particular requirements of tryptophan for different fish species and life stages hold great research value. The utilization of natural sources of tryptophan in fish feed is essential for maintaining optimal growth, health, and performance in aquaculture (Teixeira et al., 2023).

The presence of tryptophan is closely tied to various aspects of fish biology, including growth, sexual behavior, reproduction, survival, and immunity. This indispensable AA, tryptophan, plays a pivotal role in regulating both the physiological as well as behavioral functions of fish (Sahu et al., 2020). Tryptophan serves as the precursor for the production of serotonin and kynurenine (Le Floch et al., 2011). Serotonin is a critical neuromodulator, plays an active role in suppressing stress, and is involved in the regulation of pituitary growth hormone secretion (Musumeci et al., 2013). On the other hand, kynurenine and its products exhibit varied biological functions (Le Floch et al., 2011). The aim of the present chapter is to consolidate the current knowledge on some aspects related to the influence of tryptophan in fish, which is of great significance in aquaculture and fisheries management. This article also discusses the importance of tryptophan sources, urging for the development of innovative and sustainable approaches to ensure consistent and reliable nutrition in aquaculture practices.

8.2 TRYPTOPHAN METABOLISM IN FISH

The study on the metabolism of tryptophan in fish is crucial in elucidating its nutritional requirements as well as physiological effects. Tryptophan is primarily metabolized through the kynurenine pathway, leading to the synthesis of biologically active compounds, such as melatonin, serotonin, and niacin as well (Hoseini et al., 2019; Sahu et al., 2020). All these compounds perform critical roles in regulating various types of physiological functions such as modulation in growth, including immune reaction as well as behavior (Hoseini et al., 2019; Sahu et al., 2020; Wei et al., 2023). AAs play crucial and unique roles in the nutrition as well as metabolism of fish. Among these AAs, tryptophan stands out as the sole AA within proteins that derives from indole, a bicyclic ring formed by the fusion of benzene and a pyrrole group. The presence of indole in life chemistry offers distinct advantages either through its potential utilization of its C11 skeleton in various metabolism processes or its use as an R residue in proteins and peptides, contributing to their structure and stability. Tryptophan undergoes metabolism to yield biologically active indole compounds that significantly impact various life functions. Additionally, the study of the metabolic pathways of tryptophan and other aromatic AAs phenylalanine and tyrosine in living organisms reveals a kind of "signature" of the cellular evolutionary stages. These pathways trace their origins from prokaryote unicellular organisms, passing through eukaryote ancestor cells and eventually lead to distinct vegetal or animal eukaryotes possessing distinct cells, subcellular compartments, and trophic traits (Tzin and Galili, 2010). In vertebrates, upon dietary intake, tryptophan can follow one among the three pathways: catabolism through the kynurenine pathway, the serotonin pathway, or the melatonin pathway (Le Floch et al., 2011). The equilibrium among these pathways is contingent on various physiological conditions and external influences. While the established routes from tryptophan to melatonin or serotonin are well documented in fish and recognized as highly preserved among vertebrates, there is notably less data on the kynurenine pathways in fish. The kynurenine pathway takes precedence in the degradation of tryptophan, accounting for nearly 95% of the excess tryptophan. Furthermore, it is considered the primary route for the removal of excess tryptophan from the organism. This process is most active in the liver, where the portal blood supply facilitates efficient absorption by hepatocytes (Fernstrom, 2016). The initial oxidation of tryptophan is carried out by one of two heme-containing oxidoreductases, indoleamine 2, 3-dioxygenase or the closely related tryptophan 2, 3-dioxygenase. The N-formylkynurenine produced undergoes conversion to kynurenine through arylformamidase. Kynurenine can then be metabolized by kynurenine aminotransferase and thereby ultimately lead to the production of the bioactive kynurenic acid. Notably, this branch of the kynurenine pathway will only be activated in the presence of tryptophan supplementation given the enzyme's high Km (Fernstrom, 2016). Alternatively, kynurenine can undergo further transformation into 3-hydroxykynurenine and 3-hydrxyanthranilic acid, both of which are biologically active compounds. Following this, the substances can undergo oxidation leading to the formation of acroleyl aminofumarate (2-amino-3-carboxymuconate semialdehyde). The subsequent metabolic processes are closely tied to the organism's absorption of dietary niacin. The study of the complex interaction concerning tryptophan metabolism and the subsequent products has offered broad knowledge of the ensuing metabolic pathways in different fish species, aiming to derive maximum benefits for fish farmers (Ahmad et al., 2021; Zhang et al., 2023).

8.3 EMPIRICAL METHODS USED FOR TRYPTOPHAN REQUIREMENT

Various methods for determining the tryptophan requirement in fishes involve a series of experimental procedures that aim to establish the species-specific dietary tryptophan values essential for optimal growth and health status in various fish species (Hoseini et al., 2019; Teixeira et al., 2023; Zhang et al., 2023). Such methods usually include both feeding trials as well as biochemical analyses to assess the physiological responses and metabolic changes associated with different levels of dietary tryptophan. Factorial approaches and models that are designed to estimate AA requirements rely on an understanding of the factors that influence these requirements. Key factors include the protein content required for body weight gain (BWG), the AA composition of the accumulated whole-body protein, the efficiency of utilizing total or digestible AAs for body protein growth, and the maintenance requirement, which accounts for inevitable AA losses. The determination of tryptophan requirements has been established in various teleost species, using the standard dose-growth response plot method (Wilson, 2002). The determination of tryptophan requirements was based on growth parameters, protein retention coefficients, and the activity of acidic protease, and these requirements were estimated through non-linear (broken-line) and polynomial regression analyses, as well as broken-line regression models.

Several common empirical approaches utilized for determining the tryptophan requirement in fishes include growth performance trials, blood biochemical analysis, nutrient balance studies, tissue and enzyme activity assays, and behavior and stress response studies. In growth performance trial experiments, fish are fed with diets containing varying levels of tryptophan while constantly monitoring their growth parameters, such as weight gain, specific growth rate (SGR), feed conversion ratio (FCR), and protein efficiency ratio (PER). By properly analyzing the growth performance data, researchers can establish the optimal tryptophan values necessary for attaining maximum growth and development in a specific fish species (Akter et al., 2023; Yaghoubi et al., 2017; Zhang et al., 2023).

Nutrient balance studies help assess the utilization and retention of dietary tryptophan by the fishes. These studies involve the quantification of nutrient intake and excretion, allowing researchers to determine the efficiency of tryptophan utilization in relation to various other dietary components (Akter et al., 2023). By evaluating nutrient balance, research workers can gain insights into the metabolic utilization of tryptophan and its contribution to overall energy utilization and growth in different fish species (Akter et al., 2023; Yaghoubi et al., 2017; Zhang et al., 2023). Blood biochemical analyses serve a critical role in the evaluation of the physiological responses of fishes to different levels of dietary tryptophan (Zhang et al., 2023). These analyses usually involve the measurement of serum tryptophan concentrations, as well as the assessment of the related metabolites and hormones, such as melatonin, serotonin, and niacin derivatives (Hoseini et al., 2019; Sahu et al., 2021; Teixeira et al., 2023; Wei et al., 2023). By scrutinizing the changes in various parameters of blood, researchers can assess the metabolic implications of altered tryptophan levels on immune function, stress response, and overall fish health status (Zhang et al., 2023).

Experiments on tissue and enzyme activity assays facilitate the assessment of tryptophan metabolism and associated enzymatic actions in specific tissues of fish, such as the brain, liver, and intestine (Devassykutty et al., 2022; Sahu et al., 2020; Sun et al., 2023; Tejpal et al., 2009; Teixeira et al., 2022; Wei et al., 2023). These assays involve the quantification of various key enzymes having a role in the metabolism of tryptophan, such as tryptophan hydroxylase, kynurenine pathway enzymes, and serotonin-metabolizing enzymes (Hoseini et al., 2019; Wei et al., 2023). By analyzing tissue-specific enzyme activities, researchers can elucidate the impact of dietary tryptophan levels on the regulation of various biochemical pathways as well as neuroendocrine functions in fishes (Machjado et al., 2019).

Behavior and stress response studies focus on evaluating the effects of varying tryptophan levels on fish behavior, stress tolerance, and overall welfare. These studies often employ behavioral observation techniques, stress response assays, and neurobehavioral assessments to analyze the impact of tryptophan on fish aggression, feeding behavior, and coping mechanisms under different environmental conditions. By integrating behavioral analyses with physiological data, researchers can ascertain the influence of tryptophan on the neuroendocrine regulation of behavior and stress responses in fishes. The study on rainbow trout showed a change in behavior and feeding because of serotonin (derived from tryptophan) (Lepage et al., 2003). The review studies elucidated that tryptophan, being a precursor of serotonin, influences fish behavior (Sahu et al., 2020). The elevation of serotonin synthesis in the brain as a result of supplemental dietary tryptophan has been associated with various behavioral patterns such as feeding, aggression, fish maturation, stress reaction, social dominance, as well as sexual behavior (Sahu et al., 2020).

8.4 SOURCES OF DIETARY TRYPTOPHAN

Natural sources of tryptophan play a vital role in meeting the dietary requirements of fishes; however, their availability, sustainability, and prices pose significant challenges for the aquaculture industry. As already stated, tryptophan is a very important AA required for the optimum growth and overall health of fishes (Hoseini et al., 2019; Teixeira 2023). Nature has endowed us with great bounties, and we have diverse natural sources of tryptophan which can be incorporated in fish feed so as to derive optimal nutritional profiles. The influence of fishmeal, fish oil, and various plant-based sources of protein in providing essential tryptophan and other AAs to meet the dietary requirements of different fish species has been studied by various workers (Jiang et al., 2022). The implications of incorporating natural tryptophan sources in formulations of aquafeed lay great emphasis on their contributions towards sustainable and balanced nutrition in aquaculture. Furthermore, it highlights the prospective challenges and chances allied with the utilization of natural sources of tryptophan, which play a great role in supporting proficient growth, health, and welfare. The integration of various natural sources of tryptophan during the formulation of fish feed is one of the fundamental steps for the maintenance of viable, economical, and saustainable aquaculture practices. The analyses of the diverse natural sources of tryptophan emphasize their importance in fulfilling the nutritional requirements of different fish species. The common natural dietary components, such as fishmeal, fish oil, and various sources of protein derived from plants, have significance in providing essential tryptophan and therefore supporting optimum growth and health in different fish species (Prabu et al., 2017). Following lysine and methionine, tryptophan stands out as one of the AAs that tends to be most deficient in a variety of plant-based fish feeds (Coloso et al., 2004; Kim and Lall, 2000). Feed ingredients abundant in tryptophan include wheat flour, soybean oilcake, barley meal, wheat gluten, cottonseed meal, blood meal, and fishmeal. In contrast, corn, hydrolyzed feather meal, meat and bone meal, as well as corn gluten, generally have lower levels of tryptophan (NRC, 2011). The varying percentage of dietary tryptophan as per NRC, 2011, is given in Table 8.1.

TABLE 8.1 Varying Percentage of Dietary Tryptophan as Per NRC, 2011

Feedstuff	Percentage of Dietary Tryptophan
Corn	0.68–0.82%
Corn gluten	0.47%
Meat / bone meal	0.51–0.63%
Hydrolyzed feather meal	0.66%
Fish meal	0.93–1.16%
Blood meal	1.14%
Cotton seed meal	1.27%
Wheat gluten	1.3%
Barley meal	1.47%
Soybean oilcake	1.48%
Wheat flour	2.13%

Fishmeal is derived from different processing mechanisms of various marine or freshwater fish species. It serves as a primary natural source of tryptophan in aquafeeds at both small-scale as well as large-scale commercial aquaculture farms (Li et al., 2009). It is known for its high-quality protein content, balanced AA profile, and prosperous concentration of various essential nutrients, including tryptophan as well. Numerous studies in the past have highlighted the importance of fishmeal in promoting optimal growth, immunity, yield, and overall performance in various fish species (Li et al., 2009). The incorporation of fishmeal in various aquafeed formulations is to ensure that there is adequate provision of dietary tryptophan in feed, which aims to support both the metabolic as well as physiological needs of fishes.

Fish oil is extracted from various fish species, which are rich in omega-3 fatty acids (Bahurmiz et al., 2017). These omega-3 fatty acids are known as one of the essential natural sources of tryptophan in fish feed. In addition to being a source of crucial fatty acids, fish oil does contribute to the overall tryptophan content in the aquafeeds, thus supporting various physiological functions as well as growth factors in fishes. Hence, it is very important to incorporate fish oil in the diets of the fish to ensure that a balanced supply of tryptophan and various other essential nutrients that are vital for promoting optimal growth, development, and overall health in aquaculture settings are being fed to the fish.

Aquafeed formulations are mainly composed of DP, and it is the most expensive component (Wilson et al., 2002). Although fishmeal produces a static output worldwide, its composition and quality vary geographically and seasonally, and it poses a concern for contamination. Fishmeal is generally considered to be one of the best protein sources for aquatic animals (Trushenski et al., 2006). During the past decades, the global aquaculture industry and demand, as well as price and competition for this valuable feedstock, have increased rapidly. As non-fishmeal protein sources for aquafeeds become more popular, it is imperative to maintain comparable feed intake levels, feed conversion efficiency, growth rate, and survival rates (Li et al., 2009). Thus, defining the ideal dietary needs for AA and characterizing substitute sources of AA and protein have taken front stage in studies on fish nutrition.

The potential limitations associated with the reliance on specific natural sources of tryptophan emphasize the need for developing strategies that are both innovative as well as sustainable, so as to ensure consistent and balanced nutrition in fish feed formulations. Also, there is a need for considering alternative sources of tryptophan and optimizing feed formulations to promote efficient tryptophan utilization and support the long-term growth and welfare of aquaculture production systems and economic benefits as well.

8.5 DIETARY TRYPTOPHAN REQUIREMENT IN FISH

Dietary tryptophan requirements for fish species exhibit variation as influenced by factors such as life stages, type of species, environmental settings, and dietary composition (Li et al., 2021; Yaghoubi et al., 2017; Zhang et al., 2023). A large amount of research has been conducted in aquaculture to optimize tryptophan levels in the commercial diets for a variety of fish species such as sockeye salmon, *Oncorhynchus nerka* (Halver, 1965); *Cyprinus carpio* (Nose et al., 1974); rainbow trout, *O. mykiss* (Kim et al., 1987; Ogino, 1980; Walton et al., 1986); Nile tilapia, *Oreochromis niloticus* (Santiago and Lovell, 1988); Indian major carp, catla, *Catla catla* (Ravi and Devaraj, 1991); milk fish, *Chanos chanos* (Borlongan and Coloso, 1993; Coloso et al., 1992); channel catfish, *Ictalurus punctatus* (Wilson et al., 1978); Indian major carp, *Labeo rohita*

(Murthy and Varghese, 1997); gilthead sea bream, *Sparus aurata* (Kaushik, 1998); African catfish, *Clarias gariepinus* (Fagbenro and Nwanna, 1999); Asian sea bass, *Lates calcarifer* (Coloso et al., 2004); Indian major carp, mrigal, *Cirrhinus mrigala* (Ahmed and Khan, 2005; Ahmed, 2009); *Labeo rohita* (Abidi and Khan, 2010); *Heteropneustes fossilis* (Ahmed, 2012); Jian carp, *Cyprinus carpio* (Tang et al., 2013); stinging catfish, *Heteropneustes fossilis* (Farhat and Khan, 2014); catla, *Catla catla* (Zehra and Khan, 2015); Asian seabass, *Lates calcarifer* (Kumar et al., 2017); Nile tilapia, *O. niloticus* (Nguyena et al., 2019; Zaminhan et al., 2018); snakehead *C. argus* (Miao et al., 2021); Spotted murrel, *Channa punctatus* (Sharf and Khan, 2023).

Suitable levels of tryptophan in fish feed are indispensable for supporting optimal growth, augmenting immune response, and enabling wide range tolerance to stress behavior (Hoseini et al., 2019; Zhang et al., 2023). Additionally, tryptophan deficiency or excess may lead to various detrimental effects on the health of fish, such as impairment in growth, compromised immunity, and altered behavioral response (Wei et al., 2023; Yaghoubi et al., 2017). Tejpal et al. (2009) evaluated the effect of L-tryptophan supplementation on growth and crowding stress in fingerlings of *Cirrhinus mrigala* and reported that groups supplemented with L-tryptophan were found to have significantly higher ($p<0.05$) PER and specific growth. The influence of dietary tryptophan in *Labeo rohita* in mitigating stress and enhancing growth was evaluated by Kumar et al. (2014). It was observed that L-TRYPTOPHAN-supplemented groups had higher growth as well as mitigated thermal stress, with a 0.72% level in the diet being optimum to decrease thermal stress even up to a temperature of 38°C. Pianesso et al. (2015) evaluated the tryptophan requirement for juvenile *Rhamdia quelen* and its effects on hepatic and plasma metabolites, digestive enzymes activity, and growth performance. They reported a highly significant ($P < 0.001$) effect of tryptophan on acidic protease activity and growth. Fish given insufficient tryptophan levels grew slower, had greater hepato-somatic indices, and had changed enzymatic and metabolic activities, suggesting gluconeogenic processes. Based on the mathematical framework employed for assessment, the projected tryptophan AA needs for maximal growth and protein retention coefficients varied between 2.5 to 3.4 g/kg. Nguyen et al. (2019) evaluated the tryptophan requirement in juvenile Nile tilapia via semi-purified diets and reported that an increase in dietary tryptophan resulted in a positive growth response in Nile tilapia (*Oreochromis niloticus*). Fish given diets with more than 0.25% tryptophan supplement exhibited substantially greater mean TGC ($P < 0.05$) compared to results obtained for the rest of the diets. Therefore, precise determination of the tryptophan requirement in different fish species is crucial for formulating effective, budget-friendly, and sustainable aquafeeds. Fish whole-body AA composition indicates their nutritional requirements (Kaushik, 1998). It is highly consistent among fish of various sizes and even across species (Kaushik, 1998; Mambrini and Kaushik, 1995; Wilson and Cowey, 1985).

Despite the relatively high consistency of tryptophan concentration in the whole body, the tryptophan requirements among different fish studies, both across different species and for specific species, indicate a broad range (Kaushik, 1998; Mambrini and Kaushik, 1995). The dietary tryptophan requirement in *Cyprinus carpio* has been estimated to be between 0.3 and 1.1% of DP, representing a 4-fold range (Dabrowski, 1981; Tang et al., 2013). In *Labeo rohita,* dietary tryptophan levels of 0.9% and 1.13% of DP (Abidi and Khan, 2010; Murthy and Varghese, 1997) were suggested as essential tryptophan levels for optimum growth. On the other hand, other workers Akhtar et al. (2013) and Ciji et al. (2015) showed that *L. rohita* given a diet containing tryptophan levels of 3.05–5.17% of dietary protein (DP) had significantly improved weight gain compared to control fish (fed with a diet containing 0.99% tryptophan). Cabanillas-Gámez et al. (2020) recommended that in totoaba (*Totoaba macdonaldi*) under high-density culture setups, a diet supplemented with tryptophan meals improved homeostasis as measured by serotonergic activity as well as plasma and hematological parameters. Under high-density circumstances, plasma cortisol rose when the fish were fed a meal with moderate tryptophan levels. Fu et al. (2021) evaluated the influence of dietary tryptophan on intestinal gene expression, intestinal microbiota, and growth in an improved triploid crucian carp. They reported that a tryptophan concentration at the rate of 6.35 g kg^{-1} resulted in a higher daily gain average, enhanced villus height, and crypt depth in the experimental fish group as compared to the control. Alteration was observed in gene expression responsible for metabolism and nutrient transport in the intestine. 16S rRNA sequencing indicated that 6.35 g kg^{-1} tryptophan amplified the number of bacteria belonging to the genus Cetobacterium, as well as the Firmicutes/Bacteroidetes ratio at the phylum level significantly ($P < 0.05$). In addition, bacterial richness indices using the Simpson index showed a significant increase ($P < 0.05$) of community evenness with 6.35 g kg^{-1} tryptophan. Thus, appropriate dietary tryptophan not only increases growth performance but also impacts the intestinal flora of improved triploid crucian carp. Cabanillas-Gámez et al. (2022) studied the influence on plasma and blood factors of striped bass *Morone saxatilis*, on exposure to acute stressors and reported that all the selected diets supplemented with tryptophan prevented changes in glucose values under hypoxic conditions as well as changes in total protein caused by handling. These findings indicate that tryptophan played an important role in maintaining homeostasis of *M. saxatilis* restoration when subjected to acute stress conditions. Khan et al. (2023) studied the influence of L-tryptophan on feed utilization, growth, cannibalism, muscle proximate composition, and survival and reported that dietary supplementation at the rate of 1.00% was highly effective in decreasing cannibalism and also increased the survival of Asian seabass nursery reared in a recirculating aquaculture system when reared at a lower stocking density. On the other hand, at a greater stocking density and 1.50% dietary tryptophan supplementation, both growth and cannibalism increased, while the survival rate decreased. The tryptophan requirement in different fish species is given in Table 8.2.

TABLE 8.2 Tryptophan Requirement in Different Fish Species

Fish Species	Tryptophan Requirement	References
Spotted murrel, *Channa punctatus*	4.81–5.5 g/kg	Sharf and Khan (2023)
Silver pompano, *Trachinotus blochii*	0.5 g / 100 g	Devassykutty et al. (2022)
Northern snakehead, *Channa argus*	0.46%	Miao et al. (2021)
Nile tilapia, *Oreochromis niloticus*	0.31%	Nguyen et al. (2019)
Nile tilapia, *Oreochromis niloticus*	3.4 g/kg	Zaminhan et al. (2018)
Asian seabass, *Lates calcarifer*	0.5%	Kumar et al. (2017)
Catla, *Catla catla*	2.1–2.5 g/kg	Zehra and Khan (2015)
catfish, *Heteropneustes fossilis*	0.24–0.27%	Farhat and Khan (2014)
Jian carp, *Cyprinus carpio* var. *Jian*	3.5 g/kg	Tang et al. (2013)
Indian catfish, *Heteropneustes fossilis*	0.32 g	Ahmed (2012)
Rohu, *Labeo rohita*	4–4.5%	Abidi and Khan (2010)
Indian major carp, mrigal	0.38 g / 100 g	Ahmed and Khan (2005)
Asian sea, *Lates calcarifer*	3.1 g/kg	Coloso et al. (2004)
Milkfish, *Chanos chanos*	3.1 g/kg	Coloso et al. (1992)
Catla, *Catla catla*	0.95%	Ravi and Devaraj (1991)

8.6 EFFECTS OF TRYPTOPHAN ON FISH HEALTH AND BEHAVIOUR

The influence of tryptophan on various physiological and behavioral aspects of fish is quite significant (Hoseini et al., 2019). Aromatic AAs (tyrosine and phenylalanine) and tryptophan may additionally influence immune system reactions by controlling the production of hormones (thyroid hormones and catecholamines), tetrahydrobiopterin (a cofactor for synthesis of nitric oxide), along with neurotransmitters (serotonin, dopamine, melatonin, and N-cetylserotonin) (Li et al., 2009; Wu, 2009). Studies have demonstrated that the role of tryptophan in modulating stress responses, aggression, and feeding behavior in different fish species has been greatly worked upon (Hoseini et al., 2019). In some salmonids, a tryptophan shortage has been linked to disturbances in the metabolism of minerals and spine abnormalities (Shanks, 1962). Tryptophan is found in relatively little concentrations within DPs, with around 1% in animal proteins as well as 1.4% in plant proteins (Rossi and Tirapegui, 2004). The diminished growth observed in fish that were fed diets with the lowest tryptophan concentrations may be attributed to disruptions in the processes of protein synthesis. As noted by Abidi and Khan (2010), the utilization of protein for growth can be hindered when diets lack a balanced composition of AAs. Ensuring a well-balanced proportion of dietary AAs is crucial to meet the body's AA requirements. Any deficiency in an EAA can negatively impact the utilization of others, hindering the process of protein synthesis. Tryptophan, in addition to being an EAA for fish, also plays a role in the synthesis of other vital compounds. It is a limiting EAA in fish silage as well as a potential fish meal replacer in fish diets (Tacon, 1993). Tyrosine and tryptophan residues have specific functions in securely fixing membrane proteins within the cell membrane. Additionally, tryptophan, along with other aromatic AAs, is essential for facilitating interactions between glycan-proteins. When combined with niacin, tryptophan can have an impact on the metabolism of lipids and proteins (Ahmed, 2010). Furthermore, tryptophan's involvement in the regulation of neurotransmitters and neuroendocrine functions highlights its potential impact on fish welfare and overall performance. Knowledge of these values is crucial in formulating specialized diets that promote both optimal growth and stress management aspects in aquaculture settings. A reduction in aggressive behavior was observed in *Gadus morhua* when treated with tryptophan for seven days (Höglund et al., 2005). A similar aggression-suppressive behavioral effect has also been detected with *Brycon cephalus* after a seven-day tryptophan administration trial, although serotonergic activity was not reported (Wolkers et al., 2012). In a similar pattern, *Epinephelus coioides* exhibited repressed aggression behavior as well as cannibalism along with brain 5HT elevation when treated with tryptophan for 10 days (Hseu et al., 2003). Likewise, different voluntary locomotory behavior in *Dicentrarchus labrax* was observed after seven days of tryptophan administration (Herrero et al., 2007). These results indicate that tryptophan does affect teleost behavior via the serotonergic system; nevertheless, there is a need to observe effects on serotonergic and behavior response of the fish after long tryptophan administration (Sahu et al., 2020; Teixeira et al., 2023). Melatonin influenced the development of testicles in Masu salmon (Amano et al., 2004). In aquaculture feeds, replacing fish meal with plant-based protein sources is a top objective. In this context, in addition to fostering an improved immunological state, dietary supplementation with EAAs, such as tryptophan, is strategically important to guarantee that the individual's nutritional demands are satisfied (Cerqueira et al., 2020). The study carried out by Cerqueira et al. (2020) looked at the advantages of tryptophan incorporation in meals using plant-based protein sources on fish development and nutritional health (Cerqueira et al., 2020). They evaluated the effects of feeding sea bream (*Sparus aurata*) diets with 20% lower (L tryptophan) and 27% higher (H tryptophan) hypothesized tryptophan needs on fish physiology, liver metabolism, and proteome. The combined findings point to a possible immune-stimulating benefit of tryptophan supplementation without sacrificing growth or feed consumption. Dietary tryptophan deficit causes caudal fin rot disease, scoliosis, cataracts, lordosis, opercular shortage, and disruption in mineral metabolism in addition to stunting fish development (Fagbenro and Nwanna, 1999; Murthy and Varghese, 1997; Poston and Rumsey, 1983; Shanks et al., 1962; Walton et al., 1984).

8.7 RECENT ADVANCES AND FUTURE PERSPECTIVES

Recent advancements in tryptophan research have provided valuable insights into its multifaceted roles in fish physiology and behavior. Novel techniques, such as transcriptomics and metabolomics, have enabled a more comprehensive understanding of tryptophan metabolism and its effects on the health of fish (Ahmad et al., 2021; Wei et al., 2023). Furthermore, the incorporation of highly advanced molecular tools and techniques, coupled with bioinformatics approaches, holds great promise for deciphering the complex interactions between tryptophan and various physiological pathways in fish (Ahmad et al., 2021; Zhang et al., 2023). Nevertheless, future research endeavors should focus on elucidating the molecular mechanisms underlying the functions of tryptophan and exploring innovative strategies to optimize its utilization in different aquafeed formulations.

8.8 CONCLUSION

Tryptophan is a critical AA that significantly influences the growth, health, and behavior of fishes. In conclusion, tryptophan is a very critical AA that significantly influences the growth, health, as well as the behavior of fishes. Understanding the specific tryptophan requirements for different fish species is essential for devising tailored nutrition strategies that can have a vital role in promoting sustainable aquaculture practices and increasing yield as well. There is a need to explore the requirement of AA tryptophan in various fishes, which is lacking. Furthermore, continued research efforts aimed at unraveling the intricate metabolic pathways and physiological functions of tryptophan will contribute to the advancement of fish nutrition and welfare that could ultimately aid in fostering the growth and development of the aquaculture industry and providing great economic benefits as well.

KEYWORDS

- **amino acid supplementation**
- **aquafeed formulation**
- **essential amino acids**
- **fish growth and health**
- **sustainable aquaculture**
- **tryptophan in fish nutrition**

REFERENCES

Abidi, S. F., & Khan, M. (2010). Dietary tryptophan requirement of fingerling Rohu, *Labeo rohita* (Hamilton), based on growth and body composition. *Journal of the World Aquaculture Society, 41*(5), 700–709.

Ahmad, I., Ahmed, I., Fatma, S., & Peres, H. (2021). Role of branched-chain amino acids on growth, physiology and metabolism of different fish species: A review. *Aquaculture Nutrition, 27*(4), 1270–1289.

Ahmed, I. (2009). Dietary total aromatic amino acid requirement and tyrosine replacement value for phenylalanine in Indian major carp, *Cirrhinus mrigala* (Hamilton) fingerling. *Journal of Applied Ichthyology, 25*(6), 719–727.

Ahmed, I. (2010). Response to the ration levels on growth, body composition, energy, and protein maintenance requirement of the Indian catfish (*Heteropneustes fossilis* Bloch, 1974). *Fish Physiology and Biochemistry, 36*(4), 1133–1143.

Ahmed, I. (2012). Dietary amino acid L-tryptophan requirement of fingerling Indian catfish, *Heteropneustes fossilis* (Bloch), estimated by growth and haemato-biochemical parameters. *Fish Physiology and Biochemistry, 38*(5), 1195–1209.

Ahmed, I., & Khan, M. A. (2005). Dietary tryptophan requirement of fingerling Indian major carp, *Cirrhinus mrigala* (Hamilton). *Aquaculture Research, 36*(7), 687–695.

Akhtar, M., Pal, A., Sahu, N., Ciji, A., Meena, D., & Das, P. (2013). Physiological responses of dietary tryptophan fed *Labeo rohita* to temperature and salinity stress. *Journal of Animal Physiology and Animal Nutrition, 97*(6), 1075–1083.

Akter, T., Hossain, A., Rabiul Islam, M., Hossain, M. A., Das, M., Rahman, M. M., & Abdel-Tawwab, M. (2023). Effects of spirulina (*Arthrospira platensis*) as a fishmeal replacer in practical diets on growth performance, proximate composition, and amino acids profile of pabda catfish (*Ompok pabda*). *Journal of Applied Aquaculture, 35*(1), 69–82.

Amano, M., Iigo, M., Ikuta, K., Kitamura, S., Okuzawa, K., Yamada, H., & Yamamori, K. (2004). Disturbance of plasma melatonin profile by high dose melatonin administration inhibits testicular maturation of precocious male masu salmon. *Zoological Science, 24*(1), 79–85.

Bahurmiz, O. M., Adzitey, F., & Ng, W. K. (2017). Nutrient and fatty acid composition of the flesh of oil sardine (*Sardinella longiceps*) and Indian mackerel (*Rastrelliger kanagurta*) from Hadhramout coast of the Arabian Sea, Yemen. *International Food Research Journal, 24*, 2387–2393.

Barik, S. (2020). The uniqueness of tryptophan in biology: Properties, metabolism, interactions and localization in proteins. *International Journal of Molecular Sciences, 21*(23), 8776.

Borlongan, I. G., & Coloso, R. M. (1993). Requirements of juvenile milkfish (*Chanos chanos* Forsskal) for essential amino acids. *The Journal of Nutrition, 123*(8), 125–132.

Cabanillas-Gámez, M., Lopez, L. M., Galaviz, M. A., True, C. D., & Bardullas, U. (2018). Effect of L-tryptophan supplemented diets on serotonergic system and plasma cortisol in *Totoaba macdonaldi* (Gilbert, 1890) juvenile exposed to acute stress by handling and hypoxia. *Aquaculture Research, 49*(3), 847–857. https://doi.org/10.1111/are.13529.

Cerqueira, M., Schrama, D., Silva, T. S., Colen, R., Engrola, S. A., Conceição, L. E., & Farinha, A. P. (2020). How tryptophan levels in plant-based aquafeeds affect fish physiology, metabolism and proteome. *Journal of Proteomics, 221*, 103782.

Ciji, A., Sahu, N. P., Pal, A. K., & Akhtar, M. S. (2015). Dietary L-tryptophan modulates growth and immuno-metabolic status of *Labeo rohita* juveniles exposed to nitrite. *Aquaculture Research, 46*(8), 2013–2024.

Coloso, R. M., Murillo-Gurrea, D. P., Borlongan, I. G., & Catacutan, M. R. (2004). Tryptophan requirement of juvenile Asian sea bass *Lates calcarifer*. *Journal of Applied Ichthyology, 20*(1), 43–47.

Coloso, R. M., Tiro, L. B., & Benitez, L. V. (1992). Requirement for tryptophan by milkfish (*Chanos chanos* Forsskal) juveniles. *Fish Physiology and Biochemistry, 10*(1), 35–41.

Dabrowski, K. (1981). Tryptophan requirement of common carp (*Cyprinus carpio* L.) fry. *Zeitschrift für Tierphysiologie, Tierernährung und Futtermittelkunde, 46*(1), 64–71.

Devassykutty, M., Ignatius, B., Vijayagopal, P., Ebeneezar, S., Sawant, P. B., Babitha Rani, A. M., Varghese, E., Sayooj, P., & Sajina, K. A. (2022). Effect of dietary tryptophan supplementation on growth, body composition and digestive enzymes activity of juvenile silver pompano *Trachinotus blochii* (Lacepede, 1801). *Indian Journal of Fisheries, 69*(1), 73–80.

Fagbenro, O., & Nwanna, L. (1999). Dietary tryptophan requirement of the African catfish, *Clarias gariepinus*. *Journal of Applied Aquaculture, 9*(1), 65–72.

Farhat, & Khan, M. A. (2014). Dietary L-tryptophan requirement of fingerling stinging catfish, *Heteropneustes fossilis* (Bloch). *Aquaculture Research, 45*(7), 1224–1235.

Fernstrom, J. D. (2016). A perspective on the safety of supplemental tryptophan based on its metabolic fates. *The Journal of Nutrition, 146*(12), 2601S–2608S.

Fu, Y., Liang, X., Li, D., Gao, H., Wang, Y., Li, W., Xu, K., & Hu, F. (2021). Effect of dietary tryptophan on growth, intestinal microbiota, and intestinal gene expression in an Improved Triploid Crucian Carp. *Frontiers in Nutrition, 8*, 676035. https://doi.org/10.3389/fnut.2021.676035.

Guillaume, J., Kaushik, S., Bergot, P., & Metailler, R. (2001). Nutrition and feeding of fish and crustaceans. Springer, London, UK, pp. 27–58.

Halver, J. E. (1965). Tryptophan requirement of chinook, sockeye and silver salmon. *Federation Proceedings, 24*, 169.

Herrero, M., Martínez, F., Míguez, J., & Madrid, J. (2007). Response of plasma and gastrointestinal melatonin, plasma cortisol and activity rhythms of European sea bass (*Dicentrarchus labrax*) to dietary supplementation with tryptophan and melatonin. *Journal of Comparative Physiology B, 177*(4), 319–326.

Höglund, E., Bakke, M. J., Øverli, Ø., Winberg, S., & Nilsson, G. E. (2005). Suppression of aggressive behaviour in juvenile Atlantic cod (*Gadus morhua*) by L-tryptophan supplementation. *Aquaculture, 249*(1–4), 525–531.

Hopkins, F. G., & Cole, S. W. (1901). *Physiology, 27*, 418–428.

Hoseini, S. M., Perez-Jimenez, A., Costas, B., Azeredo, R., & Gesto, M. (2019). Physiological roles of tryptophan in teleosts: Current knowledge and perspectives for future studies. *Reviews in Aquaculture, 11*(1), 3–24.

Hseu, J. R., Lu, F. I., Su, H. M., Wang, L. S., Tsai, C. L., & Hwang, P. P. (2003). Effect of exogenous tryptophan on cannibalism, survival and growth in juvenile grouper, *Epinephelus coioides*. *Aquaculture, 218*(1–4), 251–263.

Jiang, Q., Zhao, Y., Zhou, X. Q., Wu, X. Y., Xu, S. X., Feng, L., & Jiang, J. (2022). Effects of dietary tryptophan on muscle growth, protein synthesis and antioxidant capacity in hybrid catfish *Pelteobagrus vachelli*♀ × *Leiocassis longirostris*♂. *British Journal of Nutrition, 127*(12), 1761–1773.

Kaushik, S. J. (1998). Whole body amino acid composition of European seabass (*Dicentrarchus labrax*), gilthead seabream (*Sparus aurata*) and turbot (*Psetta maxima*) with an estimation of their indispensable amino acid requirement profiles. *Aquatic Living Resources, 11*(5), 355–358.

Khan, M. S., Salin, K. R., Yakupitiyage, A., Tsusaka, T. W., Nguyen, L. T., & Siddique, M. A. M. (2023). L-Tryptophan mitigates cannibalism and improves growth of Asian seabass, *Lates calcarifer* reared in a RAS system. *Aquaculture Journal, 3*(3), 168–180.

Kim, J. D., & Lall, S. P. (2000). Amino acid composition of whole-body tissue of Atlantic halibut (*Hippoglossus hippoglossus*), yellowtail flounder (*Pleuronectes ferruginea*) and Japanese flounder (*Paralichthys olivaceus*). *Aquaculture, 187*(3, 4), 367–373.

Kim, K. I., Kayes, T. B., & Amundson, C. L. (1987). Effect of dietary tryptophan levels on growth, feed/gain, carcass composition and liver glutamate dehydrogenase activity in rainbow trout (*Salmo gairdneri*). *Comparative Biochemistry and Physiology Part B: Comparative Biochemistry, 88*(4), 737–741.

Kumar, P., Kailasam, M., Sethi, S. N., Sukumaran, K., Biswas, G., Subburaj, R., Thiagarajan, G., Ghoshal, T. K., & Vijayan, K. K. (2017). Effect of dietary L-tryptophan on cannibalism, growth and survival of Asian seabass, *Lates calcarifer* (Bloch, 1790) fry. *Indian Journal of Fisheries, 64*(3), 333–340.

Le Floch, N., Otten, W., & Merlot, E. (2011). Tryptophan metabolism, from nutrition to potential therapeutic applications. *Amino Acids, 41*(5), 1195–1205.

Lepage, O., Vílchez, I. M., Pottinger, T. G., & Winberg, S. (2003). Time course of the effect of dietary L-tryptophan on plasma cortisol levels in rainbow trout *Oncorhynchus mykiss*. *Journal of Experimental Biology, 206*(19), 3589–3599.

Li, P., Mai, K., Trushenski, J., & Wu, G. (2009). New developments in fish amino acid nutrition: Towards functional and environmentally oriented aqua feeds. *Amino Acids, 37*(1), 43–53.

Li, X., Zheng, S., & Wu, G. (2021). Nutrition and functions of amino acids in fish. In *Amino Acids in Nutrition and Health: Amino Acids in The Nutrition of Companion, Zoo and Farm Animals* (pp. 133–168).

Machado, M., Azeredo, R., Domingues, A., Fernandez-Boo, S., Dias, J., Conceição, L. E. C., & Costas, B. (2019). Dietary tryptophan deficiency and its supplementation compromises inflammatory mechanisms and disease resistance in a teleost fish. *Scientific Reports, 9*, 7689.

Mambrini, M., & Kaushik, S. (1995). Indispensable amino acid requirements of fish: Correspondence between quantitative data and amino acid profiles of tissue proteins. *Journal of Applied Ichthyology, 11*(4), 240–247.

Miao, S., Chang, E., Han, B., Zhang, X., Liu, X., Zhou, Z., & Zhou, Y. (2021). Dietary tryptophan requirement of northern snakehead, *Channa argus* (Cantor, 1842). *Aquaculture, 542*, 736904.

Moraes, G., & De Almeida, L. C. (2020). Nutrition and functional aspects of digestion in fish. In B. Baldisserotto, E. C. Urbinati, & J. E. P. Cyrino (Eds.), *Biology and Physiology of Freshwater Neotropical Fish* (pp. 251–271). Academic Press. https://doi.org/10.1016/B978-0-12-815872-2.00011-7.

Murthy, H. S., & Varghese, T. (1997). Dietary tryptophan requirement for the growth of rohu, *Labeo rohita*. *Journal of Applied Aquaculture*, *7*, 71–79.

Musumeci, G., Trovato, F. M., Avola, R., Imbesi, R., & Castrogiovanni, P. (2013). Serotonin/growth hormone/insulin-like growth factors axis on pre- and post-natal development: A contemporary review. *OA Anatomy*, *1*, 1–7. https://doi.org/10.13172/2052-7829-1-2-520.

Nguyen, L., Salem, S. M., Salze, G. P., Dinh, H., & Davis, D. A. (2019). Tryptophan requirement in semi-purified diets of juvenile Nile tilapia *Oreochromis niloticus*. *Aquaculture*, 502, 258–267.

Nongonierma, A. B., & FitzGerald, R. J. (2015). Milk proteins as a source of tryptophan-containing bioactive peptides. *Food Function*, 6, 2115–2127.

Nose, T., Arai, S., Lee, D. L., & Hashimoto, Y. (1974). A note on amino acids essential for growth of young carp. *Bulletin of the Japanese Society for the Science of Fish*, 40, 903–908.

NRC. (2011). *Nutrient Requirements of Fish and Shrimp*. National Academy Press.

Ogino, C. (1980). Requirements of carp and rainbow trout for essential amino acids. *Bulletin of the Japanese Society for the Science of Fish*, *46*, 171–174.

Pianesso, D., Neto, J. R., Da Silva, L. P., Goulart, F. R., Adorian, T. J., Mombach, P. I., Loureiro, B. B., Dalcin, M. O., Rotili, D. A., & Lazzari, R. (2015). Determination of tryptophan requirements for juvenile silver catfish (*Rhamdia quelen*) and its effects on growth performance, plasma and hepatic metabolites and digestive enzymes activity. *Animal Feed Science and Technology*, *210*, 172–183.

Poston, H. A., & Rumsey, G. L. (1983). Factors affecting dietary requirement and deficiency signs of L-tryptophan in rainbow trout. *Journal of Nutrition*, *113*, 2568–2577.

Prabu, E., Felix, S., Felix, N., Ahilan, B., & Ruby, P. (2017). An overview on significance of fish nutrition in aquaculture industry. *International Journal of Fisheries and Aquatic Sciences*, *5*, 349–355.

Ravi, J., & Devaraj, K. V. (1991). Quantitative essential amino acid requirements for growth of catla, *Catla catla* (Hamilton). *Aquaculture*, *96*, 281–291.

Riegel, R. E. (2012). *Bioquímica* (5th rev. ed.). Unisinos.

Rossi, L., & Tirapegui, J. (2004). Implicações do sistema serotoninérgico no exercício físico. *Arquivos Brasileiros de Endocrinologia & Metabologia*, *48*, 227–233. http://dx.doi.org/10.1590/S0004-27302004000200004.

Sahu, S., Ngasotter, S., Mog, M., Tesia, S., Sharma, S., & Waikhom, B. D. D. (2020). A review on physiological, behavioural and metabolic role of dietary tryptophan in fish. *International Journal of Chemical Studies*, *8*, 2411–2417. https://doi.org/10.22271/chemi.2020.v8.i3ai.9571.

Santiago, C. B., & Lovell, R. T. (1988). Amino acid requirements for growth of Nile tilapia. *Journal of Nutrition*, *118*, 1540–1546.

Shanks, W. E., Gahimer, G. D., & Halver, J. E. (1962). The indispensable amino acids for rainbow trout. *Progressive Fish-Culturist*, *24*, 68–73.

Sharf, Y., & Khan, M. A. (2023). Dietary tryptophan requirement of fingerling *Channa punctatus* (Bloch) based on growth, haematological parameters, intestinal enzymes, non-specific immune response, and antioxidant capacity. *Aquaculture*, *562*, 738745.

Sun, Y., Huang, H., Li, B., Su, L., Deng, J., & Cao, Z. (2023). Effects of dietary lysine level on the growth performance, protein metabolism, and antioxidant status in *Hemibagrus wyckioides* juveniles. *Journal of the World Aquaculture Society*, *54*, 1317–1336.

Tacon, A. G. J. (1993). Feed ingredients for warmwater fish: Fish meal and other processed feedstuffs. *FAO Fisheries Circular* No. 856. Food and Agriculture Organization of the United Nations, Rome, Italy.

Tang, L., Feng, L., Sun, C. Y., Chen, G. F., Jiang, W. D., Hu, K., Liu, Y., Jiang, J., Li, S. H., Kuang, S. Y., & Zhou, X. Q. (2013). Effect of tryptophan on growth, intestinal enzyme activities and TOR gene expression in juvenile Jian carp (*Cyprinus carpio* var. Jian): Studies *in vivo* and *in vitro*. *Aquaculture*, *412*, 23–33.

Teixeira, C., Pedrosa, R., Castro, C., Magalhães, R., Matos, E., Oliva-Teles, A., & Pérez-Jiménez, A. (2023). Dietary tryptophan supplementation implications on performance, plasma metabolites, and amino acid catabolism enzymes in meagre (*Argyrosomus regius*). *Fishes*, *8*, 1–12.

Tejpal, C. S., Pal, A. K., Sahu, N. P., Kumar, J. A., Muthappa, N. A., Vidya, S., & Rajan, M. G. (2009). Dietary supplementation of L-tryptophan mitigates crowding stress and augments the growth in *Cirrhinus mrigala* fingerlings. *Aquaculture*, *293*, 272–277.

Trushenski, J. T., Kasper, C. S., & Kohler, C. C. (2006). Challenges and opportunities in finfish nutrition. *North American Journal of Aquaculture*, *68*, 122–140.

Tzin, V., & Galili, G. (2010). The biosynthetic pathways for shikimate and aromatic amino acids in *Arabidopsis thaliana*. *The Arabidopsis Book*, American Society of Plant Biologists. https://doi.org/10.1199/tab.0132.

Walton, M. J., Cowey, C. B., Coloso, R. M., & Adron, J. W. (1986). Dietary requirements of rainbow trout for tryptophan, lysine and arginine determined by growth and biochemical measurements. *Fish Physiology and Biochemistry*, *2*, 161–169.

Wei, Y., Cui, X., Zhou, Z., Ma, Q., Xu, H., & Liang, M. (2023). Growth, cannibalism, and 5-TH metabolism in pufferfish (*Takifugu obscurus*♀× *Takifugu rubripes*): The role of graded levels of dietary tryptophan. *Aquaculture Nutrition*, 1–11.

Wen, H., Feng, L., Jiang, W., Liu, Y., Jiang, J., Li, S., & Zhou, X. (2014). Dietary tryptophan modulates intestinal immune response, barrier function, antioxidant status and gene expression of TOR and Nrf2 in young grass carp (*Ctenopharyngodon idella*). *Fish & Shellfish Immunology*, *40*, 275–287.

Wilson, R. P. (2002). Protein and amino acids. In J. E. Halver & R. W. Hardy (Eds.), *Fish nutrition* (3rd ed., pp. 144–179). Elsevier Science.

Wilson, R. P., Allen, O. W., Robinson, E. H., & Poe, W. E. (1978). Tryptophan and threonine requirements of fingerling channel catfish. *Journal of Nutrition*, *108*, 1595–1599.

Wilson, R., & Cowey, C. (1985). Amino acid composition of whole-body tissue of rainbow trout and Atlantic salmon. *Aquaculture*, *48*, 373–376.

Wolkers, C. P. B., Serra, M., Hoshiba, M. A., & Urbinati, E. C. (2012). Dietary L-tryptophan alters aggression in juvenile matrinxa *Brycon amazonicus*. *Fish Physiology and Biochemistry*, *38*, 819–827.

Wu, G. (2009). Amino acids: Metabolism, functions, and nutrition. *Amino Acids*, *37*, 1–17.

Wu, G. (2013). *Amino Acids: Biochemistry and Nutrition*. CRC Press.

Wu, G., Bazer, F. W., Dai, Z., Li, D., Wang, J., & Wu, Z. (2014). Amino acid nutrition in animals: Protein synthesis and beyond. *Annual Review of Animal Biosciences*, *2*, 387–417.

Yaghoubi, M., Torfi Mozanzadeh, M., Marammazi, J. G., Safari, O., & Gisbert, E. (2017). Effects of dietary essential amino acid deficiencies on the growth performance and humoral immune response in silvery-black porgy (*Sparidentex hasta*) juveniles. *Aquaculture Research*, *48*, 5311–5323.

Zaminhan, M., Michelato, M., Furuya, V. R. B., Boscolo, W. R., Araújo, F. E., Cruz, T. P., Urbich, A. V., & Furuya, W. M. (2018). Total and available tryptophan requirement of Nile tilapia, *Oreochromis niloticus*, fingerlings. *Aquaculture Nutrition*, *24*, 1553–1562.

Zehra, S., & Khan, M. A. (2015). Dietary tryptophan requirement of fingerling *Catla catla* (Hamilton) based on growth, protein gain, RNA/DNA ratio, haematological parameters and carcass composition. *Aquaculture Nutrition*, *21*, 690–701.

Zhang, X., Chang, E., Fu, Y., Liu, X., Xu, J., Wu, Y., & Miao, S. (2023). Tryptophan can alleviate the inhibition in growth and immunity of tilapia (GIFT *Oreochromis* spp.) induced by high dietary soybean meal level. *Aquaculture Reports*, *31*, 101646.

CHAPTER 9

Dietary Histidine Requirements of Different Fish Species

SAIMA ANDLEEB,[1,*] IRFAN AHMAD,[1] ASIM IQBAL BAZAZ,[2] AZRA SHAH,[1] ISHTIYAQ AHMAD,[1,3] and MOHD ASHRAF RATHER[1]

[1]*Division of Fish Genetics and Biotechnology, Faculty of Fisheries, Sher-e-Kashmir University of Agricultural Sciences and Technology of Kashmir (SKUAST–Kashmir), Rangil, Ganderbal, Jammu and Kashmir, India*

[2]*Division of Fisheries Resource Management, Faculty of Fisheries, Sher-e-Kashmir University of Agricultural Sciences and Technology of Kashmir (SKUAST–Kashmir), Rangil, Ganderbal, Jammu and Kashmir, India*

[3]*Department of Zoology, Fish Nutrition Research Laboratory, University of Kashmir, Srinagar, Jammu and Kashmir, India*

*Corresponding author

ABSTRACT

The management of fisheries and aquaculture is significantly impacted by the histidine requirements of different fish species. This study delves into the specific histidine necessities of different fish species with the aim of optimizing their growth, health, and overall performance within aquaculture systems. The research has yielded invaluable insights into the diversity of histidine requirements among these species, underscoring the significance of customized nutritional approaches to meet their distinct dietary requirements. Meeting histidine requirements is a fundamental step in promoting the progression of sustainable aquaculture practices, safeguarding the well-being of a variety of fish species, and upholding the production of top-quality seafood to meet the demands of a growing

Fish Nutrition: Protein and Amino Acid Requirements. Imtiaz Ahmed, Ishtiyaq Ahmad, and Shabihul Fatma Sayed (Eds.)

DOI: 10.1201/9781998511570-9

global population. Ongoing research in this field holds the potential to further enhance our comprehension and refine the effectiveness and sustainability of fish farming practices.

9.1 INTRODUCTION

Histidine plays a crucial role as an essential amino acid (EAA) in the synthesis of hemoglobin in both aquatic and terrestrial animals. It is vital for various physiological processes, including growth, tissue formation, repair, osmoregulation, and the protection of nerve cells through the maintenance of myelin sheaths (Ahmad, 2013). Additionally, it serves as a significant antioxidant and buffer in the organs of different fish species, preventing acidosis in white muscle caused by anaerobic metabolism (Gao et al., 2016; Wade and Tucker, 1998). Histidine, along with its imidazole derivatives, contributes to the taste and texture of aquaculture seafood, and its dietary supplementation can enhance sensory attributes, such as flavor (Ogata, 2002). Dietary intake of histidine has been associated with increased intramuscular histidine levels and pH, along with reduced muscle gapping in post-mortem Atlantic cod (Forde-Skjaevik et al., 2006). The positive impact of histidine on zinc uptake in rainbow trout and the improvement of intestinal digestive enzyme activities in juvenile Jian carp with dietary zinc supplementation have also been reported (Glover et al., 2003; Tan et al., 2011). Histidine plays a role in stress mitigation by influencing antioxidant status. It enhances superoxide scavenging ability, acting as an antioxidant in the lens of the eye (Bjerkas and Sveier, 2004; Boldyrev and Severin, 1990). Furthermore, histidine has been demonstrated to prevent oxidative stress-induced inflammation in human intestinal epithelial cells (Son et al., 2005). Furthermore, dietary histidine positively influences the activities of glutamate pyruvate transaminase (GPT) in both hepatopancreas and muscle, potentially due to its connection to the synthesis of carnosine, a significant metabolic product of histidine (Stifel et al., 1971; Zhao et al., 2012).

Histidine holds essential status as an amino acid (AA) in mammals, fish, and poultry, as it is not synthesized *de novo* and must be acquired through the diet (Wu et al., 2009). In these organisms, a deficiency in histidine leads to a reduction in body weight, and maintaining nitrogen balances becomes challenging on a histidine-depleted diet. The research by Kriengsinyos et al. (2002) suggests that histidine deficiency may not impact nitrogen balance when total protein intake exceeds the recommended level of 0.75 g/kg. However, histidine deficiency does result in decreased AA oxidation and reduced protein turnover. Considering these aspects of protein metabolism, histidine is deemed indispensable for healthy adults. Some studies also emphasize the essential role of histidine for ruminants (Onodera et al., 2003). Red meat and fish have high levels of histidine, but its concentration varies among fish species, ranging from histidine-rich fish with dark muscle to histidine-poor fish with white muscle. In the bodies of various species, histidine exists in different forms, including histidine-containing protein, free L-histidine, N(alpha)-acetylhistidine (NAH) and histidine-containing dipeptides such as anserine and carnosine (HRC).

9.2 METABOLIC PROCESSES OF HISTIDINE

Histidine requirements vary across species, with humans expressing them as a daily intake per kilogram of body weight. In contrast, for many other species, lysine serves as the reference AA, known as the first limiting AA, and the requirements for other AAs, including histidine, are often expressed in relation to lysine requirements. The histidine ratio (% of histidine/lysine) varies, according to the Food and Agriculture Organization of the United Nations (FAO): 37 for common carp, 34 for tilapia, 37 for common carp, 33 for salmonids, and 39 for trout.

Histidine follows diverse metabolic pathways. It can be methylated to yield 1-methyl or 3-methyl histidine, or transaminase can convert it to imidazole-pyruvic acid, which can then be reduced to produce imidazole-lactic acid. Histidine can also combine with β-alanine to create carnosine and anserine, substances involved in protecting against oxidative stress (Taylor et al., 1991). Alternatively, it can undergo decarboxylation to form histamine or undergo irreversible degradation. This irreversible, non-oxidative deamination involves histidine ammonia-lyase, also known as histidase, a cytoplasmic enzyme, resulting in the formation of trans-urocanic acid and ammonia, with the removal of the α-amino group of L-histidine (Mehler and Tabor, 1953). This step represents the primary aspect of histidine degradation studied in mammals and bacteria.

Urocanic acid undergoes hepatic metabolism, where urocanate hydratase converts it to 4,5-dihydro-4-oxo-5-imidazolepropanoic acid. Subsequently, it transforms into formimino-glutamate (FIGLU) and N5-formimino-tetrahydrofolate. Notably, in the epidermis, urocanate hydratase is absent, and urocanic acid represents the final product of histidine catabolism. This substance builds up in the epidermis and acts as a natural sunscreen to protect the skin from UV ray damage (Kinuta et al., 1994). Histidase, the rate-limiting enzyme in histidine degradation, is found exclusively in the liver and epidermis (Zannoni and La Du, 1963). Histidase deficiency has been reported in some cases. In such instances, the principal step in histidine catabolism involves transamination, resulting in the accumulation of imidazole-pyruvate, imidazole-lactate, and imidazole-acetate (Stifel and Herman, 1971). The expression of histidase can be influenced by dietary factors, with high-protein or AA-imbalanced diets stimulating its expression (Torres et al., 1999). Histidase activity increases proportionally with the rise in protein content in the diet. Additionally, studies indicate that estrogen can modulate histidase. It's noteworthy that the decarboxylation of histidine to histamine by histidine decarboxylase is a common metabolic conversion in all animal species except fish.

9.3 INFLUENCE OF HISTIDINE LEVELS AND SUPPLEMENTATION IN FISH

9.3.1 IMPACT OF HISTIDINE IN FISH GROWTH

In a study with grass carp, for two weeks, young individuals weighing between 279.1 and 685.4 g were given a diet supplemented with different amounts of histidine (2.0, 3.7, 5.9, 7.9, 9.8, and 12.2 g/kg diet). The determined dietary histidine requirement for optimal growth performance was 7.63 g/kg, as growth performance declined above this level (Jiang, 2016). In catfish, the histidine requirement was established at 3.7 ± 0.1 g/kg in terms of the dry diet or 15.4 g/kg when represented as a proportion of dietary protein (DP). Serum-free histidine levels did not increase until the dietary requirement was met, and muscle carnosine was undetectable in the catfish (Wilson et al., 1980). Nile tilapia juveniles, when fed diets with varying histidine levels (4.2, 5.4, 7.1, 8.9, 9.8, and 11.5 g/kg dry diet), achieved optimal final weight, weight increase, feed conversion ratio (FCR), PER, and net consumption of protein at 8.9 g/kg of histidine. However, fish fed

7.1 and 9.8 g/kg of histidine showed higher whole-body protein content. Unlike mammals and birds, fish can generate new muscle fibers throughout their lifetime, with muscle growth influenced by myogenic regulatory factors. According to the study, fish fed 9.8 and 11.5 g/kg of histidine had higher myogenin expression than fish fed 4.2 to 7.1 g/kg, indicating that dietary histidine positively affects muscle growth (Michelato et al., 2017).

9.3.2 HISTIDINE'S BUFFERING FUNCTION IN FISH

Fish possess a substantial quantity of free histidine and histidine-related compounds (HRC). The buffering capacity of HRC was investigated by Abe et al. (1985) in Pacific blue marlin (*Makaira nigricans*) and the rainbow trout (*Salmo gairdneri*). Their findings indicated that the contribution of HRC to total cellular buffering ranged from 62% in white marlin muscle to 7% in red muscle of trout. These findings imply that the buffering capacity of HRC is dependent on the fish species and the muscle type. The other primary buffers in red muscle include phosphate, protein, and taurine. Histidine buffering capacities may function through the presence of N(alpha)-acetylhistidine (NAH), which, in contrast to endothermic birds and mammals, is only present in large quantities in the brain and lenses of ectothermic vertebrates, such as amphibians, ray-finned fish, and reptiles. Histidine N-acetyltransferase (HISAT; EC 2.3.1.33) is known to produce NAH from histidine and acetyl-CoA, although the gene encoding HISAT is still unknown. Notably, 13 species, including five anabantids, seven cichlids, and one catfish, showed significant levels of NAH in their skeletal muscles (>1 μmol/g). The highest NAH level (10.37 μmol/g) was detected in *Betta splendens* (Siamese fighting fish). Additionally, NAH contents in the tissues of *Oreochromis niloticus* (Nile tilapia), *Trichogaster trichopterus* (three spot gourami), *Mikrogeophagus ramirezi* (ram cichlid), *Kryptopterus bicirrhis* (glass catfish), and *Parachromis managuensis* (Guapote tiger) ranged from 3.17 to 6.16 micromol/g. On the other hand, NAH was found in little amounts (<0.25 μmol/g) in the skeletal muscles of four species of reptiles and five species of amphibians (Abe, 2000).

9.4 FUNCTION OF HISTIDINE IN PREVENTING CATARACTS IN FISH

Another way that histidine functions is its ability to reduce the occurrence and intensity of cataracts in fish, a significant concern in the aquaculture industry. This observation was made following a cataract outbreak in rapidly growing Atlantic salmon that were fed diets lacking blood meal, a rich histidine source (Breck et al., 2003). The biomolecule N(alpha)-acetylhistidine (NAH) is notably present in the lens, and its concentrations closely resemble the amounts of histidine in food consumed by Atlantic salmon.

Dietary histidine emerges as a crucial factor in preventing cataracts, with its positive effects linked to elevated histidine levels and the accumulation of NAH in the lens, providing buffering and antioxidant properties (Bjerkas, 2006).

Studies on cultured lenses revealed a significant reduction in NAH concentration under oxidative stress conditions, demonstrating the influence of histidine enrichment on the innate antioxidative defense system through increased expression of glutaredoxin in the media (Remo et al., 2011). NAH in the lens has also been associated with maintaining lens-water homeostasis, a vital aspect in both freshwater and seawater environments. After NAH is exported, a particular acylase cleaves histidine in the ocular fluid. The lens actively absorbs the released histidine and resynthesizes it into NAH, acting as a molecular water pump to keep the lens highly dehydrated, thereby preventing cataract formation (Remo et al., 2014).

Dietary histidine levels have been shown to either prevent or slow the formation of cataracts. Histidine plays a critical function in regulating the concentration of NAH in the lens. The effect of diet histidine on the severity and prevalence of cataracts has been investigated in a number of research studies. Trosse et al. (2009), for example, demonstrated a negative correlation between NAH content and cataract scores in adult Atlantic salmon given dietary histidine in three separate dosages for a duration of four months. Similarly, lens NAH concentrations were positively associated with histidine consumption, and cataract prevalence and severity were inversely correlated with dietary histidine concentration in salmon smolts fed varying doses of histidine for a 13-week period. The scientists came to the conclusion that 14.4 g of histidine/kg of diet was the ideal amount to consume in order to reduce the chance of developing cataracts (Remo et al., 2014).

Waagbo et al. (2010) conducted a noteworthy experiment to examine the impact of feeding times and food composition high in histidine on the severity of cataract development. Severe cataracts were observed between

July and September, and the amount of histidine fed during the first and second periods had a clear correlation with the severity. The authors came to the conclusion that supplementing with histidine just prior to and during the early stages of cataract formation could reduce the risk of cataract development.

9.5 HISTIDINE LEVELS IN FEED COMPONENTS AND DIETARY REQUIREMENTS

As histidine is an indispensable AA, the primary source for its incorporation into the diets of various cultivable fish species is through dietary inclusion. The histidine content in different feed ingredients ranges from 0.89% to 3.54% on a protein basis, as documented by the National Research Council (NRC) in 2011. Ingredients such as fish meal containing rice protein concentrate (2.02%), pea protein concentrate (2.49%), salmon byproduct (3.51%), solvent extracted (1.93%), soybean meal (SB) protein concentrate (2.50%), wheat, soft grain (2%), wheat flour (3.54%), rice bran (2.61%), rice bran with germ, white fishmeal (2.16%), rice polishing (2.15%), wheat bran (2.14%), rice broken with polishing (2.57%), and meat and bone meal (2%) exhibit high histidine content on a protein basis. Conversely, poultry byproduct meal (1.90%), poultry byproduct meal, feed-grade low (1.88%), whey dried (1.68%), oats (1.63%), gelatin (0.89%), and hydrolyzed poultry feather meal (0.84%) have lower histidine content on a protein basis. The effective utilization of these protein sources in fish feeds relies on appropriate histidine supplementation to meet the specific requirements.

Histidine requirements have been established for several cultivable fish species, ranging between 0.9% and 4.8% of DP (Table 9.1). Reported dietary histidine requirements vary widely among different species, leading to some contradictions in the available data. For instance, established dietary histidine requirements include 1.54% for channel catfish *Ictalurus punctatus* (Wilson et al., 1980), 1.6% for Chum salmon *Oncorhynchus keta* (Akiyama et al., 1985), 1.7% for gilthead sea bream *Sparus auratus* (Kaushik, 1998), 1.1% for rainbow trout *Oncorhynchys mykiss* (Rodehutscord et al., 1997), 2.1% for Mrigal carp *Cirrhinus mrigala* (Ahmed and Khan, 2005), 1.0% for African catfish *Clarias gariepinus* (Khan and Abidi, 2009), 3.51% for Stinging catfish *Heteropneustes fossilis*

TABLE 9.1 Dietary Histidine Requirements for Diverse Commercially Cultivated Fish Species, as Outlined by KHAN in 2018

Species	Crude Protein Content Expressed as a Percentage of the Dry Diet)	Standards for Determining Requirements (in grams)	Dietary Protein Percentage Needed for Histidine Requirements	Weight of Fish (g)	Protein Content in the Diet	References
Anguilla japonica	30	–	2.1	–	–	Arai et al. (1979)
Chanos chanos	40	WG; FE; SR	2	≤ 8.0	CS; GL; FM	Borlongan and Coloso (1993)
Clarias gariepinus	40	AWG; FCR; PER; S-GR; PRE	1.5	0.2	CS; GL; AA	Khan and Abidi (2009).
Cirrhinus mrigala	40	LWG; FCR; PER	2.1	0.6	CS; GL; AA	Ahmed and Khan (2005)
Catla	40	–	2.5	–	CS; GL; AA	Ravi and Devaraj (1991)
	33	AWG; FCR; PG	2.06	0.65	CS; GL; AA	Zehra and Khan (2014)
Clarias macrocephalus x Clarias gariepinus	–	–	1.2	–	–	Unprasert (1994)
Cyprinus carpio	32	SGR; FE; PER	2.3	8.7	FM; GL; AA	Zhao et al. (2012)
	48	WG	2.1	1.7	–	Ogino (1980)
	40	–	4.8	–	–	Klein and Halver (1970)
Ctenopharyngodon idella	38	WG; SGR; FCR; FI	3.2	3.68	CS; GL; AA	Gao et al. (2016)
Dicentrarchus labrax	62	–	1.64	–	–	Kaushik (1998)
Heteropneustes fossilis	38	AWG; PG; FCR	2.48	6.6	CS; GL; AA	Farhat and Khan (2013)
	43	LWG; FCR; PER; PRE	3.5	5.2	CS; GL; AA	Khan and Abidi (2014)
	40	LWG; FCR; PER	1.35	4.6	CS; GL; AA	Ahmed (2013)

TABLE 9.1 *(Continued)*

Species	Crude Protein Content Expressed as a Percentage of the Dry Diet)	Standards for Determining Requirements (in grams)	Dietary Protein Percentage Needed for Histidine Requirements	Weight of Fish (g)	Protein Content in the Diet	References
Ictalurus punctatus	24	WG	1.5	200	CS; GL; AA	Wilson et al. (1980)
Labeo rohita	40	WG; SGR; FCR	2.25	–	CS; GL; AA	Murthy and Varghese (1995)
	40	LWG; SGR; FCR; PER	2.25	3.5	CS; GL; AA	Abidi and Khan (2004)
Megalobrama amblycephala	33	WG; SGR; FCR SR	3.6	23.3	FM; CS; GL; AA	Arop et al. (2018)
Onchorhynchus tshawytscha	40	GAIN; FEED/GAIN	1.8	2.9	CS; GL; AA	Arai and Ogata (1993)
Oncorhynchus kisutch	40	GAIN; FEED/GAIN	1.8	3.1	CS; GL; AA	Arai and Ogata (1993)
	–	–	0.9	–	–	Akiyama and Arai (1993)
Onchorhynchus keta	40	WG	1.6	1.5	CS; GL; AA	Akiyama et al. (1985)
	–	–	1.6	–	–	Santiago and Lovell (1988)
Oreochromis niloticus	26	WG; FCR; PER	3.1	4.8	RC; MBM; PBM; FEM	Michelato et al. (2017)
	34	WG; PD	1.1.2	40	WG; AA	Rodehutscord et al. (1997)
	28	WG	1.7	0.08	CS; GL; AA	Scott et al. (1996)
Oncorhynchus mykiss	40	–	1.6	–	–	Klein and Halver (1970)
Pseudosciaena crocea	44	WG; FE; SR	2.0	6	FM; CG; GL	Li et al. (2014)
Salmo salar	40	LWG; FCR; SGR	2.0	62	CS; CG	Rollin (1999)
	45	–	4.0	–	–	Moon and Gatlin (1991)

TABLE 9.1 *(Continued)*

Species	Crude Protein Content Expressed as a Percentage of the Dry Diet)	Standards for Determining Requirements (in grams)	Dietary Protein Percentage Needed for Histidine Requirements	Weight of Fish (g)	Protein Content in the Diet	References
Sciaenops ocellatus	35	–	1.7	–	–	Gaber et al. (2018)
	35	–	1.6	–	–	–
	35	RWG; FER; PER; PR; PH	1.6	0.98	RDM; AA	Peachey et al. (2018)
Scophthalmus maximus	70	–	1.5	–	–	Kaushik (1998)
Sparus aurata	57	–	1.7	–	–	–
	40:36	LWG; DDP	3.5	2.2	FM; SB; WB	Kaushik (1998)

SGR: specific growth rate; RE: protein retention efficiency; LWG: live weight gain; WG: weight gain; FCR: feed conversion ratio; PER: protein efficiency ratio; PFE: feed efficiency; DDP: daily deposition; SR: survival; WGM: wheat gluten meal; FEM: feather meal; RC: rice; SB: soybean meal; FM: fish meal; xWB: wheat bran; PBM: poultry by product meal; RE: protein retention efficiency; CS: casein; GL: gelatin; AA: amino acids mixture; RDM: red drum meal; PH; plasma histidine.

(Farhat and Khan, 2013), 3.6% for Singhi, *Heteropneustes fossilis* (Khan and Abidi, 2014), 2.38% for *Cyprinus carpio*, 2.1% for *C. carpio* (Nose, 1979), 3.2% for juvenile Grass carp *Ctenopharyngodon idella* (Gao et al., 2015), 3.1% for blunt snout bream *Megalobrama amblycepahala* (Arop et al., 2018), and 3.1% for Nile tilapia *Oreochromis niloticus* (Michelato et al., 2017), all expressed as a percentage of DP.

9.6 INDICATORS OF INSUFFICIENCY AND ADVERSE EFFECTS

Insufficient balance of histidine in feeds can lead to the development of deficiency signs and toxicity. A deficiency in dietary histidine is associated with diminished growth, cataracts, increased incidence of lordosis, anorexia, allergic reactions, reduced gastric acid secretion, and heightened stress response (Yoshikawa et al., 2014). Fish fed diets deficient in histidine also exhibit lower feed intake, possibly because, after ingesting an AA-imbalanced diet, the animals recognize the deficiency and respond by reducing their feed intake (Maurin et al., 2006). The ability of histidine to mitigate cataracts is attributed to the synthesis of N-acetyl histidine in the lens. Peachey et al. (2018) observed significantly higher histidine levels in the lenses of fish fed an adequate histidine diet compared to those with histidine deficiency. Furthermore, excessive histidine has been found to inhibit the African catfish (*Clarias gariepinus*) growth (Khan and Abidi, 2009), Indian major carp *C. catla* (Ravi and Devaraj, 1991), and *Labeo rohita* (Murthy and Varghes, 1995). This might be attributed to the disruption of AA balance leading to tissue toxicity (Mertz, 1972) or widespread necrosis in the hepatopancreatic epithelial cells (Recodo, 1991), ultimately influencing fish growth. Elevated levels of free histidine could result in increased histamine production, and excess histamine may induce proinflammatory cytokines, potentially affecting erythrocytes (Jiang et al., 2016). Erythrocyte osmotic fragility was notably higher in fish fed a histidine-sufficient diet compared to those on a histidine-deficient diet. Erythrocyte osmotic fragility, indicating the amount of hemolysis under osmotic stress, serves as a measure of erythrocyte quality and membrane integrity. Faster erythrocyte breakdown was observed in cases of improper membrane function (Robbins, 1984). In stinging catfish (Farhat and Khan, 2013) and grass carp (Gao et al., 2016), dietary histidine significantly influenced erythrocyte fragility. According to Arop et al. (2018), as feeding histidine levels rose to 9.9 g/kg, the relative gene expression of TOR mRNA in the liver of blunt snout bream increased, declining thereafter as dietary histidine levels continued to rise. Similar trends were observed in the gill of grass carp fed with graded levels of histidine (Jiang et al., 2016).

9.7 INFLUENCE ON BODY COMPOSITION

Supplementing with histidine has been noted to impact carcass quality, a crucial concern for aquaculturists as it directly influences the yield of the final product. Consequently, considerable research has been directed towards enhancing carcass quality and optimizing nutrient utilization for muscle growth. Given that protein is a major and essential component of tissues, there is a priority to increase both the quantity and quality of protein in tissues. Histidine supplementation has demonstrated a positive effect on the protein content of the entire body of fish (Zhao et al., 2012). Numerous studies have shown that the protein content in the entire body increases with higher dietary histidine levels but decreases significantly when histidine levels surpass the optimal requirement (Ahmed, 2013; Khan and Abidi, 2009, 2014; Michelato et al., 2017).

However, in certain fish species such as yellow croaker (Li et al., 2014), grass carp (Gao et al., 2016), singhi (Khan and Abidi, 2014), and red drum (Peachey et al., 2018), dietary histidine levels did not significantly affect whole-body composition and muscle composition. Additionally, histidine supplementation has been documented to lower body fat in a number of fish species (Farhat and Khan, 2013). Given that histidine deficiency leads to reduced growth and cataracts, it is imperative to ensure the optimal inclusion of dietary histidine to enhance fish growth performance (Gaber et al., 2016).

9.8 SPREAD OF FREE L-HISTIDINE AND RELATED DIPEPTIDES THROUGHOUT FISH TISSUES

Since Lukton and Olcott's initial work in 1958, numerous studies have investigated the presence of dipeptides in fish muscle. A typical distribution of these dipeptides in fish and whale muscles, with substantial amounts of dipeptides, has been predominantly identified in the past decade.

High-speed marine pelagic swimmers characterized by myoglobin-rich dark flesh, such as tuna, sardine, and mackerel, exhibit significant quantities of free L-histidine in their white muscle—a unique feature not observed in other examined animals (Konosu et al., 1974; Lukton and Olcott, 1958; Suyama and Suzuki, 1975). The L-histidine content in the white muscle ranges from 15 to 94 μmol/g in dark-fleshed fishes, 4 to 20 μmol/g in 'intermediate' fishes, and 0.07 to 1 μmol/g in white-fleshed species. Carp muscle, although white, stands out as an exception with a relatively high concentration of L-histidine. However, the level in red muscle of these fishes is significantly lower than that of white muscle.

Alongside L-histidine, various HRCs like carnosine, anserine, and balenine are found in the muscles of different fishes and vertebrates. In contrast to L-histidine, the content of these HRCs is not influenced by muscle color but displays high species specificity (Abe, 1983; Crush, 1970). Although these dipeptides may be broadly distributed in fish muscles, the occurrence of balenine may be limited. Balenine concentrations are generally low in most fish muscles but remarkably high in the skeletal muscle of certain whales, with no systematic differences observed between baleen and toothed whales.

Histidine-related peptide concentrations in fish red muscle are generally low, ranging from one-tenth to one-quarter of their concentration in white muscle. Regarding dipeptide distribution, concentrations in white muscle are closely linked to the swimming ability of fish species. While there is no significant difference in the levels of these compounds among dorsal or ventral muscle portions, dorsal muscle tends to show slightly higher concentrations (Abe, 1983). Individual differences in the concentration of these compounds within a given species are small, yet substantial differences between regional populations may exist, as evidenced by histidine in rainbow trout (*Oncorhynchus mykiss*) and anserine in skipjack tuna (*Katsuwonus pelamis*).

In general, fishes belonging to the same suborder group exhibit similar distribution patterns of these compounds, regardless of family, as observed in *Salmonidae*, *Cyprinidae*, and *Anguillidae* (Abe, 1983). Fishes with less pronounced swimming ability, however, contain much smaller amounts of these compounds in their white muscle. In contrast to the abundance of these compounds in fish skeletal muscle, their levels are extremely low in cardiac muscle and smooth muscles, such as the stomach and intestine. In skipjack tuna, minimal amounts of dipeptides are present in non-muscular tissues, including the liver, brain, gill, kidney, spleen, pyloric caecum, and ovary (Abe, 1983). Blood levels of these compounds are also notably low, well below the concentrations observed in skeletal muscle tissue. Consequently, the distribution of these compounds in fish tissues strongly suggests their role in enhancing swimming function and performance.

Among all examined fish species, skipjack tuna (*Katsuwonus pelamis*) and Pacific blue marlin (*Makaira nigricans*) represent extremes. Live skipjack tuna (1–2 kg body weight) captured off Hawaii exhibited substantial amounts of L-histidine and anserine, along with a lesser amount of carnosine in their white muscle. The average L-histidine level was 93.5 μmol/g, reaching a maximum of 105 μmol/g—the highest value recorded for all species studied thus far. Anserine was similarly abundant, with a maximum of about 69.1 μmol/g and an average concentration of 51.1 μmol/g (Abe, 1986). The combined total of these compounds approached 150 μmol/g in white muscle. In contrast, Pacific blue marlin (50–100 kg body weight) caught off Hawaii during the Annual Billfish Tournament exhibited the highest anserine content among all animals studied, with an average value of 105 μmol/g and a maximum of 120 μmol/g (Abe, 1985). The combined total of histidine and related compounds in marlin white muscle exceeded 120 μmol/g.

However, the levels in the red muscle of both species amounted to 15–30% of those found in the white muscle. Fish white muscle, known for its high anaerobic nature, is thought to be utilized primarily during burst swimming. In contrast, red muscle is employed during generally slower-speed, sustainable swimming (Hochachka, 1980; Tsukamoto, 1984). Tuna species are considered the 'ultimate teleosts' due to their ability to achieve high burst swimming speeds and maintain muscle temperatures as much as 10°C above ambient temperatures. The ability of tunas and marlins to engage in burst swimming is sustained by elevated capacities for anaerobic glycolysis in white muscle, represented by the activity of lactate dehydrogenase (LDH), which in tunas is the highest found in any animal (Guppy and Hochachka, 1978). During burst swimming, significant amounts of lactate, nearly 100 μmol/g muscle, accumulate as an end product in tuna white muscle (Guppy et al., 1979). Therefore, the highest concentrations of L-histidine and related dipeptides in the anaerobic white muscle of tunas and billfishes may indicate their contribution to the extreme ability for anaerobic burst swimming. An in-depth analysis of the data suggests a positive correlation between the concentration of these dipeptides and the ability of a given species to perform anaerobic exercises such as burst swimming, sprint running, flying, and diving.

9.9 MARINE MAMMALS

The muscles of these creatures have less than 1 μmol/g of free L-histidine but a considerable quantity of one of the dipeptides. Notably, whale muscle is distinctive due to its substantial balenine content, ranging from 20 to 80 μmol/g (Suyama et al., 1977). Whales typically exhibit β values comparable to those of warm-bodied fishes. Marine diving mammals, characterized by a relatively high concentration of carnosine in their skeletal muscle, show higher average β values than terrestrial mammals. While LDH activity does not exhibit significant differences between marine and terrestrial mammals, it closely correlates with β. Furthermore, in these diving mammals, there is a high connection between the β value and the concentration of muscle myoglobin (Castellini and Somero, 1981). Consequently, HRC buffering may serve to prevent a decline in muscle pH during extended breath-hold diving, especially when oxygen stores are depleted and substantial amounts of lactate and protons are generated.

9.10 CONCLUSION

The histidine requirement of different fish species holds significant importance in the management of aquaculture and fisheries. It is crucial to comprehend the unique histidine demands of diverse fish species in order to enhance their growth, overall health, and performance within aquaculture systems. This investigation has yielded valuable insights into the varying histidine requirements among these species, underscoring the necessity for customized nutritional approaches to cater to their specific dietary necessities. Addressing these histidine requirements can play a vital role in promoting sustainable advancements in aquaculture practices, safeguarding the well-being of various fish species, and ensuring the production of top-quality seafood for an expanding global population. Ongoing research in this field will further refine our understanding and enhance the effectiveness and sustainability of fish farming methods.

KEYWORDS

- **aquaculture feed formulation**
- **essential amino acids**
- **fish growth and health**
- **histidine in fish nutrition**
- **species-specific dietary requirements**
- **sustainable aquaculture practices**

REFERENCES

Abe, H. (1983). Distribution of free L-histidine and its related compounds in marine fishes. *Bulletin of the Japanese Society of Scientific Fisheries, 49*, 1683–1687.

Abe, H. (1995). Histidine-related compounds. In P. W. Hochachka & T. P. Mommsen (Eds.), *Biochemistry and Molecular Biology of Fishes: Vol. 4. Metabolic Biochemistry* (pp. 309–333). Elsevier.

Abe, H. (2000). Role of histidine-related compounds as intracellular proton buffering constituents in vertebrate muscle. *Biochemistry (Moscow), 65*(7), 757–765.

Abe, H., Brill, R. W., & Hochachka, P. W. (1986). Metabolism of L-histidine, carnosine, and anserine in skipjack tuna. *Physiological Zoology, 59*, 439–450.

Abe, H., Dobson, G. P., Hoeger, U., & Parkhouse, W. S. (1985). Role of histidine-related compounds to intracellular buffering in fish skeletal muscle. *American Journal of Physiology, 249*, R449–R454.

Abidi, S. F., & Khan, M. A. (2004). Dietary histidine requirement of fingerling Indian major carp, *Labeo rohita* (Hamilton). *Aquaculture Nutrition, 13*(6), 424–430.

Ahmed, I. (2013). Dietary amino acid L-histidine requirement of fingerling Indian catfish, *Heteropneustes fossilis* (Bloch), estimated by growth and whole-body protein and fat composition. *Journal of Applied Ichthyology, 29*(3), 602–609.

Ahmed, I., & Khan, M. A. (2005). Dietary histidine requirement of fingerling Indian major carp, *Cirrhinus mrigala* (Hamilton). *Aquaculture Nutrition, 11*(5), 359–366.

Akiyama, T., & Arai, S. (1993). Amino acid requirements of chum salmon fry and supplementation of amino acids to diet. In M. R. Collie & J. P. McVey (Eds.), *Proceedings of the 20th U.S. Japan Symposium on Aquaculture Nutrition* (pp. 35–48). UJNR, Department of Commerce.

Akiyama, T., Arai, S., & Murai, T. (1985). Threonine, histidine and lysine requirements of chum salmon fry. *Nippon Suisan Gakkaishi, 51*(4), 635–639.

Arai, S., & Ogata, H. (1993). Quantitative amino acid requirements of fingerling coho salmon. In M. R. Collie & J. P. McVey (Eds.), *Proceedings of the 20th U.S. Japan Symposium on Aquaculture Nutrition* (pp. 19–28). UJNR, Department of Commerce.

Arai, S., Nose, T., & Hashimoto, Y. (1979). Amino acid essential for the growth of eels, *Anguilla anguilla* and *A. japonica*. *Nihon Suisan Gakkaishi*, *38*(7), 753–759.

Arop, O. M. W., Liang, H., Ge, X., Ren, M., Habte-Tsion, H. M., et al. (2018). Dietary histidine requirement of juvenile blunt snout bream (*Megalobrama amblycephala*). *Aquaculture Nutrition*, *24*(3), 1122–1132.

Bjerkås, E., & Sveier, H. (2004). The influence of nutritional and environmental factors on osmoregulation and cataracts in Atlantic salmon (*Salmo salar* L.). *Aquaculture*, *235*(1–4), 101–122.

Bjerkås, E., Breck, O., & Waagbø, R. (2006). The role of nutrition in cataract formation in farmed fish. *CAB Reviews: Perspectives in Agriculture, Veterinary Science, Nutrition and Natural Resources*, *1*, 1–10.

Boldyrev, A. A., & Severin, S. E. (1990). The histidine-containing dipeptides, carnosine and anserine: Distribution, properties and biological significance. *Advances in Enzyme Regulation*, *30*, 175–188.

Borlongan, I. G., & Coloso, R. M. (1993). Requirements of juvenile milkfish (*Chanos chanos* Forsskål) for essential amino acids. *Journal of Nutrition*, *123*(1), 125–132.

Breck, O., Bjerkås, E., Campbell, P., Arnesen, P., Haldorsen, P., & Waagbø, R. (2003). Cataract preventative role of mammalian blood meal, histidine, iron and zinc in diets for Atlantic salmon (*Salmo salar* L.) of different strains. *Aquaculture Nutrition*, *9*, 341–350.

Castellini, M. A., & Somero, G. N. (1981). Temperature-induced changes in aerobic and anaerobic metabolism in muscle tissue of fast- and slow-swimming fish. *Journal of Comparative Physiology B*, *143*(2), 191–198.

Crush, K. G. (1970). Carnosine and related substances in animal tissues. *Comparative Biochemistry and Physiology*, *34*, 3–30.

Dickson, K. A., & Somero, G. N. (1987). Temperature-related muscle potential and performance in tunas. *Physiological Zoology*, *60*(6), 699–706.

Dizon, A. E., Brill, R. W., & Yuen, H. S. H. (1978). Correlations between environment, physiology, and activity and the effects on thermoregulation in skipjack tuna. In G. D. Sharp & A. E. Dizon (Eds.), *The Physiological Ecology of Tunas* (pp. 233–259). Academic Press.

Farhat, & Khan, M. A. (2013). Effects of varying levels of dietary L-histidine on growth, feed conversion, protein gain, histidine retention, haematological and body composition in fingerlings stinging catfish, *Heteropneustes fossilis* (Bloch). *Aquaculture*, *404, 405*, 130–138.

Forde-Skjaervik, O., Skjaervik, O., Morkore, T., Thomassen, M. S., & Rorvik, K. A. (2006). Dietary influence on quality of farmed Atlantic cod (*Gadus morhua*): Effect on glycolysis and buffering capacity in white muscle. *Aquaculture*, *252*(2–4), 409–420.

Gaber, M. M., Salem, M. E. S., Zaki, M. A., & Nour, A. M. (2016). Amino acid requirements of gilthead bream (*Sparus aurata*) juveniles. *World Journal of Environmental and Technology*, *4*(3B), 18–24.

Gao, H. N., Hu, H., Zheng, N., & Wang, J. Q. (2015). Leucine and histidine independently regulate milk protein synthesis in bovine mammary epithelial cells via mTOR signaling pathway. *Journal of Zhejiang University Science B*, *16*(6), 560–572.

Gao, Y. J., Jian, Y., Chen, X. Q. L., Yang, H. J., Li, X. F., et al. (2016). Effects of graded levels of histidine on growth performance, digested enzymes activities, erythrocyte osmotic fragility and hypoxia-tolerance of juvenile grass carp *Ctenopharyngodon idella. Aquaculture, 452*, 388–394.

Glover, C. N., Bury, N. R., & Hogstrand, C. (2003). Zinc uptake across the apical membrane of freshwater rainbow trout intestine is mediated by high affinity, low affinity, and histidine-facilitated pathways. *Biochimica et Biophysica Acta (BBA)–Bio Membranes, 1614*(2), 211–219.

Guppy, M., & Hochachka, P. W. (1978). Controlling the highest lactate dehydrogenase activity known in nature. *American Journal of Physiology, 234*(1), R136–R140.

Guppy, M., Hulbert, W. C., & Hochachka, P. W. (1979). Metabolic sources of heat and power in tuna muscles. II. Enzyme and metabolite profiles. *Journal of Experimental Biology, 82*, 303–320.

Hochachka, P. W. (1980). *Living Without Oxygen* (pp. 85–94). Harvard University Press.

Hochachka, P. W., & Somero, G. N. (1984). *Biochemical Adaptation* (pp. 337–348). Princeton University Press.

Jiang, W. D., Feng, L., Qu, B., Wu, P., Kuang, S. Y., et al. (2016). Changes in integrity of the gill during histidine deficiency or excess due to depression of cellular anti-oxidative ability, induction of apoptosis, inflammation and impairment of cell-cell tight junctions related to Nrf2: TOR and NF-κB signaling in fish. *Fish & Shellfish Immunology, 56*, 111–122.

Jiang, W. D., Qu, B., Feng, L., Jiang, J., Kuang, S. Y., Wu, P., Tang, L., Tang, W. N., Zhang, Y. A., Zhou, X. Q., et al. (2016). Histidine prevents Cu-induced oxidative stress and the associated decreases in mRNA encoding tight junction proteins in the intestine of grass carp (*Ctenopharyngodon idella*). *PLoS ONE, 11*(6), e0157001.

Kaushik, S. J. (1998). Whole body amino acid composition of European seabass (*Dicentrarchus labrax*), gilthead seabream (*Sparus aurata*) and turbot (*Psetta maxima*) with an estimation of their IAA requirement profiles. *Aquatic Living Resources, 11*(5), 355–358.

Khan, M. A., & Abidi, S. F. (2009). Optimum histidine requirement of fry African catfish, *Clarias gariepinus* (Burchell). *Aquaculture Research, 40*(9), 1000–1010.

Khan, M. A., & Abidi, S. F. (2014). Dietary histidine requirement of Singhi, *Heteropneustes fossilis* fry (Bloch). *Aquaculture Research, 45*(8), 1341–1345.

Kinuta, M., Ubuka, T., Yao, W. B., Zhao, Y. Q., & Shimizu, H. (1994). Isolation of S-[2-carboxy-1-(1H-imidazol-4-yl) ethyl]-3-thiolactic acid, a new metabolite of histidine, from normal human urine and its formation from S-[2-carboxy-1-(1H-imidazol-4-yl) ethyl]cysteine. *Biochemical Journal, 297*, 475–478.

Klein, R. G., & Halver, J. E. (1970). Nutrition of salmonid fishes: Arginine and histidine requirements of chinook and coho salmon. *Journal of Nutrition, 100*(9), 1105–1109.

Konosu, S., Watanabe, K., & Shimizu, T. (1974). Distribution of nitrogenous constituents in the muscle extracts of eight species of fish. *Bulletin of the Japanese Society of Scientific Fisheries, 40*, 909–915.

Kriengsinyos, W., Rafii, M., Wykes, L. J., Ball, R. O., & Pencharz, P. B. (2002). Long-term effects of histidine depletion on whole-body protein metabolism in healthy adults. *Journal of Nutrition, 132*(11), 3340–3348.

Li, Y., Cheng, Z., Kangsen, I., & Ai, Q. M. (2014). Dietary histidine requirement for juvenile large yellow croaker, *Pseudosciaena crocea* R. *The Israeli Journal of Aquaculture–Bamidgeh, 66*, 1–7.

Lukton, A., & Olcott, H. S. (1958). Content of free imidazole compounds in the muscle tissue of aquatic animals. *Food Research, 23*, 611–618.

Maurin, A. C., Bruhat, A., Jousse, C., Cherasse, Y., & Fafournoux, P. (2006). Cellular adaptation to amino acid availability: Mechanism involved in the regulation of gene expression. *Bulletin de l'Académie Vétérinaire de France, 159*(3), 319–326.

Mehler, A. H., & Tabor, H. (1953). Deamination of histidine to form urocanic acid in liver. *Journal of Biological Chemistry, 201*, 775–784.

Mertz, E. T. (1972). The protein and amino acid needs. In J. E. Halver (Ed.), *Fish Nutrition* (pp. 105–143). Academic Press.

Michelato, M., Zaminhan, M., Boscolo, W. R., Nogaroto, V., Vicari, M., Artoni, R. F., Furuya, V. R. B., & Furuya, W. M. (2017). Dietary histidine requirement of Nile tilapia juveniles based on growth performance, expression of muscle-growth-related genes and hematological responses. *Aquaculture, 467*, 63–70.

Moon, H. Y., & Gatlin, D. M., III. (1991). Total sulfur amino acid requirement of juvenile red drum, *Sciaenops ocellatus*. *Aquaculture, 95*(1, 2), 97–106.

Murthy, H., & Varghese, T. (1995). Arginine and histidine requirements of the Indian major carp, *Labeo rohita* (Hamilton). *Aquaculture Nutrition, 1*(4), 235–239.

National Research Council (NRC). (2011). *Nutrient Requirements of Fish and Shrimp*. National Academies Press.

Nose, T. (1979). Summary report on the requirements of essential amino acids for carp. In J. E. Halver & K. Tiews (Eds.), *Finfish Nutrition and Fish Feed Technology* (pp. 145–156).

Ogata, H. Y. (2002). Muscle buffering capacity of yellowtail fed diets supplemented with crystalline histidine. *Journal of Fish Biology, 61*(6), 1504–1512.

Ogino, C. (1980). Requirements of carp and rainbow trout for essential amino acids. *Bulletin of the Japanese Society of Scientific Fisheries, 46*(2), 171–174.

Onodera, R. (2003). Essentiality of histidine in ruminant and other animals including human beings. *Asian-Australasian Journal of Animal Sciences, 16*, 445–454.

Peachey, B. L., Scott, E. M., & Gatlin, D. M. III. (2018). Dietary histidine requirement and physiological effects of dietary histidine deficiency in juvenile red drum *Sciaenops ocellatus*. *Aquaculture, 483*, 244–251.

Ravi, J., & Devaraj, K. V. (1991). Quantitative essential amino acid requirements for growth of catla, *Catla catla* (Hamilton). *Aquaculture, 96*(3–4), 281–291.

Recodo, A. G. (1991). *Histophysiology Effects of Various Histidine Levels in The Diet on The Hepatopancreas of Penaeus monodon* (Master's thesis). Institute of Biology, College of Science, U. P. Diliman, Quezon City, Philippines. 104 pp.

Remø, S. C., Hevrøy, E. M., Olsvik, P. A., Fontanillas, R., Breck, O., & Waagbø, R. (2014). Dietary histidine requirement to reduce the risk and severity of cataracts is higher than the requirement for growth in Atlantic salmon smolts, independently of the dietary lipid source. *British Journal of Nutrition, 111*, 1759–1772.

Remø, S. C., Olsvik, P. A., Torstensen, B. E., Amlund, H., Breck, O., & Waagbø, R. (2011). Susceptibility of Atlantic salmon lenses to hydrogen peroxide oxidation *ex vivo* after being fed diets with vegetable oil and methylmercury. *Experimental Eye Research, 92*, 414–424.

Robbins, S. L., Cotran, R. S., & Kumar, V. (1984). *Pathologic Basis of Disease*. W. B. Saunders.

Rodehutscord, M., Becker, A., Pack, M., & Pfeffer, E. (1997). Response of rainbow trout (*Oncorhynchus mykiss*) to supplements of individual essential amino acids in a semi-purified diet, including an estimate of the maintenance requirement for essential amino acids. *Journal of Nutrition, 127*(6), 1166–1175.

Rollin, X. (1999). *Critical Study of Indispensable Amino Acids Requirements of Atlantic salmon (Salmo salar L.) fry* (Doctoral dissertation). Université catholique de Louvain, Louvain, Belgium.

Santiago, C. B., & Lovell, R. T. (1988). Amino acid requirements for growth of Nile tilapia. *Journal of Nutrition, 118*(12), 1540–1546.

Scott, L., Lall, S. P., & Anderson, D. M. (1996). Dietary histidine requirement of Atlantic salmon (*Salmo salar*) reared in sea water. *CFIA Eastern Nutrition Conference Proceedings*, Dartmouth/Halifax, Canada. 259 pp.

Son, D. O., Satsu, H., & Shimizu, M. (2005). Histidine inhibits oxidative stress and TNF-α-induced interleukin-8 secretion in intestinal epithelial cells. *FEBS Letters, 579*(21), 4671–4677.

Stevens, E. D., Lam, H. M., & Kendall, J. (1974). Vascular anatomy of the counter-current heat exchanger of skipjack tuna. *Journal of Experimental Biology, 61*, 145–153.

Stifel, F. B., & Herman, R. H. (1971). Histidine metabolism. *American Journal of Clinical Nutrition, 24*, 207–217.

Suyama, M., & Suzuki, H. (1975). Nitrogenous constituents in the muscle extracts of marine elasmobranchs. *Bulletin of the Japanese Society of Scientific Fisheries, 41*, 787–790.

Suyama, M., Suzuki, T., & Yamamoto, A. (1977). *Journal of Tokyo University of Fisheries, 63*, 189–196.

Tan, L. N., Feng, L., Liu, Y., Jiang, J., Jiang, W. D., et al. (2011). Growth, body composition and intestinal enzyme activities of juvenile Jian carp (*Cyprinus carpio* var. Jian) fed graded levels of dietary zinc. *Aquaculture Nutrition, 17*(3), 338–345.

Taylor, R. G., Levy, H. L., & McInnes, R. R. (1991). Histidase and histidinemia: Clinical and molecular considerations. *Molecular Biology and Medicine, 8*, 101–116.

Torres, N., Beristain, L., Bourges, H., & Tovar, A. R. (1999). Histidine-imbalanced diets stimulate hepatic histidase gene expression in rats. *Journal of Nutrition, 129*, 1979–1983.

Trosse, C., Waagbø, R., Breck, O., Stavrum, A. K., Petersen, K., & Olsvik, P. A. (2009). Genome-wide transcription analysis of histidine-related cataract in Atlantic salmon (*Salmo salar* L.). *Molecular Vision, 15*, 1332–1350.

Tsukamoto, I. C. (1984). Contribution of the red and white muscles to the power output required for swimming by the yellowtail. *Bulletin of the Japanese Society of Scientific Fisheries, 50*, 2031–2042.

Unprasert, N. G. (1994). *An Evaluation of The Use of Ideal Protein Concept to Estimate Essential Amino Acid Requirements of the Clarias hybrid (Clarias macrocephalus × Clarias gariepinus)* (Master's thesis). Mississippi State University, Mississippi State, MS.

Waagbø, R., Trosse, C., Koppe, W., Fontanillas, R., & Breck, O. (2010). Dietary histidine supplementation prevents cataract development in adult Atlantic salmon, *Salmo salar* L., in seawater. *British Journal of Nutrition, 104*, 1460–1470.

Wade, M. A., & Tucker, H. N. (1998). Antioxidant characteristics of L-histidine. *Journal of Nutritional Biochemistry, 9*(6), 308–315.

Waters, V., & Fierstine, H. L. (1964). Measurement of swimming speeds of yellowfin tuna and wahoo. *Nature, 202*, 208–209.

Wilson, R. P., Poe, W. E., & Robinson, E. H. (1980). Leucine, isoleucine, valine and histidine requirements of fingerling channel catfish. *Journal of Nutrition, 110*(4), 627–633.

Wu, G. (2009). Amino acids: Metabolism, functions, and nutrition. *Amino Acids, 37*, 1–17.

Yoshikawa, T., Nakamura, T., Shibakusa, T., Sugita, M., Naganuma, F., et al. (2014). Insufficient intake of L-histidine reduces brain histamine and causes anxiety-like behaviors in male mice. *Journal of Nutrition, 144*(10), 1637–1641.

Zannoni, V. G., & La Du, B. N. (1963). Determination of histidine α-deaminase in human stratum corneum and its absence in histidinaemia. *Biochemical Journal, 88*, 160–162.

Zehra, S., & Khan, M. A. (2014). Dietary histidine requirement of fingerling *Catla catla* (Hamilton) based on growth, protein gain, histidine gain, RNA/DNA ratio, hematological indices and carcass composition. *Aquaculture Research, 47*(4), 1028–1039.

Zhao, B., Feng, L., Liu, Y., Kuang, S. Y., Tang, L., et al. (2012). Effects of dietary histidine levels on growth performance, body composition and intestinal enzymes activities of juvenile Jian carp (*Cyprinus carpio* var. Jian). *Aquaculture Nutrition, 18*(2), 220–232.

CHAPTER 10

Leucine Requirement of Different Fish Species

BHAVESH CHOUDHARY,[1,*] VEDIKA V. KANOJE,[2] SUPARNA DEB,[1] and VINAYAKA CHOUDHARY[3,4]

[1]*College of Fisheries, Central Agricultural University (Imphal), Lembucherra, Agartala, Tripura, India*

[2]*College of Fisheries, Dr. Balasaheb Sawant Konkan Krishi Vidyapeeth ((DBSKKV, Dapoli)), Shirgaon, Ratnagiri, Maharashtra, India*

[3]*Vidya Bhawan Rural Institute, Mohanlal Sukhadia University, Udaipur, Rajasthan, India*

[4]*Regional Institute of Education (NCERT, New Delhi), Ajmer, Rajasthan, India*

*Corresponding author

ABSTRACT

Accurate estimates of the maintenance requirement of amino acids (AAs) are important for feed formulation. Branched-chain amino acids (BCAAs) serve not only as substrates for protein synthesis but also play a role in regulating protein synthesis. BCAAs regulate various key signaling pathways. Leucine is believed to be an indispensable AA in most aquatic species because its carbon skeleton cannot be synthesized *in vivo*. It functions as a major regulator in body composition, growth performance, mTOR activation, and metabolism at multiple levels. Leucine plays a crucial part in investigations, which have consistently emphasized the significance of dietary quantities of BCAAs in the development and growth of fish. Based on the existing

Fish Nutrition: Protein and Amino Acid Requirements. Imtiaz Ahmed, Ishtiyaq Ahmad, and Shabihul Fatma Sayed (Eds.)

DOI: 10.1201/9781998511570-10

evidence, it appears that fish need slightly greater quantities of BCAAs throughout the early stages of life compared to later stages. The optimum dietary leucine requirement of fishes can be revealed by various trial data and analysis of feed conversion ratio (FCR), protein efficiency ratio (PER), specific growth rate (SGR), final weight (FW), weight gain rate (WG), leucine retention efficiency, and body protein deposition (BPD). This chapter provides full information on leucine metabolism, the empirical method used for leucine requirements, the dietary leucine requirement of different fish species, natural or synthetic leucine sources, the interaction of leucine with other AAs, the ideal protein concept, important functions in the fish body, nutritional disorder of leucine deficiency, and factors affecting leucine requirement. The optimum dietary leucine requirement has been determined for various fish species, showing a wide range of values ranging from 11.5 to 91.0 g/kg of dietary protein (DP) to date.

10.1 INRODUCTION

Supplemental feed containing all of the necessary nutrients required by the organism has become an integral component of every successful culture system. Feed is being developed rich in vital nutrients for direct consumption of fish in more extensive and semi-intensive activities, although natural food resources remain the predominant and optimum source of nourishment (Ahmed and Khan, 2004). Supplementary feeds are utilized as an alternate source of dietary elements that meet the nutritional demands of the particular species in intensive culture, when the role of natural creatures is not important in terms of stocking density. Because dietary protein (DP) is one of the most expensive substances in fish feed, it is also regarded as the first source of nitrogen waste products entering the culture system (Rahimnejad and Lee, 2013). As a result, optimizing DP levels and enhancing nutrition retention by fish might not only assist reduce level of nitrogen in their surroundings, but also have a favorable influence on total aquaculture expenditure (Thoman et al., 1999). However, understanding the protein requirements of fish is useless unless we also know the essential amino acid (EAA) needs of the particular fish. Proteins are made up of bundles of AAs that not only function as protein building blocks but also as critical controllers of significant metabolic pathways (Jobgen et al., 2006; Meijer, 2003). Out of all the AAs needed by fish for their many functions in the body, 10 have been identified as important for fish growth and development (NRC, 2011;

Twibell et al., 2003; Wilson, 1985) and that requires an exogenous source in the form of feed that fulfills these AA demands (NRC, 2011). A shortage of these EAA may result in lower growth and poor feed efficacy (Wilson and Halver, 1986), as well as an influence on the general physiology of fish. As a result, meeting the necessary AA needs of farmed fish is critical to achieving optimal fish development. Branched chain amino acids (BCAAs), such as leucine, isoleucine, and valine, are necessary AAs for animals, including fish. They not only serve as building blocks for tissue protein but also have functional roles in fish, including protein synthesis, insulin release, and inhibition of the process of breaking down proteins into smaller components (Jobgen et al., 2006; Meijer, 2003; Wu, 2013).

Growth enhancement in fish is mostly determined by the outcomes of protein synthesis and deposition (Hochachka and Mommsen, 1995). Protein synthesis is controlled by variations in peptide chain initiation caused by changes in the rate of mRNA translation (Rhoads et al., 2007; Tang et al., 2013). Because it is an activator of the target of rapamycin (TOR), leucine (Leu) works as a functional AA that promotes the production of muscle proteins and inhibits proteolysis in both mammals and fish (Ahmad et al., 2021; Nakashima et al., 2007). It has the ability to upregulate pathways affecting muscle protein synthesis through both insulin-dependent and independent pathways. Leu is also directly involved in the activation of mTOR in muscle and helps stimulate the downstream phosphorylation of p70S6 kinase and 4E-BP1 that signal RNA translation and protein synthesis (Zhao et al., 2020). Leu has been suggested to play an important role in regulating the rapamycin (TOR) targeting pathway, gluconeogenesis, and adipogenesis in rainbow trout hepatocytes (Lansard et al., 2009, 2010). However, its nutritional imbalance (deficiency or excess) can cause severe biochemical dysfunction in a type of growth inhibition present in numerous aquaculture species (Abidi and Khan, 2007; Ahmed and Khan, 2006; Deng et al., 2014; Farhat and Khan, 2014; Fernstrom, 2005). The need of Leu for many fish species has been studied from various parts of the world, and the complete evidence has been previously studied (Ahmed and Khan, 2006); later on, many studies are being added, such as yellow croaker, *Pseudosciaena crocea* (Yan et al., 2010); *Ctenopharyngodon idella* (Deng et al., 2014); stinging catfish, *Heteropneustes fossilis* (Farhat and Khan, 2014); Catla (Zehra and Khan, 2015); *Megalobrama amblycephala* (Ren et al., 2015); *Trachinotus ovatus* (Tan et al., 2016); *Oreochromis niloticus* (Gan et al., 2016; Zehra and

Khan, 2017); *Sciaenops ocellatus* (Castillo and Gatlin, 2018); *Carassius auratus* (Zou et al., 2018); *Epinephelus fuscoguttatus* × *E. lanceolatus* (Zhou et al., 2019); *Epinephelus coioides* (Niu et al., 2021) *Micropterus salmoides* (Yang et al., 2022); *Channa punctatus* (Sharf et al., 2023). While the Leu requirement of rainbow trout has been previously shown by Ogino (1980) and NRC (1993), these studies did not establish a correlation between growth data and hematobiochemical parameters, as well as the relative expression of TOR and 4E-BP levels. This correlation is crucial for stimulating protein synthesis.

The nutritional value of DP in fish is influenced by their AA composition and bio-availability (Wilson and Cowey, 1985; Wilson and Poe, 1985). The integral quantifiable AA (IDAA) requirements of different fish species have been studied for optimal growth and feed utilization, cost-effective feed formulations, and desired carcass quality (NRC, 2011). Diets with a balanced IDAA composition have great nutritional value for fish, especially in the early stages of life (Ostaszewska et al., 2010). AA is used not only for protein synthesis but also as an energy source for fish larvae and their other life stages (Rønnestad and Conceição, 2005). According to Dabrowski et al. (2003), a meal containing synthetic dipeptides can boost rainbow trout growth during the early stages of life. In a methionine AA study, Luo et al. (2005) found that marine grouper given an intact protein control diet gained weight noticeably more than fish fed a crystalline AA diet, and they came to the conclusion that crystalline AA may be utilized less effectively than AA obtained from an intact protein. According to several studies (Cowey and Walton, 1988; Dabrowski et al., 2003, 2007; Murai et al., 1987; Rønnestad et al., 2000), the ineffectiveness of free AAs in fish is thought to be caused by their quicker absorption and subsequent degradation than complete proteins. According to an alternate theory put out by Zarate and Lovell (1997), more free AA is lost during filtering than is bound AA in the aqueous medium prior to consuming it. Therefore, crystalline AAs have been used in nutrient embedding or microencapsulation techniques to slow the release of nutrients and AAs, which in turn reduces the rate of nutrient absorption (De La Higuera et al., 1998; Villamar and Langdon, 1993) and the leaching rate of free AAs from feed in contact with water (Segovia-Quintero and Reigh, 2004).

Leucine (Leu) is one of the AAs that is involved in a variety of physiological functions in fish, including growth, maintenance, and metabolic processes (Tan et al., 2016). After absorption, Leu is converted to alpha-keto-isocaproate (α-KIC), which is subsequently oxidized and decarboxylated to produce acetyl-CoA, allowing the TCA cycle to take over and carry out the vital biological processes (Zhang et al., 2017; Zhou et al., 2019). In addition to promoting protein synthesis, Leu plays a critical role in maintaining blood glucose levels, nitrogen stabilization, somatotrophic hormones, energy metabolism, body fat deposition, and hemoglobin levels (Abidi and Khan, 2007; Ahmed and Khan, 2006; Li et al., 2011; Norton and Layman, 2006). Since it is necessary for the synthesis of protein in muscle tissue, leucine (Leu) is crucial for the growth and health of fish. It is a ketogenic AA that stands out from other BCAAs in that it can trigger the release of insulin from pancreatic islet cells (Panten et al., 1974). Growth retardation can result from severe metabolic malfunction caused by Leu deficiency. According to Conceicao et al. (1997), one of the limiting AAs controlling the turbot larvae's rate of development is Leucine. This may be true for other flatfish species, and an essential need has been reported for only a limited number of farmed fish species, including Chinook salmon (Chance et al., 1964), common carp (Nose, 1979), channel catfish (Wilson et al., 1980), Japanese eel (Nose, 1979), Nile tilapia (Santiago and Lovell, 1988), chum salmon (Akiyama, 1987), lake trout (Hughes et al., 1983), and the Indian major carp (Abidi and Khan, 2007).

It is one of the most prevalent AAs in high-quality protein diets and may be found in all proteins (Wu 2013a; Li et al., 2011b). In addition to placental cells (Kim 2007; Kim et al., 2011a, b; Li et al., 2007), mammary epithelial cells (Lei et al., 2012a), and enterocytes (Rhoads and Wu 2009), leucine has a special signaling function in skeletal muscle and adipose tissue (Li et al., 2011a). Being one of the most prevalent AAs, it is frequently found in meals with high-quality proteins. Leu encourages energy metabolism to fuel protein synthesis while inhibiting protein breakdown (glucose uptake, mitochondrial biogenesis, and fatty acid oxidation).

10.2 LEUCINE METABOLISM

Leucine, isoleucine, and valine, which are classified as branched-chain amino acids (BCAA), constitute approximately one-third of muscle protein. Among these, leucine has undergone the most extensive investigation because of its increased oxidation rate compared to isoleucine or valine. Leucine has the ability to enhance the production of proteins in muscle and is strongly linked to the release of substances that are used to create glucose, such as alanine, from muscle (Mero, 1999) (Figure 10.1). Leucine metabolism is a process by which AA leucine is broken down and utilized by the body. It is one of the 10 EAAs required for protein synthesis and various biological functions. The breakdown of leucine occurs primarily in the liver and muscle tissues. It undergoes a series of enzymatic reactions, leading to the formation of acetyl-CoA and acetoacetate. These products can then enter into the citric acid cycle and ultimately contribute to energy production in the form of ATP (Ahmad et al., 2021).

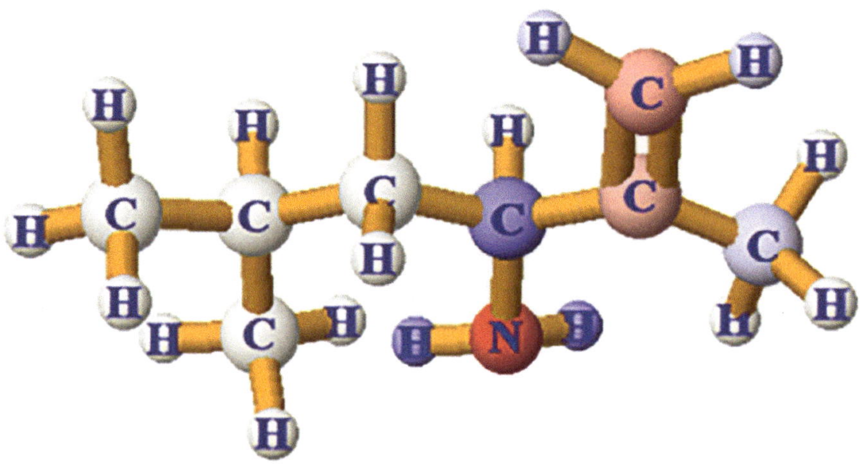

FIGURE 10.1 Leucine molecule structure.

It also plays a vital function in the process of protein synthesis and muscle maintenance (Jürss and Bastrop, 1995). Fish use leucine for several purposes, including growth and maintenance of muscle tissues, immune function, and energy production. The rate of leucine metabolism can vary depending on fish species, its age, size, and environmental factors (Duan et al., 2016). According to the theory of Krebs (1972), AA catabolism is typically studied through changes in the concentrations of the enzymes involved in these pathways and changes in their affinities for their substrates (Cowey, 1980; Van Waarde, 1983; Walton and Cowey, 1981; Zebian and Creach, 1979). The relative relevance of each of these processes may also be determined by understanding the rates of AA oxidation and incorporation into proteins.

Leucine metabolism involves a series of enzymatic reactions that occur within cells of an organism. Some general overviews involved in the breakdown of leucine are:

1. **Uptake of Leucine:** It obtained through the diet and catabolism (breakdown by a large particle) of the various protein in the digestive system of fish. After the digestion of the various AAs the leucine and other AAs is absorbed into the bloodstream of fishes from digestive system.
2. **Transport:** The digested leucine is transported from bloodstream into the cells of various tissues in the fish body. Where it will undergo into the metabolism through the metabolic processes.
3. **Trans-Amination:** Inside the cells, leucine undergoes various transamination, where it transfers its amino group to alpha-ketoglutarate, forming alpha-ketoisocaproate (α-KIC) and glutamate.
4. **Deamination:** α-KIC undergoes oxidative decarboxylation, leading to removal of the amino group as ammonia, resulting in isovaleryl-CoA and propionyl-CoA.
5. **Branched-Chain Keto Acid Dehydrogenase Complex (BCKDH):** Isovaleryl-CoA then enters BCKDH complex, which further metabolizes it to produce acetyl-CoA and propionyl-CoA.
6. **Fate of Metabolites:** Acetyl-CoA can enter the citric acid cycle or Krebs cycle to be used for energy production, while propionyl-CoA can be converted to succinyl-CoA and eventually enter the same cycle.
7. **Utilization and Storage:** Acetyl-CoA and other intermediates generated in the citric acid cycle can be used to produce ATP, the primary energy currency of cells, or be utilized for various biosynthetic processes, including protein synthesis.

   It is important to note that leucine metabolism doesn't involve only the catabolic pathway, but anabolic processes for protein synthesis and other cellular components also take place in the body of fish. The regulation of leucine metabolism is tightly controlled by various factors, such as dietary intake, hormonal balance of AAs, and energy utilization, and so forth. If leucine metabolism does not occur in the fish body, it can have several detrimental effects on its health and overall well-being. Some potential effects of impaired leucine metabolism are:

1. **Protein Synthesis and Muscle Growth:** Leucine is a key AA in promoting protein synthesis, particularly in muscle tissues. Impaired leucine metabolism could result in reduced growth, muscle atrophy, and decreased overall muscle function. This can lead to weakened swimming ability and compromised predator or prey interaction.
2. **Energy Production:** It involves energy production through conversion to acetyl-CoA and entry into the citric acid cycle. Without proper leucine metabolism, the fish might have difficulty generating sufficient energy for essential cellular processes, leading to fatigue, decreased activity levels, and overall reduced vitality (National Research Council (1993) Nutrient requirements of fish. National Academies Press).
3. **Ammonia Accumulation:** The breakdown of AAs like leucine produces ammonia as a by-product. Impaired leucine metabolism could lead to ammonia accumulation, which is toxic to fish. High ammonia levels can damage the fish's gills, disrupt osmoregulation, and lead to various health issues, including ammonia poisoning.
4. **Neurological Function:** Leucine also involved in neurotransmitter synthesis and brain function. A disruption in leucine metabolism might affect neurological processes, potentially leading to altered behavior, impaired nervous system function, and cognitive abnormalities.
5. **Immune System:** Leucine are necessary for proper immune system function. Impaired leucine metabolism could weaken the fish's ability to mount an effective immune response against pathogens, making it more susceptible to infectious diseases.
6. **Metabolic Disorders:** If leucine metabolism is severely compromised, it could contribute to metabolic disorders and imbalances, disrupting the delicate equilibrium of AA and other metabolites within the fish's body.

10.3 EMPIRICAL METHODS USED FOR LEUCINE REQUIREMENT

Determining the leucine requirements of fishes typically involves several empirical methods (Figure 10.2) that involve experimentation and observation (D'mello, 1982). The main objectives of the use of empirical methods in any organism are to help researchers, scientists, and nutritionists establish or identify the optimal leucine requirements for particular species and life stages. These empirical methods are often conducted in controlled laboratory or aquaculture settings.

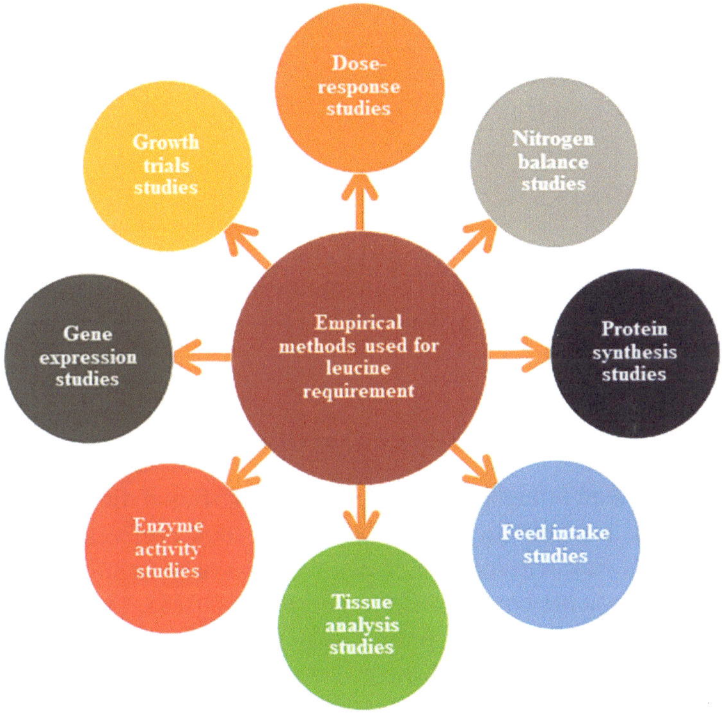

FIGURE 10.2 Empirical methods used for leucine requirement in different fish species.

Some common methods are:

10.3.1 DOSE-RESPONSE STUDIES

These involve feeding the fish with different levels of leucine and observing the growth, health, and other relevant parameters. By analyzing the responses to different dietary leucine levels, researchers can determine the minimum amount of leucine required for optimal growth and performance. Kaushik et al. reported in 1988 that patterns of changes in free AA concentrations in plasma, muscle, or liver in rainbow trout were not consistent enough to estimate AA needs according to dose-response analysis.

10.3.2 NITROGEN BALANCE STUDIES

It involves measuring the intake and excretion of nitrogen in fish fed diets with different leucine levels. The aim is to find the dietary leucine level at which nitrogen intake and output are balanced, indicating that the fish's protein metabolism is in equilibrium.

10.3.3 PROTEIN SYNTHESIS STUDIES

Researchers may study the rate of protein synthesis in fish by measuring the incorporation of labeled leucine into newly synthesized proteins. This helps determine the optimal level of leucine required to support efficient protein synthesis, which is important for growth and tissue maintenance.

10.3.4 FEED INTAKE STUDIES AND GROWTH TRIALS

Studying the feed intake behavior of fish when offered diets with different leucine levels can provide information about their preferences and intake patterns. This can help identify if fish tend to consume more or less feed based on the leucine content (Zehra and Khan, 2015). Controlled growth trials are conducted where fish are fed diets containing different levels of leucine, and their growth performance is monitored over a specific period. This helps establish the leucine level that results in the highest growth rate and best feed conversion ratio (FCR).

10.3.5 TISSUE ANALYSIS

Analyzing the nutrient content in the various fish tissues under different dietary conditions can provide insights into leucine utilization and requirements. Tissue analysis helps researchers understand how fish allocate leucine for different physiological processes.

10.3.6 ENZYME ACTIVITY AND GENE EXPRESSION

The person who finds out the requirement may also study the activity of enzymes involved in leucine metabolism and monitor the expression of relevant genes. Changes in enzyme activity and gene expression patterns can provide information about the fish's response to varying leucine levels.

10.4 DIETARY LEUCINE REQUIREMENT

It is important to note that leucine requirements can vary based on factors such as species, size, age, environmental conditions, and dietary composition. Various factors have the potential to change the optimum dietary requirement of any kind of fish or other animals. Optimum dietary leucine levels have been determined for various fish species, showing a wide range of values ranging from 11.5 to 91.0 g/kg of DP (Table 10.1).

TABLE 10.1 Optimum Leucine Requirement of Different Fish Species

Sl. No.	Species	Fry/Fingerling/ Juvenile	Leucine Requirement (g Kg^{-1})	References
1.	*Carassius auratus gibelio*	Juvenile	54.28	Zou et al. (2018)
2.	*Cyprinus carpio*	Fingerling	33.0	–
3.	*Catla catla*	Fry	37.0	Ravi and Devaraj (1991)
4.	*Catla catla*	Fingerling	47.57–47.87	Zehra and Khan (2015)
5.	*Oncorhynchus tshawytscha*	Fingerling	39.02	Chance et al. (1964)
6.	*Cirrhinus mrigala*	Juvenile	43.30	Benakappa and Varghese (2003)
7.	*Cirrhinus mrigala*	Fingerling	38.50	Ahmed and Khan (2006)
8.	*Ctenopharyngodon idella*	Juvenile	42.43	Deng et al. (2014)
9.	*Heteropneustes fossilis*	Fingerling	43.40	Farhat and Khan (2014)
10.	*Ictalurus punctatus*	Fingerling	35.0	Wilson et al. (1980)
11.	*Labeo rohita*	Fingerling	37.5–39.2	Abidi and Khan (2007)
12.	*Megalobrama amblycephala*	Juvenile	42.4–47.4	Ren et al. (2015)
13.	*Oncorhynchus mykiss*	Juvenile	44.0	Ogino (1980)
14.	*Oreochromis niloticus*	Juvenile	43.1	Gan et al. (2016)
15.	*Oreochromis niloticus*	Fingerling	44.5–46.0	Zehra and Khan (2017)
16.	*Pseudosciaena crocea*	Juvenile	67.9	Yan et al. (2010)
17.	*Salvelinus namaycush*	Juvenile	27.4–36.6	Hughes et al. (1983)
18.	*Sciaenops ocellatus*	Juvenile	42.43	Castillo and Gatlin (2018)
19.	*Trachinotus ovatus*	Juvenile	71.5–80.2	Tan et al. (2016)

TABLE 10.1 *(Continued)*

Sl. No.	Species	Fry/Fingerling/ Juvenile	Leucine Requirement (g Kg^{-1})	References
20.	*Oncorhynchus mykiss*	–	91.0	Choo et al. (1991)
21.	*Labeo rohita*	–	46.0	Murthy and Varghese (1997a)
22.	*Oreochromis niloticus*	Juvenile	33.9	Santiago and Lovell (1988)
23.	*Chanos chanos*	Juvenile	51.1	Borlongan and Colosa (1993)
24.	*Labeo rohita*	Fingerling	11.5–12.0	Giri et al. (2015)
25.	*Mylopharyngodon piceus*	Juvenile	54.75–70.75	Wu et al. (2017)
26.	*Pseudosciaena crocea*	Juvenile	67.9	Li et al. (2010)
27.	*Epinephelus fuscoguttatus* ♀ × *Epinephelus lanceolatus* ♂	Juvenile	60.7	Zhou et al. (2019)
28.	*Oncorhynchus mykiss*	Fingerling	38.77	Ahmad et al. (2021)
29.	*Sciaenops ocellatus*	Juvenile	42.43	(Castillo and Gatlin, 2018)
30.	*Micropterus salmoides*	Juvenile	58.51	Yang et al. (2022)
31.	*Epinephelus coioides*	Juvenile	88.88	Niu et al. (2021)
32.	*Channa punctatus*	Fingerling	44.44	Sharf et al. (2023)
33.	*Pelteobagrus vachelli* × *Leiocassis longirostris*	–	73.04	Zhao (2020)

10.5 SOURCES OF DIETARY LEUCINE (NATURAL & SYNTHETIC)

10.5.1 NATURAL SOURCES

Leucine is available in many natural food sources. Some of the main natural food-derived sources having a high amount of dietary leucine are given in Table 10.2.

TABLE 10.2 Natural Sources of Dietary Leucine

Natural Sources of Dietary Leucine				
Vital wheat gluten	Seaweed (dried)	Hemp seeds	Pumpkin seeds	Chicken
Cottage cheese	Beef	Milk	Chia seeds	Sesame seeds
Cashew, Oats	Flaxseed	Pistachio nuts	Brazil nuts	Walnut
Hazelnut	Pine nuts	Soybeans	Buckwheat groats	Black beans
Navy beans	Kidney beans	Lentils	Adzuki beans	Chickpeas
Macadamia nut	Pecans	Fava bean	Rye grain	Mung bean
Lima beans	Peanuts	Almonds	Pork	

10.5.2 SYNTHETIC SOURCES

Synthetic dietary sources of leucine are supplements or additives that contain leucine in a concentrated form. These sources are often used in animal nutrition, including aquaculture, to provide an additional and controlled amount of leucine in the diet. Some common synthetic dietary sources of leucine include:

1. **L-Leucine:** L-Leucine is the pure, biologically active form of leucine. It's available as a standalone AA supplement and is commonly used to boost the leucine content of animal feeds.
2. **Leucine Peptides:** Leucine peptides are short chains of AAs containing leucine. These peptides can be derived from various protein sources and can provide a source of leucine with potentially improved absorption and utilization.
3. **Amino Acid Mixtures:** Some formulations provide a balanced mixture of EAAs, including leucine, in specific ratios. These mixtures are designed to mimic the AA profile of high-quality protein sources and can be used to optimize the AA composition of animal feeds.
4. **Feed Additives:** In aquaculture, certain feed additives may contain leucine along with other nutrients or compounds that support growth, immune function, or other physiological processes.
5. **Functional Amino Acids:** These are specialized products (Table 10.2) that contain specific AAs, including leucine, in forms that can have beneficial effects beyond basic nutrition. They are often used to target specific health or performance goals in animals.

It's important to note that while these synthetic dietary sources can be useful for providing targeted nutrition, they should be used with caution and in consultation with animal nutrition experts. Over-supplementation or imbalanced AA ratios can have unintended consequences on animal health and performance. It is generally best to incorporate synthetic sources of leucine into a well-balanced diet that includes natural protein sources to ensure comprehensive and appropriate nutrition.

10.6 INTERACTION OF LEUCINE WITH OTHER AMINO ACIDS

10.6.1 BRANCHED-CHAIN AMINO ACIDS (BCAAS) SYNERGY

Leucine is a BCAA, and it interacts synergistically with the other two BCAAs, isoleucine, and valine. Together, they promote muscle protein synthesis and are essential for muscle maintenance and recovery.

10.6.2 PROTEIN SYNTHESIS

Leucine acts as a signaling molecule to initiate protein synthesis through the mammalian target of rapamycin (mTOR) pathway. This pathway is essential for muscle growth and adaptation to exercise. Other AAs, especially EAAs, also contribute to this process by providing the necessary building blocks for protein synthesis.

10.6.3 NUTRIENT TRANSPORT

AAs are transported into cells through specific transporters, and competition for these transporters can occur between different AAs. Leucine may compete with other large neutral AAs, such as phenylalanine, tyrosine, tryptophan, and isoleucine for cellular uptake.

10.6.4 ENERGY PRODUCTION

Leucine can be catabolized into acetyl-CoA, a key molecule in the citric acid cycle (Krebs cycle) that generates energy through cellular respiration. It can interact with other AAs, such as isoleucine, in this metabolic process.

10.6.5 INTERACTIONS IN PROTEIN STRUCTURES

In the context of protein folding and tertiary structure, leucine's hydrophobic properties can play a role in stabilizing protein cores alongside other hydrophobic AAs.

10.6.6 NEUROTRANSMITTER SYNTHESIS

Leucine is involved in the synthesis of important neurotransmitters, such as glutamate and gamma-aminobutyric acid (GABA), which are essential for brain function.

10.6.7 GLUCONEOGENESIS REGULATION

Leucine can influence gluconeogenesis, the process by which the body produces glucose from non-carbohydrate sources. It can enhance the conversion of certain AAs, such as alanine, into glucose in the liver.

10.7 LEUCINE AND IDEAL PROTEIN CONCEPT

10.7.1 IDEAL PROTEIN CONCEPT

An ideal protein is one that offers the precise ratio of AAs required for optimum efficiency and maximal growth (e.g., in terms of size, carcass weight, yield, and body composition). It is possible to achieve the necessary AA demands while using less protein in the food source by creating the optimum protein-based feed. In order to drastically lower production costs, boost farm profitability, and decrease the harvest of wild fish used as fishmeal, protein utilization efficiency must be maximized, and the amount needed in feed must be reduced. By incorporating the optimal protein concept into fish feed formulations, as it has been successfully done for poultry and pig feed, nitrogen pollution from fish production may also be decreased. Because fish, like other animals, are not completely effective in utilizing what they ingest, the ideal protein concept in feed preparation necessitates that the necessary AA needs of the fish be supplied using digestible AAs rather than total AAs (Miles et al., 2007).

The EAA need of fish has been demonstrated to correspond with the EAA pattern of that fish's whole-body tissue when the quantitative AA requirements of a fish species are unknown (Akiyama et al., 1997; Ngamsnae et al., 1999; Tacon, 1989; Wilson and Poe, 1985). Using a model of the ideal protein, Moon and Gatlin (1991), Wilson (1991), and Brown (1995) offered the prediction of the EAA needs of fish.

The necessary AAs that make up protein, rather than protein itself, are what fish need in order to survive. In other words, rather than protein itself, fish require crucial AAs from their diet. The AAs that make up protein are released during digestion and are either taken up by the body as single AAs or in the form of di- and tripeptides, which are shorter chains of AAs. All living things employ a set of around 20 known AAs as the foundation for their proteins. Fish cannot synthesize the 10 necessary AAs or can only do so at a pace that is insufficient for biological demands; thus, they must be supplemented in the diet through external sources. Fish can synthesize several AAs at a rate that is sufficient for cellular protein synthesis, making them non-essential or dietary AAs. Body cells need both essential and non-EAAs for the synthesis of proteins. Utilizing nutrient-dense feeds can help save expenses and decrease nitrogenous waste generated by aquaculture facilities. In comparison to the amount of energy taking up space in a feed pellet, a nutrient-dense diet will provide a considerable amount of a certain nutrient or nutrients (Miles et al., 2007).

The ideal protein would be one in which a fish diet has precisely the right quantity of each key AA needed for a certain fish species (mainly for the synthesis of lean muscle tissue). No lack or surplus of AAs will result from the use of this ideal protein. Additionally, modifications to the food's AA composition would be done in a way that preserved the relative AA ratio. In other words, the idea of the perfect protein is based on the premise that AAs are necessary and must be given in a certain ratio to one another. AAs will be effectively used for maintenance, health, and the synthesis of new structural proteins (i.e., lean fillets) in farmed fish given such a diet, leading to optimal feed efficiency and growth (Miles et al., 2007).

10.7.2 WHY IDEAL PROTEIN CONCEPT IS NEEDED?

Typically, extra protein is added to feeds for one of two explanations: either the protein is poorly digested, necessitating the addition of additional protein to fulfill AA requirements, or excessive protein is supplied because the precise amounts of EAAs needed are unknown. A considerable margin of safety is provided by the surplus protein, reducing the likelihood that the diet may be deficient in key AAs. In an effort to fulfill the need for one or more of the EAAs that are limited in supply, it is not practical or required to raise the total protein content of a diet to the point where excessive levels of several AAs are included. Based on the digestible AA content of the feed components and a suitable protein, a diet should be created (Miles et al., 2007).

Fish excrete extra nitrogen in the form of ammonia, which contributes significantly to water pollution, which might have a severe effect on the ecosystem. The ideal situation would be to develop a low-protein diet that would reduce the amount of nitrogen excreted while simultaneously fulfilling all requirements for vital AAs because each species of fish and the various proteins within each species have their own unique AA makeup. Since commercially accessible synthetic EAAs (such as methionine, lysine, and threonine) are now readily available and are effectively used by workers (Miles et al., 2007).

10.7.3 IDEAL PROTEINS CONCEPT USED BY AQUACULTURE FEEDS INDUSTRY

The AA profiles of a fish species' whole-body tissue closely match its nutritional needs (Akiyama et al., 1997; Ngamsnae et al., 1999; Tacon 1989; Wilson and Poe 1985), and several studies have used the ideal protein concept to predict the EAA needs of fish. Because it offers a perfect ratio of AAs and reduces nitrogen excretion, the optimal protein concept might be crucial for fish nutrition. The optimal ratio of EAAs to one another will remain the same, even while the absolute demand of a specific AA will alter since dietary AA requirements are influenced by a variety of dietary, genetic, and environmental variables (Pack and Rodehutscord 1995). Since lysine is predominantly used for body protein accretion and is not complicated by other metabolic pathways for maintenance or synthesis, it is chosen as the reference AA. However, digestible values should be used to represent optimum AA ratios.

10.8 PHYSIOLOGICAL AND NUTRITIONAL FUNCTIONS OF LEUCINE

- Dietary Leu dramatically improved the density and diameter of muscle fibers in fish, demonstrating that it has favorable effects on muscle growth and development that are consistent with the findings for C2C12 (Areta et al., 2014).
- IGFs influence how fish skeletal muscle develops (Rius-Francino et al., 2011).

- Ingesting optimum Leu increased the expression of IGF-I in the liver (Zhou et al., 2019).
- The development of fish muscle is a complex and carefully regulated process that involves the proliferation and differentiation of myoblasts (Johnston et al., 1999).
- To enhance muscle development in fish, IGF-I promotes myoblast differentiation and proliferation (Coolican et al., 1997; Montserrat et al., 2012; Rius-Francino et al., 2011).
- Leu is regarded as a crucial nutrition control element involved in muscle protein synthesis. The ratio of tissue RNA to protein may be a good indicator of a fish's ability to synthesize proteins (Houlihan et al., 1989).
- According to Lang et al. (2005) Leu-induced TOR phosphorylation may not need the PI3K/AKT signaling pathway.
- The density and hardness of muscle fiber both considerably enhanced with dietary leucine intake (Mang et al., 2022). It confirmed the general rule that muscle hardness and fish muscle fiber density are positively connected and showed that consuming more leucine may enhance the texture of blunt snout bream meat (Lin et al., 2016).
- Leucine has been shown to encourage their proliferation in the muscle cells of mammals (Chen et al., 2013).
- A leucine (ketogenic amino acid) has the ability to oxidize and release energy. The essential enzyme for its oxidative metabolism is branched-chain -ketoacid dehydrogenase complex (BCKDH). Increasing the leucine concentration can initiate the metabolism of BCKDH, activate its function, and create metabolites that are used to produce energy (Duan et al., 2016).
- A crucial energy metabolism sensor in an animal's body is AMP-activated protein kinase (AMPK), which may be triggered by energy and subsequently activate the downstream energy sensor, Sirtuin 1 (Sirt1). This signaling system, known as AMPK/Sirt1, controls the gene expression or activity of downstream molecules and can control several bodily metabolic functions (Cantó et al., 2010; Hardie 2007).
- Another crucial component of the intestinal barrier is the intestine's immune system (Zhao et al., 2020). Immune parameters include complement, immune globulin, immune-related enzymes (LZM

and ACP), and immune globulin, and have been thought to be essential tools for examining the intestinal immunological barrier in fish (Li et al., 2018; Liang et al., 2020). LZM is involved in the fish's innate immune response as a first line of defense against microbial invasion (Harikrishnan et al., 2010; Ramasamy et al., 2010; Saurabh et al., 2008; Uribe et al., 2011).

- Important lysosomal hydrolytic enzymes ACP and AKP help the body defend itself against external infections and microbial invasion (Rajalakshmi et al., 2005).
- In fish Blunt snout bream (*Megalobrama amblycephala*), Liang et al. (2018a) found increases in plasma C3 immunoglobulin M (IgM) concentrations and reductions in the expression of genes for pro-inflammatory proteins in the hepatopancreas.
- Improvements in intestinal morphology, growth, and protein deposition in the Golden pompano *Trachinotus ovatus*, according to Tan et al. (2016).
- Increases in immune defense effector mRNA levels or enzyme activities, as well as non-specific immunity, are observed in Black carp *Mylopharyngodon piceus*.
- Oxidative stress is also decreased Wu et al. (2017).

10.9 NUTRITIONAL DISORDERS OF LEUCINE DEFICIENCY

There is not a lot of knowledge on how AA deficiency affects the IGF system in fish. Very few studies on how lysine or leucine deficits affect *in vitro*-cultured fish muscle cells. The majority of studies conducted to date investigate the *in vivo* consequences of AA shortage or supplementation in fish diets, and multiple studies have demonstrated the significance of maintaining the necessary amount of the various EAA, depending on the species, in aquafeeds for appropriate development and health (Davies et al., 1997; Deng et al., 2010; Siwicki et al., 2003; Van et al., 2014; Yang et al., 2011).

Regarding the effects of AA deficits on the gene expression of IGF system members, the predominant consequence was a reduction in IGF-I expression at days 4 and 8, and an increase in IGF-II expression at day 8 with lysine deficiency. In the absence of lysine, IGF-IRb gene expression also drastically reduced at day 8. Various data suggest that lysine shortage may be impairing IGFs' ability to do their job in these muscle cells.

According to research by Rolland et al. (2015), rainbow trout's liver transcript levels of IGF-I, but not IGF-II, increased linearly as dietary methionine levels increased, and Atlantic salmon saw a similar outcome when dietary lysine levels increased (Hevrøy et al., 2007). When leucine, methionine, and lysine were stimulated *in vitro* to regulate the expression of genes involved in intermediary metabolism in rainbow trout hepatocytes, Lansard et al. (2010) found that leucine had effects related to a pool of AA, whereas lysine had only modest effects. Hevrøy et al. (2007) found that high lysine consumption increased the levels of muscle IGF-II mRNA in Atlantic salmon by a factor of seven *in vivo*. Results suggested that lysine may have a significant local anabolic regulator role in regulating the growth of muscle tissue in gilthead sea bream (García et al., 2014; Jiménez et al., 2013; Montserrat et al., 2007, b; Rius-Francino et al., 2011; Vélez et al., 2014).

Lack of lysine or leucine had no effect on muscle IGFBP-4 or IGFBP-5 in an investigation by Azizi et al. (2016). According to Hevrøy et al. (2007), liver IGFBP-1b was discovered to be down-regulated in response to decreased lysine consumption, which may be related to the fish's catabolic condition. However, Bower and Johnston (2010a) showed enhanced expression of both IGFBP-5 isoforms after AA stimulation, suggesting the development of an anabolic environment. However, in the experiment (Azizi et al., 2016), the deficit of lysine AA only reduced the gene expression of IGF-IRb in day 8 cells, indicating that both IGF-IRs responded differently to lysine restriction in myocytes. All of these studies suggest that a more drastic restriction, such as total AA suppression like during fasting, may be necessary and that a single AA shortage, either lysine or leucine, is not sufficient to decide substantial alterations on the gene expression of IGFBPs and IGF-IRa in these cells.

Particularly in the case of lysine, AA deficiency clearly affected the growth of myocytes. In the absence of lysine, there is a definite down-regulation in the expression of the genes for MyoD2 at day 4 and Myf5, myogenin, PCNA, and MHC at day 8. Although the absence of leucine did not significantly vary, it displayed comparable tendencies. Deficits in lysine and leucine appear to slow down cell division and muscle differentiation, which is corroborated by the discernible decline in PCNA protein expression (Azizi et al., 2016). According to research by Velez et al. (2014), AA treatment alone or in conjunction with IGF-I increased PCNA and myogenin gene expression in cultured myocytes from gilthead sea bream. Leucine did not influence the expression of any of the genes examined, and its failure to clearly demonstrate diminished myogenesis in the culture images suggests that the two AAs have different functions and different mechanisms of action. In order to control the proliferation of myocytes, leucine may thus have minimal effects working at a post-transcriptional level. Lower amounts of MyoD paralogs were found in fasting salmon myocytes than in cells that had been treated with AA, according to Bower and Johnston (2010b). During leucine starvation in C2C12 myoblasts, Averous et al. (2012) also noted an up-regulation of Myf5 mRNA and protein level and a reduction in MyoD protein level but not mRNA. MHC protein expression was elevated during differentiation in the existence of leucine. In a study by Averous et al. (2012) with murine cells, however, MHC protein expression was not present during leucine starvation. However, research (Azizi et al., 2016) found that lysine had a greater impact on myocyte growth than leucine, with a reduction in the expression of the Myf5, myogenin, PCNA, and MHC genes expression.

Future studies on the importance of this specific AA in the diets of particular species will be triggered by the AA deficiency discovered in fish. The *in vivo* investigation in rainbow trout by Alami-Durante et al. (2010) reveals that variations in dietary EAA have considerable impacts on fish myogenesis, and all researchers agreed that AA has important effects on myocyte function.

10.10 FACTORS AFFECTING LEUCINE REQUIREMENT

Several factors can affect the leucine requirements of fishes. These factors influence the amount of leucine that an animal needs for optimal growth, health, and performance. Some of the key factors affecting leucine requirements include (Figure 10.3):

10.10.1 SPECIES AND LIFE STAGE

Different species of animals have varying metabolic rates, growth patterns, and physiological needs. Leucine requirements can vary significantly between species, and they can also change as animals transition through different life stages, such as larval, juvenile, and adult stages.

FIGURE 10.3 Factors affecting leucine requirement in different fishes.

10.10.2 GROWTH RATE

Animals that are growing rapidly generally require more nutrients, including leucine, to support the synthesis of new tissues. So, we can say faster growth rates typically correspond to higher leucine requirements.

10.10.3 DIET COMPOSITION

The overall AA composition of the diet, including the balance of other EAAs, can influence the leucine requirements. An imbalance in AAs may lead to decreased leucine utilization.

10.10.4 PROTEIN CONTENT

Diets with higher protein content often require higher levels of EAAs, like leucine, to maintain the AA balance. Low-protein diets might have relatively lower leucine requirements.

10.10.5 ENERGY INTAKE

The energy intake of animals is closely tied to their metabolism and growth. Adequate energy supply is important for utilizing leucine effectively for various physiological processes.

10.10.6 ENVIRONMENTAL CONDITIONS

Environmental stressors such as temperature fluctuations, water quality, and disease challenges can impact the metabolic rate of animals. Stressful conditions may increase leucine requirements due to the higher demand for energy and tissue repair.

10.10.7 HEALTH AND IMMUNE STATUS

Animals under immune challenge or experiencing health issues may have altered nutrient requirements, including leucine. Immune responses and tissue repair processes can increase the demand for AAs.

10.10.8 GENETICS

Genetic factors can influence an animal's metabolism, growth potential, and nutrient utilization efficiency. Different genetic strains or breeds might have varying leucine requirements.

10.10.9 FEEDING FREQUENCY AND FEEDING REGIMEN

Feeding frequency and meal size can impact nutrient utilization. Animals fed more frequently might have better AA utilization compared to those fed infrequently.

10.10.10 AQUACULTURE SYSTEM

The type of aquaculture system (e.g., pond, recirculating system) can influence leucine requirements. Water quality, stocking density, and other factors inherent to the system can affect nutrient needs.

10.10.11 FEED INGREDIENTS

The AA composition of the feed ingredients employed in formulating the diet can influence leucine requirements. Ingredients that are rich in leucine might decrease the overall requirement. Due to the complexity of these factors, determining accurate leucine requirements requires a combination of empirical studies, growth trials, and careful observation to optimize the nutritional composition of animal diets.

10.11 FUTURE PERSPECTIVES

To fulfill the protein demand of the world, the nutritionist/aquaculturist will provide information about the optimum nutritional diet for particular fish in the future. In the era of development of new innovative technology and advancement in the current technology of aquaculture, systems will aim to find out the proper nutritional requirement of the particular fish species, and their relations are very much needed by the aquaculture nutritionists/researchers. To know the optimum requirement of the protein/AAs for particular fish species, which is an economically important fish for any region, scientists will give more effort in the future in the development and upgradation of the fish nutrition and aquaculture department.

10.12 CONCLUSION

To maximize development, protein accretion, and fish health, the dietary content of BCAAs must be adjusted in accordance with the needs of the fish. Overall, dietary leucine levels may enhance the nutritional composition of different fishes' muscles, including their meat protein, lipid, collagen, fatty acid, and AA compositions, and may also enhance the texture of their fillets by enhancing their ability to hold water and myofiber density. Leucine supplementation improved whole-body AA retention, increased growth and feed utilization, decreased body condition indicators, and encouraged protein synthesis. The optimum dietary leucine requirement has been determined for various fish species, showing a wide range of values ranging from 11.5 to 91.0 g/kg of DP to date.

KEYWORDS

- **aquafeed formulation**
- **branched-chain amino acids**
- **essential amino acids**
- **leucine in fish nutrition**
- **protein synthesis and metabolism**
- **sustainable aquaculture**

REFERENCES

Abidi, S. F., & Khan, M. A. (2007). Dietary leucine requirement of fingerling Indian major carp, *Labeo rohita* (Hamilton). *Aquaculture Research, 38*(5), 478–486.

Ahmad, I., Ahmed, I., & Dar, N. A. (2021). Effects of dietary leucine levels on growth performance, hematobiochemical parameters, liver profile, intestinal enzyme activities and target of rapamycin signalling pathway related gene expression in rainbow trout, *Oncorhynchus mykiss* fingerlings. *Aquaculture Nutrition, 27*(6), 1837–1852.

Ahmad, I., Ahmed, I., Fatma, S., & Peres, H. (2021). Role of branched chain amino acids on growth, physiology and metabolism of different fish species: A review. *Aquaculture Nutrition*. https://doi.org/10.1111/anu.13267.

Ahmed, I., & Khan, M. A. (2004). Dietary lysine requirement of fingerling Indian major carp, *Cirrhinus mrigala* (Hamilton). *Aquaculture, 235*(1–4), 499–511. https://doi.org/10.1016/j.aquaculture.2003.12.009.

Ahmed, I., & Khan, M. A. (2006). Dietary branched-chain amino acid valine, isoleucine and leucine requirements of fingerling Indian major carp, *Cirrhinus mrigala* (Hamilton). *British Journal of Nutrition, 96*(3), 450–460. https://doi.org/10.1079/BJN20061845.

Akiyama, T. (1987). *Studies on the Essential Amino Acids and Scoliosis Caused by Tryptophan Deficiency of Chum Salmon Fry* (Ph.D. dissertation). University of Kyushu, Japan.

Akiyama, T., Oohara, I., & Yamamoto, T. (1997). Comparison of essential amino acid requirements with A/E ratio among fish species. *Fisheries Science, 63*(6), 963–970.

Alami-Durante, H., Wrutniak-Cabello, C., Kaushik, S. J., & Médale, F. (2010). Skeletal muscle cellularity and expression of myogenic regulatory factors and myosin heavy chains in rainbow trout (*Oncorhynchus mykiss*): Effects of changes in dietary plant protein sources and amino acid profiles. *Comparative Biochemistry and Physiology Part A: Molecular & Integrative Physiology, 156*(4), 561–568. https://doi.org/10.1016/j.cbpa.2010.04.015.

Areta, J. L., Hawley, J. A., Ye, J., Chan, M. H. S., & Coffey, V. G. (2014). Increasing leucine concentration stimulates mechanistic target of rapamycin signaling and cell growth in C2C12 skeletal muscle cells. *Nutrition Research, 34*(11), 1000–1007. https://doi.org/10.1016/j.nutres.2014.09.011.

Averous, J., Gabillard, J. C., Seiliez, I., & Dardevet, D. (2012). Leucine limitation regulates *myf5* and *myoD* expression and inhibits myoblast differentiation. *Experimental Cell Research, 318*(2), 217–227. https://doi.org/10.1016/j.yexcr.2011.10.015.

Azizi, S., Nematollahi, M. A., Mojazi Amiri, B., Vélez, E. J., Lutfi, E., Navarro, I., Capilla, E., & Gutiérrez, J. (2016). Lysine and leucine deficiencies affect myocytes development and IGF signaling in gilthead sea bream (*Sparus aurata*). *PLOS ONE, 11*(1), e0147618. https://doi.org/10.1371/journal.pone.0147618.

Benakappa, S., & Varghese, T. J. (2003). Isoleucine, leucine and valine requirement of juvenile Indian major carp, *Cirrhinus mrigala* (Bloch, 1975). *Acta Ichthyologica et Piscatoria, 33*(2), 161–172.

Borlongan, I. G., & Coloso, R. M. (1993). Requirements of juvenile milkfish (*Chanos chanos* Forsskal) for essential amino acids. *The Journal of Nutrition, 123*(1), 125–132. https://doi.org/10.1093/jn/123.1.125.

Bower, N. I., & Johnston, I. A. (2010a). Transcriptional regulation of the IGF signaling pathway by amino acids and insulin-like growth factors during myogenesis in Atlantic salmon. *PLOS ONE, 5*(5), e11100. https://doi.org/10.1371/journal.pone.0011100.

Bower, N. I., & Johnston, I. A. (2010b). Paralogs of Atlantic salmon myoblast determination factor genes are distinctly regulated in proliferating and differentiating myogenic cells. *American Journal of Physiology-Regulatory, Integrative and Comparative Physiology, 298*(6), R1615–R1626. https://doi.org/10.1152/ajpregu.00114.2010.

Brown, P. B. (1995). Using whole-body amino acid patterns and quantitative requirements to rapidly develop diets for new species such as striped bass (*Morone saxatilis*). *Journal of Applied Ichthyology, 11*(3, 4), 342–346.

Cantó, C., Jiang, L. Q., Deshmukh, A. S., et al. (2010). Interdependence of AMPK and SIRT1 for metabolic adaptation to fasting and exercise in skeletal muscle. *Cell Metabolism, 11*(3), 213–219. https://doi.org/10.1016/j.cmet.2010.02.006.

Castillo, S., & Gatlin, D. M. III. (2018). Dietary requirements for leucine, isoleucine and valine (branched-chain amino acids) by juvenile red drum *Sciaenops ocellatus*. *Aquaculture Nutrition, 24*(3), 1056–1065. https://doi.org/10.1111/anu.12644.

Chance, R. E., Mertz, E. T., & Halver, J. E. (1964). Nutrition of salmonid fishes. Isoleucine, leucine, valine and phenylalanine requirements of Chinook salmon and interrelations between isoleucine and leucine for growth. *The Journal of Nutrition, 83*(2), 177–185.

Chen, X. L., Huang, Z. Q., Chen, D. W., Yang, T., & Liu, G. (2013). MicroRNA-27a is induced by leucine and contributes to leucine-induced proliferation promotion in C2C12 cells. *International Journal of Molecular Sciences, 14*(7), 14076–14084.

Choo, P. S., Smith, T. K., Cho, C. Y., & Ferguson, H. W. (1991). Dietary excesses of leucine influence growth and body composition of rainbow trout. *The Journal of Nutrition, 121*(12), 1932–1939.

Conceição, L. E. C., Meeren, T., Verreth, J. A. J., Evjen, M. S., Houlihan, D. F., & Fyhn, H. J. (1997). Amino acid metabolism and protein turnover in larval turbot (*Scophthalmus maximus*) fed natural zooplankton or *Artemia*. *Marine Biology, 129*, 255–265.

Coolican, S. A., Samuel, D. S., Ewton, D. Z., McWade, F. J., & Florini, J. R. (1997). The mitogenic and myogenic actions of insulin-like growth factors utilize distinct signaling pathways. *Journal of Biological Chemistry, 272*(10), 6653–6662. https://doi.org/10.1074/jbc.272.10.6653.

Cowey, C. B. (1980). Protein metabolism in fish. In P. J. Buttery & D. B. Lindsay (Eds.), *Protein Deposition in Animals* (pp. 271–288). Butterworths.

Cowey, C. B., & Walton, M. J. (1988). Studies on the uptake of (14C) amino acids derived from both dietary (14C) protein and dietary (14C) amino acids by rainbow trout, *Salmo gairdneri* Richardson. *Journal of Fish Biology, 33*, 293–305.

D'Mello, J. P. F. (1982). A comparison of two empirical methods of determining amino acid requirements. *World's Poultry Science Journal, 38*(2), 114–119.

Dabrowski, K., Arslan, M., Terjesen, B. F., & Zhang, Y. (2007). The effects of dietary indispensable amino acid imbalances on feed intake: Is there a sensing of deficiency and neural signaling present in fish? *Aquaculture, 268*, 136–142.

Dabrowski, K., Lee, K. J., & Rinchard, J. (2003). The smallest vertebrate, teleost fish, can utilize synthetic dipeptide-based diets. *Journal of Nutrition, 133*(12), 4225–4229.

Davies, S. J., Morris, P. C., & Baker, R. T. M. (1997). Partial substitution of fish meal and full-fat soya bean meal with wheat gluten and influence of lysine supplementation in diets for rainbow trout, *Oncorhynchus mykiss* (Walbaum). *Aquaculture Research, 28*, 317–328.

De la Higuera, M., Garz6n, A., Hidalgo, M. C., Perag6n, J., Cardenete, G., & Lupiáñez, J. A. (1998). Influence of temperature and dietary-protein supplementation either with free or coated lysine on the fractional protein-turnover rates in the white muscle of carp. *Fish Physiology and Biochemistry, 18*(1), 85–95.

Deng, D. F., Dominy, W., Ju, Z. Y., Koshio, S., Murashige, R., & Wilson, R. P. (2010). Dietary lysine requirement of juvenile Pacific threadfin (*Polydactylus sexfilis*). *Aquaculture, 308*(1), 44–48.

Deng, Y. P., Jiang, W. D., Liu, Y., Jiang, J., Kuang, S. Y., Tang, L., Wu, P., Zhang, Y. A., Feng, L., & Zhou, X. Q. (2014). Differential growth performance, intestinal antioxidant status and relative expression of Nrf2 and its target genes in young grass carp (*Ctenopharyngodon idella*) fed with graded levels of leucine. *Aquaculture, 434*, 66–73. https://doi.org/10.1016/j.aquaculture.2014.07.026.

Duan, Y. H., Li, F. N., Li, Y. H., et al. (2016). The role of leucine and its metabolites in protein and energy metabolism. *Amino Acids, 48*(1), 41–51.

Farhat, L., & Khan, M. A. (2014). Response of fingerling stinging catfish, *Heteropneustes fossilis* (Bloch), to varying levels of dietary leucine in relation to growth, feed conversion, protein utilization, leucine retention and blood parameters. *Aquaculture Nutrition, 20*(3), 291–302. https://doi.org/10.1111/anu.12077.

Fernstrom, J. D. (2005). 4th amino acid assessment workshop: Branched-chain amino acids and brain function. *Journal of Nutrition, 135*(6 Suppl), 1539S–1546S. https://doi.org/10.1093/jn/135.6.1539S.

Gan, L., Zhou, L. L., Li, X. X., & Yue, Y. R. (2016). Dietary leucine requirement of juvenile Nile tilapia, *Oreochromis niloticus*. *Aquaculture Nutrition, 22*(5), 1040–1046. https://doi.org/10.1111/anu.12353.

García de la Serrana, D., Codina, M., Capilla, E., Jiménez-Amilburu, V., Navarro, I., Du, S. J., & Gutiérrez, J. (2014). Characterization and expression of myogenesis regulatory factors during *in vitro* myoblast development and *in vivo* fasting in the gilthead sea bream (*Sparus aurata*). *Comparative Biochemistry and Physiology Part A: Molecular & Integrative Physiology, 167*, 90–99. https://doi.org/10.1016/j.cbpa.2013.10.020.

Giri, S. S., Sen, S. S., Chi, C., Kim, H. J., Yun, S., Park, S. C., & Sukumaran, V. (2015). Effect of dietary leucine on growth parameters and expression of antioxidant, immune, and inflammatory genes in the head kidney of *Labeo rohita* fingerlings. *Veterinary Immunology and Immunopathology, 167*, 36–43. https://doi.org/10.1016/j.vetimm.2015.07.004.

Hardie, D. G. (2007). AMP-activated/SNF1 protein kinases: Conserved guardians of cellular energy. *Nature Reviews Molecular Cell Biology, 8*(10), 774–785.

Harikrishnan, R., Balasundaram, C., & Heo, M. S. (2010). *Lactobacillus sakei* BK19 enriched diet enhances the immunity status and disease resistance to streptococcosis infection in kelp grouper, *Epinephelus bruneus*. *Fish & Shellfish Immunology, 29*(6), 1037–1043.

Hevrøy, E. M., El-Mowafi, A., Taylor, R. G., Olsvik, P. A., Norberg, B., & Espe, M. (2007). Lysine intake affects gene expression of anabolic hormones in Atlantic salmon, *Salmo salar*. *General and Comparative Endocrinology, 152*(1), 39–46.

Hochachka, P. W., & Mommsen, T. P. (Eds.). (1995). *Biochemistry and Molecular Biology of Fishes* (Vol. 2). Elsevier.

Houlihan, D. F., Hall, S. J., & Gray, C. (1989). Effects of ration on protein turnover in cod. *Aquaculture, 79*(1), 103–110. https://doi.org/10.1016/0044-8486(89)90450-X.

Hughes, S. G., Rumsey, G. L., & Nesheim, M. C. (1983). Dietary requirements for essential branched-chain amino acids by lake trout. *Transactions of the American Fisheries Society, 112*(5), 812–817.

Jiménez-Amilburu, V., Salmerón, C., Codina, M., Navarro, I., Capilla, E., & Gutiérrez, J. (2013). Insulin-like growth factors effects on the expression of myogenic regulatory factors in gilthead sea bream muscle cells. *General and Comparative Endocrinology, 188*, 151–158. https://doi.org/10.1016/j.ygcen.2013.02.033.

Jobgen, W. S., Fried, S. K., Fu, W. J., Meininger, C. J., & Wu, G. (2006). Regulatory role for the arginine-nitric oxide pathway in metabolism of energy substrates. *Journal of Nutritional Biochemistry, 17*(9), 571–588. https://doi.org/10.1016/j.jnutbio.2005.12.001.

Johnston, I. A. (1999). Muscle development and growth: Potential implications for flesh quality in fish. *Aquaculture, 177*(1–4), 99–115. https://doi.org/10.1016/S0044-8486(99)00072-1.

Jürss, K., & Bastrop, R. (1995). Amino acid metabolism in fish. In P. W. Hochachka & T. P. Mommsen (Eds.), *Biochemistry and Molecular Biology of Fishes* (Vol. 4, pp. 159–189). Elsevier.

Kaushik, S. J., Fauconneau, B., Terrier, L., & Gras, J. (1988). Arginine requirement and status assessed by different biochemical indices in rainbow trout (*Salmo gairdneri* R.). *Aquaculture, 70*(1), 75–95.

Kim, J. Y., Burghardt, R. C., Wu, G., Johnson, G. A., Spencer, T. E., & Bazer, F. W. (2011a). Select nutrients in the ovine uterine lumen: VII. Effects of arginine, leucine, glutamine, and glucose on trophectoderm cell signaling, proliferation, and migration. *Biology of Reproduction, 84*(1), 62–69.

Kim, J. Y., Burghardt, R. C., Wu, G., Johnson, G. A., Spencer, T. E., & Bazer, F. W. (2011b). Select nutrients in the ovine uterine lumen: IX. Differential effects of arginine, leucine, glutamine and glucose on interferon tau, ornithine decarboxylase and nitric oxide synthase in the ovine conceptus. *Biology of Reproduction, 84*(6), 1139–1146.

Kim, S. W., Mateo, R. D., Yin, Y. L., & Wu, G. Y. (2007). Functional amino acids and fatty acids for enhancing production performance of sows and piglets. *Asian-Australasian Journal of Animal Sciences, 20*(2), 295–306.

Krebs, H. A. (1972). Some aspects of the regulation of fuel supply in omnivorous animals. *Advances in Enzyme Regulation, 10*, 397–420.

Lang, C. H., & Frost, R. A. (2005). Endotoxin disrupts the leucine-signaling pathway involving phosphorylation of mTOR, 4E-BP1, and S6K1 in skeletal muscle. *Journal of Cellular Physiology, 203*(1), 144–155. https://doi.org/10.1002/jcp.20207.

Lansard, M., Panserat, S., Plagnes-Juan, E., Seiliez, I., & Skiba-Cassy, S. (2010). Integration of insulin and amino acid signals that regulate hepatic metabolism-related gene expression in rainbow trout: Role of TOR. *Amino Acids, 39*(3), 801–810. https://doi.org/10.1007/s00726-010-0533-3.

Lansard, M., Panserat, S., Seiliez, I., Polakof, S., Plagnes-Juan, E., Geurden, I., Médale, F., Kaushik, S., Corraze, G., & Skiba-Cassy, S. (2009). Hepatic protein kinase B (Akt)–target of rapamycin (TOR)-signaling pathways and intermediary metabolism in rainbow trout (*Oncorhynchus mykiss*) are not significantly affected by feeding plant-based diets. *British Journal of Nutrition, 102*(11), 1564–1573. https://doi.org/10.1017/S000711450999095X.

Lei, J., Feng, D. Y., Zhang, Y. L., Zhao, F. Q., Wu, Z. L., San Gabriel, A., Fujishima, Y., Uneyama, H., & Wu, G. (2012). Nutritional and regulatory role of branched-chain amino acids in lactation. *Frontiers in Bioscience, 17*, 2725–2739.

Li, F., Yin, Y., Tan, B., Kong, X., & Wu, G. (2011). Leucine nutrition in animals and humans: mTOR signaling and beyond. *Amino Acids, 41*(5), 1185–1193. https://doi.org/10.1007/s00726-011-0983-2.

Li, F., Yin, Y., Tan, B., Kong, X., & Wu, G. (2011a). Leucine nutrition in animals and humans: mTOR signaling and beyond. *Amino Acids, 41*(5), 1185–1193.

Li, P., Yin, Y. L., Li, D. F., Kim, S. W., & Wu, G. Y. (2007). Amino acids and immune function. *British Journal of Nutrition, 98*(2), 237–252.

Li, S. A., Jiang, W. D., Feng, L., Liu, Y., Wu, P., Jiang, J., Kuang, S. Y., Tang, L., Tang, W. N., Zhang, Y. A., et al. (2018). Dietary myo-inositol deficiency decreased intestinal immune function related to NFκB and TOR signaling in the intestine of young grass carp (*Ctenopharyngodon idella*). *Fish & Shellfish Immunology, 76*, 333–346.

Li, X. L., Rezaei, R., Li, P., & Wu, G. (2011b). Composition of amino acids in feed ingredients for animal diets. *Amino Acids, 40*(4), 1159–1168.

Li, Y., Ai, Q., Mai, K., Xu, W., Cheng, Z., & He, Z. (2010). Dietary leucine requirement for juvenile large yellow croaker *Pseudosciaena crocea* (Richardson, 1846). *Journal of Ocean University of China, 9*(3), 371–375.

Liang, D., Yang, Q., Tan, B., Dong, X., Chi, S., Liu, H., & Zhang, S. (2020). Dietary vitamin A deficiency reduces growth performance, immune function of intestine, and alters tight junction proteins of intestine for juvenile hybrid grouper (*Epinephelus fuscoguttatus* ♀ × *Epinephelus lanceolatus* ♂). *Fish & Shellfish Immunology, 107*, 346–356.

Liang, H., Mokrani, A., Ji, K., Ge, X., Ren, M., Xie, J., Liu, B., Xi, B., & Zhou, Q. (2018). Dietary leucine modulates growth performance, Nrf2 antioxidant signaling pathway and immune response of juvenile blunt snout bream (*Megalobrama amblycephala*). *Fish & Shellfish Immunology, 73*, 57–65.

Lin, W. L., Yang, X. Q., Li, L. H., et al. (2016). Effect of ultrastructure on changes of textural characteristics between crisp grass carp (*Ctenopharyngodon idellus* c. et v.) and grass carp (*Ctenopharyngodon idellus*) inducing heating treatment. *Journal of Food Science, 81*(2), E404–E411.

Luo, Z., Liu, Y., Mai, K., Tian, L., Yang, H., Tan, X., & Liu, D. (2005). Dietary l-methionine requirement of juvenile grouper *Epinephelus coioides* at a constant dietary cystine level. *Aquaculture, 249*(1–4), 409–418.

Meijer, A. J. (2003). Amino acids as regulators and components of nonproteinogenic pathways. *Journal of Nutrition, 133*(6), 2057S–2062S. https://doi.org/10.1093/jn/133.6.2057S.

Mero, A. (1999). Leucine supplementation and intensive training. *Sports Medicine, 27*(6), 347–358. https://doi.org/10.2165/00007256-199927060-00001.

Miles, R., & Chapman, F. (2007). Concept of Ideal Protein in formulation of aquaculture feeds: FA144/FA144, 3/2007. *EDIS*. https://doi.org/10.32473/edis-fa144-2007.

Montserrat, N., Capilla, E., Navarro, I., & Gutiérrez, J. (2012). Metabolic effects of insulin and IGFs on gilthead sea bream (*Sparus aurata*) muscle cells. *Frontiers in Endocrinology, 3*, 55. https://doi.org/10.3389/fendo.2012.00055.

Montserrat, N., Gabillard, J. C., Capilla, E., Navarro, M. I., & Gutiérrez, J. (2007). Role of insulin, insulin-like growth factors, and muscle regulatory factors in the compensatory growth of the trout (*Oncorhynchus mykiss*). *General and Comparative Endocrinology, 150*(3), 462–472.

Montserrat, N., Sánchez-Gurmaches, J., García de la Serrana, D., Navarro, I., & Gutiérrez, J. (2007). IGF-I binding and receptor signal transduction in primary cell culture of muscle cells of gilthead sea bream: Changes throughout *in vitro* development. *Cell and Tissue Research, 330*(3), 503–513.

Moon, H. Y., & Gatlin III, D. M. (1991). Total sulfur amino acid requirement of juvenile red drum, *Sciaenops ocellatus*. *Aquaculture, 95*(1, 2), 97–106.

Murai, T., Ogata, H., Hirasawa, Y., Akiyama, T., & Nose, T. (1987). Portal absorption and hepatic uptake of amino acids in rainbow trout force-fed complete diets containing casein or crystalline amino acids. *Bulletin of the Japanese Society of Scientific Fisheries, 53*(10), 1847–1859.

Murthy, H. S., & Varghese, T. J. (1997). Dietary requirement of the Indian major carp, *Labeo rohita*, for the essential amino acid leucine. *Mysore Journal of Agricultural Sciences, 31*(4), 348–351.

Nakashima, K., Yakabe, Y., Ishida, A., Yamazaki, M., & Abe, H. (2007). Suppression of myofibrillar proteolysis in chick skeletal muscles by alpha-ketoisocaproate. *Amino Acids, 33*(4), 499–503. https://doi.org/10.1007/s00726-006-0404-0.

National Research Council (NRC). (1993). *Nutrient Requirements of Fish*. National Academy Press.

National Research Council (NRC). (2011). *Nutrient Requirements of Fish* (pp. 1–392). National Academy Press.

National Research Council (NRC). (2011). *Nutrient Requirements of Fish and Shrimp* (p. 376). National Academy Press.

Ngamsnae, P., Silva, D. E. S. S., & Gunasekera, R. M. (1999). Arginine and phenylalanine requirement of juvenile silver perch *Bidyanus bidyanus* and validation of the use of body amino acid composition for estimating individual amino acid requirement. *Aquaculture Nutrition, 5*(3), 173–180.

Niu, X., Qian, X., Feng, H., Yi, K., Li, D., Chen, W., & Ye, J. (2021). Growth and metabolic responses of grouper juveniles (*Epinephelus coioides*) fed diets containing varying levels of leucine. *Aquaculture, 534*, 736281.

Norton, L. E., & Layman, D. K. (2006). Leucine regulates translation initiation of protein synthesis in skeletal muscle after exercise. *Journal of Nutrition, 136*(2), 533S–537S. https://doi.org/10.1093/jn/136.2.533S.

Nose, T. (1979). Summary report on the requirements of essential amino acids for carp. In K. Tiews & J. E. Halver (Eds.), *Finfish Nutrition and Fish Feed Technology* (pp. 145–156). Heenemann GmbH.

Ogino, C. (1980). Requirements of carp and rainbow trout for essential amino acids. *Bulletin of the Japanese Society of Fisheries Science, 46*(2), 171–174. https://doi.org/10.2331/suisan.46.171.

Ostaszewska, T., Dabrowski, K., Kamaszewski, M., Grochowski, P., Verri, T., Rzepkowska, M., & Wolnicki, J. (2010). The effect of plant protein-based diet supplemented with dipeptide or free amino acids on digestive tract morphology and PepT1 and PepT2 expressions in common carp (*Cyprinus carpio* L.). *Comparative Biochemistry and Physiology Part A, 157*(2), 158–169.

Pack, M., & Rodehutscord, M. (1995). Extending the ideal protein concept to fish nutrition. *Feed Mix, 3*, 23–25.

Panten, U., Christians, J., Kriegstein, E., Poser, W., & Hasselblatt, A. (1974). Studies on the mechanism of l-leucine- and alpha-ketoisocaproic acid-induced insulin release from perifused isolated pancreatic islets. *Diabetology, 10*(2), 149–154.

Rahimnejad, S., & Lee, K. J. (2013). Dietary valine requirement of juvenile red sea bream *Pagrus major*. *Aquaculture, 416–417*, 212–218. https://doi.org/10.1016/j.aquaculture.2013.09.026.

Rajalakshmi, S., & Mohandas, A. (2005). Copper-induced changes in tissue enzyme activity in a freshwater mussel. *Ecotoxicology and Environmental Safety, 62*(2), 140–143.

Ramasamy Harikrishnansupa, C. B. S. M., & Heosupc Sup, M. S. (2010). Effect of a mixed herb-enriched diet on the innate immune response and disease resistance of *Paralichthys olivaceus* against *Philasterides dicentrarchi* infection. *Journal of Aquatic Animal Health, 22*(4), 235–243.

Ravi, J., & Devaraj, K. V. (1991). Quantitative essential amino acid requirements for growth of catla, *Catla catla* (Hamilton). *Aquaculture, 96*(3–4), 281–291. https://doi.org/10.1016/0044-8486(91)90158-4.

Ren, M., Habte-Tsion, H. M., Liu, B., Miao, L., Ge, X., Xie, J., Liang, H., Zhou, Q., & Pan, L. (2015). Dietary leucine level affects growth performance, whole body composition, plasma parameters and relative expression of TOR and TNF-α in juvenile blunt snout bream, *Megalobrama amblycephala*. *Aquaculture, 448*, 162–168. https://doi.org/10.1016/j.aquaculture.2015.06.008.

Rhoads, J. M., Cori, B. A., Harrell, R., Niu, X. M., Gatlin, L., Phillips, O., Blikslager, A., Moeser, A., Wu, G., & Odle, J. (2007). Intestinal ribosomal p70S6K signaling is increased in piglet rotavirus enteritis. *American Journal of Physiology-Gastrointestinal and Liver Physiology, 292*(3), G913–G922. https://doi.org/10.1152/ajpgi.00468.2006.

Rius-Francino, M., Acerete, L., Jiménez-Amilburu, V., Capilla, E., Navarro, I., & Gutiérrez, J. (2011). Differential effects on proliferation of GH and IGFs in sea bream (*Sparus aurata*) cultured myocytes. *General and Comparative Endocrinology, 172*(1), 44–49. https://doi.org/10.1016/j.ygcen.2011.03.024.

Rolland, M., Dalsgaard, J., Holm, J., Gómez-Requeni, P., & Skov, P. V. (2015). Dietary methionine level affects growth performance and hepatic gene expression of GH-IGF system and protein turnover regulators in rainbow trout (*Oncorhynchus mykiss*) fed plant protein-based diets. *Comparative Biochemistry and Physiology Part B, 181*, 33–41. https://doi.org/10.1016/j.cbpb.2014.11.009.

Rønnestad, I., & Conceição, L. E. C. (2005). Aspects of protein and amino acids digestion and utilization by marine fish larvae. In J. M. Starck & T. Wang (Eds.), *Physiological and Ecological Adaptations to Feeding in Vertebrates* (pp. 389–416). Science Publishers.

Rønnestad, I., Conceicao, L. E. C., Aragai, C., & Dinis, M. T. (2000). Free amino acids are absorbed faster and assimilated more efficiently than protein in postlarval Senegal sole (*Solea senegalensis*). *Journal of Nutrition, 130*(11), 2809–2812.

Santiago, C. B., & Lovell, R. T. (1988). Amino acid requirements for growth of Nile tilapia. *Journal of Nutrition, 118*(12), 1540–1546.

Saurabh, S., & Sahoo, P. K. (2008). Lysozyme: An important defense molecule of fish innate immune system. *Aquaculture Research, 39*(3), 223–239.

Segovia-Quintero, M. A., & Reigh, R. C. (2004). Coating crystalline methionine with tripalmitin-polyvinyl alcohol slows its absorption in the intestine of Nile tilapia, *Oreochromis niloticus*. *Aquaculture, 238*(1–4), 355–367.

Sharf, Y., & Khan, M. A. (2023). Dietary leucine requirement of fingerling *Channa punctatus* (Bloch) based on growth, feed conversion and leucine retention efficiency, hematological parameters, antioxidant and intestinal enzyme activities. *Amino Acids, 55*(3), 451–468. https://doi.org/10.1007/s00726-023-03240-1.

Siwicki, A. K., Morand, M., Fuller, J., Nissen, S., Goryczko, K., Ostaszewski, P., et al. (2003). Influence of feeding the leucine metabolite beta-hydroxy-beta-methylbutyrate (HMB) on the non-specific cellular and humoral defense mechanisms of rainbow trout (*Oncorhynchus mykiss*). *Journal of Applied Ichthyology, 19*(1), 44–48.

Tacon, A. G. (1989). *Nutrición y Alimentación De Peces Y Camarones Cultivados: Manual De Capacitación* (pp. 1–572). Organización de las Naciones Unidas para la Agricultura y la Alimentación.

Tan, X., Lin, H., Huang, Z., Zhou, C., Wang, A., Qi, C., & Zhao, S. (2016). Effects of dietary leucine on growth performance, feed utilization, non-specific immune responses and gut morphology of juvenile golden pompano *Trachinotus ovatus*. *Aquaculture, 465*, 100–107.

Tang, L., Feng, L., Sun, C. Y., Chen, G. F., Jiang, W. D., Hu, K., Liu, Y., Jiang, J., Li, S. H., Kuang, S. Y., & Zhou, X. Q. (2013). Effect of tryptophan on growth, intestinal enzyme activities and TOR gene expression in juvenile jian carp (*Cyprinus carpio* var. Jian): Studies *in vivo* and *in vitro*. *Aquaculture, 412–413*, 25–33. https://doi.org/10.1016/j.aquaculture.2013.07.002.

Thoman, E. S., Davis, D. A., & Arnold, C. R. (1999). Evaluation of growout diets with varying protein and energy levels for red drum (*Sciaenops ocellatus*). *Aquaculture, 176*(4), 343–353. https://doi.org/10.1016/S0044-8486(99)00118-0.

Twibell, R. G., Griffin, M. E., Martin, B., Price, J., & Brown, P. B. (2003). Predicting dietary essential amino acid requirements for hybrid striped bass. *Aquaculture Nutrition, 9*(6), 373–381. https://doi.org/10.1046/j.1365-2095.2003.00266.x.

Uribe, C., Folch, H., Enriquez, R., & Moran, G. (2011). Innate and adaptive immunity in teleost fish: A review. *Veterinární Medicína, 56*(10), 486–503.

Van Nguyen, M., Rønnestad, I., Buttle, L., Van Lai, H., & Espe, M. (2014). Imbalanced lysine to arginine ratios reduced performance in juvenile cobia (*Rachycentron canadum*) fed high plant protein diets. *Aquaculture Nutrition, 20*(1), 25–35.

Van Waarde, A. (1983). Aerobic and anaerobic ammonia production by fish. *Comparative Biochemistry and Physiology Part A, 74*(4), 675–684.

Vélez, E. J., Lutfi, E., Jiménez-Amilburu, V., Riera-Codina, M., Capilla, E., Navarro, I., et al. (2014). IGF-I and amino acids effects through TOR signaling on proliferation and differentiation of gilthead sea bream cultured myocytes. *General and Comparative Endocrinology, 205*, 296–304. https://doi.org/10.1016/j.ygcen.2014.05.024.

Villamar, D. F., & Langdon, C. (1993). Delivery of dietary components to larval shrimp (*Penaeus vannamei*) by means of complex microcapsules. *Marine Biology, 115*(4), 635–642.

Walton, M. J., & Cowey, C. B. (1981). Distribution and some kinetic properties of serine catabolizing enzymes in rainbow trout *Salmo gairdnerii*. *Comparative Biochemistry and Physiology Part B, 68B*(1), 147–150.

Wang, M., Guo, H., Huang, Y., Liu, W., Wang, X., Xiao, K., Xiong, W., Hua, H., Li, X., & Jiang, G. (2022). Dietary leucine supplementation improves muscle fiber growth and development by activating AMPK/Sirt1 pathway in blunt snout bream (*Megalobrama amblycephala*). *Aquaculture Nutrition*, Article ID 7285851, 16 pages. https://doi.org/10.1155/2022/7285851.

Wilson, P. (1993). Amino acid nutrition of fish: A new method of estimating requirement values. In Proceedings of the 20th US-Japan Symposium of Aquaculture Nutrition (pp. 49–54). NOAA, Department of Commerce.

Wilson, R. P. (1985). Amino acid and protein requirements of fish. In C. B. Cowey, A. M. Mackie, & J. G. Bell (Eds.), *Nutrition and Feeding in Fish* (pp. 1–15). Academic Press.

Wilson, R. P., & Cowey, C. B. (1985). Amino acid composition of whole-body tissue of rainbow trout and Atlantic salmon. *Aquaculture, 48*(4), 373–376.

Wilson, R. P., & Halver, J. E. (1986). Protein and amino acid requirements of fishes. *Annual Review of Nutrition, 6*, 225–244. https://doi.org/10.1146/annurev.nu.06.070186.001301.

Wilson, R. P., & Poe, W. E. (1985). Relationship of whole body and egg essential amino acid patterns to amino acid requirement patterns in channel catfish, *Ictalurus punctatus*. *Comparative Biochemistry and Physiology Part B, 80*(2), 385–388.

Wilson, R. P., Poe, W. E., & Robinson, E. H. (1980). Leucine, isoleucine, valine and histidine requirements of fingerling channel catfish. *Journal of Nutrition, 110*(3), 627–633.

Wu, C., Chen, L., Lu, Z., Chu, Y., Li, L., Wang, M., Zhang, G., Zhang, M., & Ye, J. (2017). The effects of dietary leucine on the growth performances, body composition, metabolic abilities and innate immune responses in black carp *Mylopharyngodon piceus*. *Fish & Shellfish Immunology, 67*, 419–428.

Wu, G. (2013). Functional amino acids in nutrition and health. *Amino Acids, 45*(3), 407–411. https://doi.org/10.1007/s00726-013-1500-6.

Wu, G. (2013a). *Amino Acids: Biochemistry and Nutrition*. CRC Press.

Wu, G. Y., Bazer, F. W., Davis, T. A., Kim, S. W., Li, P., Rhoads, J. M., Satterfield, M. C., Smith, S. B., Spencer, T. E., & Yin, Y. L. (2009). Arginine metabolism and nutrition in growth, health and disease. *Amino Acids, 37*(1), 169–175.

Yang, H., Rahman, M. M., Li, X., Sharifuzzaman, S. M., & Leng, X. (2022). Dietary leucine requirement of juvenile largemouth bass (*Micropterus salmoides*) based on growth, nutrient utilization and growth-related gene analyses. *Aquaculture, 555*, 738207.

Yang, S. D., Liu, F. G., & Liou, C. H. (2011). Assessment of dietary lysine requirement for silver perch (*Bidyanus bidyanus*) juveniles. *Aquaculture, 312*(1–4), 102–108.

Zarate, D. D., & Lovell, R. T. (1997). Free lysine (L-lysine·HCl) is utilized for growth less efficiently than protein-bound lysine (soybean meal) in practical diets by young channel catfish (*Ictalurus punctatus*). *Aquaculture, 159*(1–4), 87–100.

Zebian, M. F., & Creach, Y. (1979). Fraction an amine libre et degradation oxydative de quelques acides amines chez la carpe (*Cyprinus carpio* L.). In J. E. Halver & K. Tiews (Eds.), *Fish Nutrition and Fish Feed Technology* (Vol. II, pp. 531–544). H. Heenemann, Berlin.

Zehra, S., & Khan, M. A. (2015). Dietary leucine requirement of fingerling *Catla catla* (Hamilton) based on growth, feed conversion ratio, RNA/DNA ratio, leucine gain, blood indices and carcass composition. *Aquaculture International, 23*(3), 577–595. https://doi.org/10.1007/s10499-014-9837-8.

Zehra, S., & Khan, M. A. (2017). Dietary leucine requirement for growth and maintenance, and its utilization efficiency for fingerling of Nile tilapia, *Oreochromis niloticus* (Linnaeus). *Borneo Journal of Marine Science and Aquaculture, 1*, 16–24.

Zhang, S., Yang, Q., Ren, M., Qiao, S., He, P., Li, D., & Zeng, X. (2016). Effects of isoleucine on glucose uptake through the enhancement of muscular membrane concentrations of GLUT1 and GLUT4 and intestinal membrane concentrations of Na+/glucose co-transporter 1 (SGLT-1) and GLUT2. *British Journal of Nutrition, 116*(4), 593–602. https://doi.org/10.1017/S0007114516002439.

Zhao, Y., Li, J. Y., Jiang, Q., Zhou, X. Q., Feng, L., Liu, Y., Jiang, W. D., Wu, P., Zhou, J., Zhao, J., & Jiang, J. (2020). Leucine improved growth performance, muscle growth, and muscle protein deposition through AKT/TOR and AKT/FOXO3a signaling pathways in hybrid catfish, *Pelteobagrus vachelli* × *Leiocassis longirostris*. *Cells, 9*(2), 327. https://doi.org/10.3390/cells9020327.

Zhao, Y., Yan, M. Y., Jiang, Q., Yin, L., & Jiang, J. (2020). Isoleucine improved growth performance, and intestinal immunological and physical barrier function of hybrid catfish *Pelteobagrus vachelli* × *Leiocassis longirostris*. *Fish & Shellfish Immunology, 109*, 20–33.

Zhou, C., Lin, H., Huang, Z., Wang, J., Wang, Y., & Yu, W. (2020). Effects of dietary leucine levels on intestinal antioxidant status and immune response for juvenile golden pompano (*Trachinotus ovatus*) involved in Nrf2 and NF-κB signaling pathway. *Fish & Shellfish Immunology, 107*, 336–345.

Zhou, Z., Wang, X., Wu, X., Gao, Y., Li, X., Dong, Y., & Yao, W. (2019). Effects of dietary leucine levels on growth, feed utilization, neuro-endocrine growth axis and TOR-related signaling expression of juvenile hybrid grouper (*Epinephelus fuscoguttatus* ♀ × *Epinephelus lanceolatus* ♂). *Aquaculture, 504*, 172–181. https://doi.org/10.1016/j.aquaculture.2019.02.005.

Zhou, Z., Wu, X., Li, X., Dong, Y., Wang, X., Mu, W., Gatlin, D. M. III, & Zhang, Y. (2020). The optimum dietary isoleucine requirement of juvenile hybrid grouper (*Epinephelus fuscoguttatus* ♀ × *Epinephelus lanceolatus* ♂). *Aquaculture Nutrition*. Advance online publication. https://doi.org/10.1111/anu.13085.

Zou, T., Cao, S., Xu, W., Han, D., Liu, H., Yang, Y., Jin, J., Zhu, X., & Xie, S. (2018). Effects of dietary leucine levels on growth, tissue protein content and relative expression of genes related to protein synthesis in juvenile gibel carp (*Carassius auratus gibelio* var. CAS III). *Aquaculture Research, 49*(6), 2240–2248. https://doi.org/10.1111/are.13682.

CHAPTER 11

Arginine Requirement of Different Fish Species

MOHAMMAD AMIN HAJAM* and SHEZA SHAKEEL MIR

Department of Biochemistry, University of Kashmir, Srinagar, Jammu and Kashmir, India

*Corresponding author

ABSTRACT

Amino acids (AAs) are essential ingredients in fish nutrition for physiological functions, optimum growth, and development. Arginine is an essential amino acid (EAA) and must be incorporated into the diet. The requirement of arginine ranges from 3% to 8% of dietary protein (DP). Among the various feed components, the arginine level ranges from 1.12% to 11.23% of the DP basis. Arginine has a range of physiological functions, including growth, osmoregulation, immunity, antioxidant effects, and ammonia detoxification. It acts as the precursor for nitric oxide (NO), polyamines, and other important metabolites. Arginine plays a role in insulin and growth hormone stimulation to maintain growth and development. It regulates homeostasis by altering the mechanism that adenosine 5'-monophosphate (AMP) activates to the activated protein kinase (AMPK) pathway, activates the target of rapamycin (TOR) signaling pathway, and hence regulates protein synthesis. The aforementioned functions of the AA arginine have been covered in this chapter to provide a complete understanding of the role of arginine and its necessity in the fish diet for optimal growth.

Fish Nutrition: Protein and Amino Acid Requirements. Imtiaz Ahmed, Ishtiyaq Ahmad, and Shabihul Fatma Sayed (Eds.)

DOI: 10.1201/9781998511570-11

11.1 INTRODUCTION

In aquaculture, fish need to be fed balanced diets containing the required amount of nutrients for proper growth and development. Protein diets are essentially important for optimum fish growth and to maintain fish physiologically and immunologically active (Hosseini et al., n.d.). Proteins constitute the most critical diet for structural development and metabolism. Amino acids (AAs) act as the building blocks of these macromolecules, which are critical molecules and play central roles in metabolic pathways, hormone synthesis, antioxidant reactions, and various physiological activities in the body (Mukhtar et al., n.d.). Fishes cannot synthesize all AAs in their body and hence are dependent on the incorporation of AAs in their diet. Arginine is one of the EAAs, and arginine requirements are to be fulfilled as the urea cycle in fish is very poor; thus, diet is an important source of arginine for sustainable fish culture (National Research Council, 2011). The evidence that supports arginine biosynthesis in fishes from citrulline in fewer fish species is not sufficient and occurs at very low rates when compared with the higher vertebrates (Buentello et al., n.d.). Among the various feed components, arginine level ranges from 1.12 to 11.23% of the dietary protein (DP) basis. Fishmeal with a crude protein content ranging from 61.3 to 72%, gelatin with a protein content of 88%, hydrolyzed poultry feather meal with a protein content of 82.9%, 84.1% crude protein content in the soybean meal (SB) protein, poultry byproduct meal with a crude protein content of 59.5%, and rice bran with a crude protein content of 13.7%, are the fish feedstuffs which contain the higher content of arginine. The feedstuffs which contain the lower arginine levels include rice bran with a protein content of 12.2%, oats with a crude protein content of 11.4%, wheat bran with a crude protein content of 15.4%, whey dried with a crude protein content of 12.2%, and wheat with a protein content of 10.2% (Jobling, 2012). The evaluation of dietary arginine needs of various fish species was evaluated using growth response curves; certain fish species displayed the broken-like response, and others displayed the polynomial response. The presence of arginine in the diet brings more versatility to feed because of its diverse physiological functions. Arginine stimulates the production of growth hormone and insulin (Floyd et al., 1966) and possibly induces the somatotropic axis in the fish character (Plisetskaya et al., 1991). The reason for the stimulation of these hormones is possibly due to the involvement of arginine in anabolic processes. Studies reveal that the effect of dietary arginine in fish contributes towards stress and cortisol responses (Aragao et al., 2008). Metabolites derived from arginine AAs strongly affect processes like tissue repair by collagen synthesis and cell replication. Arginine is involved in the process of nitrogen metabolism and polyamine synthesis and is considered one of the versatile AAs. The presence of the amidino group on the arginine AA provides this group for the synthesis of creatine. Phosphocreatine acts as the major energy reservoir of high-energy phosphate for the regeneration of ATP in the muscles. Arginine acts as the precursor of another molecule, nitric oxide (NO). The biosynthesis of nitric oxide is catalyzed by the enzyme nitric oxide synthase (NOS), for which arginine acts as the only substrate. Nitric oxide has many biological activities including anti-oxidative activity, ammonia detoxification, and brown adipose tissue development (Liang et al., 2018). Arginine in fish helps in improving specific and non-specific immunity by regulating the level of cytokines released into the blood such as C3 and IgM (Chen et al., 2015). AA arginine regulates protein synthesis by activating the signaling pathway of the mammalian target of rapamycin (mTOR) through modulation of the AMP-activated protein kinase (AMPK) pathway (Chen et al., 2015).

11.2 ARGININE REQUIREMENT IN VARIOUS FISHES

Fish diets should possess an arginine concentration between 3.2% to 8% to maintain proper health (Hoseini et al., 2020). The requirement of arginine in the diet increases with age because the activity of various enzymes gets reduced with age, which includes pyrroline-5-carboxylate synthase, ornithine transcarboxylase, and carbamoyl phosphate synthase (P5C, OTC, and CPS III, respectively) (Andersen et al., 2016). The variations in arginine requirements vary with fish species. Initial studies reported that fishes have relatively similar arginine requirements, from 1.2%–1.5% of diet dry matter and 3.7%–3.9% of crude protein (CP), but exceptionally in salmon species of the Pacific Ocean, which has 2.04% diet dry matter and 5.4% requirement of crude protein (Ball et al., 2007). But later on, studies revealed that carnivorous species have higher diet requirements of more than 2%–6% of DM and CP, respectively (Zhou et al., 2012). This difference might be because of experimental treatment differences, protein quality differences, species differentiation, and other biological causes. It makes sense that the requirements for arginine are higher in carnivorous fish species, as they have higher requirements for protein deposition and activation of the TOR signaling pathway than omnivorous species because arginine is critically important in synthesizing the proteins required for structural tissues and also in activating the TOR signaling pathway as a signal. Dietary requirements not only vary with fish species, but fish size has a direct correlation with arginine requirements, and it has been estimated that arginine requirements decrease with increased fish size. In one study carried out on gibel carp (*Carassius auratus gibelio*), the dietary requirement of bigger fish was 1.29% of DM and 4.2% of CP, while in smaller fish it was 1.64% DM and 5.3% of CP (Tu et al., 2015). Generally, during an experimental duration of 8 or 12 weeks of developmental stages, fish with lower initial body weights always exhibit better relative growth rates (Table 11.1).

TABLE 11.1 Requirements of Arginine in Different Fish Species

Fish Species	Crude Protein (% Age of Dry Diet)	Arginine Requirement (% of Dietary Protein)	References
Acipenser schrenckii♀ x *A. baerii*♂	40	6.2	Wang et al. (2017)
Clarias gariepinus x *C. macrocephalus*	40	4.5–5.0	Singh and Khan (2007)
Carassis auratus gabelio	31	5.3	Tu et al. (2015)
Catla Catla	31	5.1	Zehra and Khan (2013)
Chanos chanos	40	5.3	Borlongan (1991)
Cirrhinus mrigala	40	4.6	Ahmed and Khan (2004)
Clarias gariepinus	40	4.5	Fagbenro et al. (1999)
Cyprinus carpio	48	4.3	Nose (1979)
Dicentrarchus labrax	46	3.9	Tibaldi et al. (1994)
Epinephelus awoara	43	6.5	Zhou et al. (2012)
Epinephelus coioides	50	6.1	Han et al. (2018)
Heteropneustes fossilis	40	4.0	Ahmed (2013)
Ictalurus punctatus	24	3.3–3.8 (glutamate added)	Buentello and Gatlin. (1999)
		4.2–5.0 (glycine added)	
Labeo rohita	40	5.8	Murthy and Varghese (1995)

TABLE 11.1 *(Continued)*

Fish Species	Crude Protein (% Age of Dry Diet)	Arginine Requirement (% of Dietary Protein)	References
Lates calcarifer	47	3.8	Murillo-Gurrea et al. (2001)
Megalobrama amblycephala	34	7.2	Ren et al. (2013)
Micropterus salmoides	45.9	4.1	Zhou et al. (2012)
Morone saxatilis x M. chrysops	33	4.4	Griffin et al. (1994)
Oncorhynchus kisutch	45	4.9–5.5	Luzzana et al. (1998)
Oncorhynchus mykiss	40	3.5–4.2	Chiu et al. (1988)
Oreochromis niloticus	28	4.8	Neu et al. (2016)
Pagrus major	50	4.5	Rahimnejad and Lee (2014)
Paralichthys olivaceus	50	4.1–4.9	Alam et al. (2002)
Pelteobagrus fulvidraco	42	6.5	Chen et al. (2016)
Perca flavescens	33	4.9	Twibell and Brown (1997)
Rachycentron canadum	46	6.2	Ren et al. (2014)
Salminus brasiliensis	43	3.3	Dairiki et al. (2013)
Salmo salar	40	4.1	Lall et al. (1994)
Sciaenops ocellatus	35	5.0	Barziza et al. (2000)
Sparus aurata	42–44	5.55	Peres and Oliva-Teles (2009)
Sparus macrocephalus	38	7.7–8.1	Zhou et al. (2010)
Trachinotus ovatus	43	6.3–6.4	Lin et al. (2015)

11.3 PHYSIOLOGICAL ROLES OF ARGININE IN FISH

11.3.1 ROLE OF ARGININE IN THE GROWTH OF FISH BY ACTIVATION OF ENDOCRINE SYSTEM

Arginine plays a role in the activation of the endocrine system of fish. The somatotropic axis is responsible for the growth of a fish. A study has shown that during malnutrition, the somatotropic axis, in turn, depends on growth hormone and insulin-like growth factor 1 (IGF-1) (Pohlenz et al., 2013). The arginine concentration of a fish diet is directly proportional to the growth of a fish, as it increases the growth hormone and IGF-1 levels (Yue et al., 2015). Since the growth of fish depends on the insulin hormone, an arginine-rich diet promotes the growth of the fish by increasing the production of insulin. This indirect link depicts the relationship of arginine with fish growth. In channel catfish, dietary arginine intake resulted in an increased expression at the transcriptional level of IGF-1 in the liver and muscle tissues, with an elevated secretion of growth hormone and IGF-I in plasma. The performance of fish growth is closely related to the levels of GH and IGF-I (Chen et al., 2012). The high arginine diet intake in largemouth bass (*Micropterus salmoides*) had considerably elevated hepatic IGF-I and pituitary GH mRNA levels. This upregulation exhibits a strong positive correlation with fish-specific growth rate (SGR) (Chen et al., 2012).

Arginine also increases the mTOR signaling pathway responsible for sensing growth. Experimental observations from different studies carried out in young grass carp (*Ctenopharyngodon idella*) have shown that high dietary levels dramatically raised the relative mRNA expression of TOR in the gill. In a similar experimental outcome, dietary arginine levels showed an increase in the relative mRNA expression of TOR and ribosome protein kinase 1 (S6K1) in blunt snout bream (Liang et al., 2016).

11.3.2 ARGININE AS OSMOREGULATOR IN SOME FISH SPECIES

Arginine acts as an important precursor for a neuropeptide hormone known as arginine-vasotocin (AVT) (Balment et al., n.d.). AVT helps in maintaining the water and salinity in the case of teleosts. It is antidiuretic and helps in conserving water. This hormone comes into play when water conservation becomes essential. Several fish physiology and behavioral functions, including circadian rhythms, seasonal biology, responses toward stress, metabolism, reproduction, osmoregulation, and cardiovascular functions, are integrated by the AVT neuroendocrine system (Balment et al., 2006). Hence, it helps fish to adapt according to the prevailing environmental conditions by maintaining salinity, as it has a coordinating role in teleosts.

11.3.3 ARGININE IMPROVES FISH IMMUNITY

Nitric oxide is one of the metabolites of arginine, and it has dual functions. One is that it can oxidize bacterial cell walls, and the second property is that it can also inactivate various mitochondrial enzymes responsible for respiratory chain transport, therefore making it difficult for bacterial cells to survive. Polyamines are important for cell division, and biosynthesis of polyamines is dependent on arginine (Xu et al., 2020). Arginine, when present in optimum quantity in the diet, maintains a good leucocyte count

(Buentello et al., 2007). Moreover, when more arginine is present, more polyamines are synthesized, and hence macrophages are stimulated to secrete more cytokine, as cytokine secretion is mediated by polyamine precursors (Zehra and Khan, 2013). Therefore, it causes inflammation in fish. In this way, arginine helps boost the immunity of fish and makes fish disease-resistant. Arginine supplementation promotes fish immune response and resistance to the disease by enhancing the production of NO. The increased supply may reduce disease resistance; the supply must be properly evaluated during the feeding practices. As per the study conducted regarding the adequate supply of arginine needed for proper immune response against infections and disease resistance, dietary arginine levels were summarized as (Table 11.2).

TABLE 11.2 Requirements of Arginine for Proper Immune Response Against Disease Resistance

Fish Species	Arginine Dosage (%of DM/% of CP)	References
Yellow grouper, *Epinephelus awoara*	2.51 to 2.83 / 5.83 to 6.58	Zhou et al. (2012).
Golden pompano, *Trachinotus ovatus*	2.65 to 2.98 / 6.16 to 6.93	Lin et al. (2015)
Jian carp, *Cyprinus carpio* var. Jian	1.27 to 1.61 / 3.74 to 4.74	(Chen et al. (2015)
Red sea bream, *Pagrus major*	2.54 to 3.08 / 5.08 to 6.16	Rahimnejad and Lee (2014)
Indian Major Carp, *Cirrhinus mrigala*	2.78/8.00	Varghese et al. (2020)
Yellow catfish, *Pelteobagrus fulvidraco*	2.26 to 2.74 / 5.02 to 6.09	Zhou et al. (2015)
Senegalese sole, *Solea senegalensis*	6.9/12.78	Costas et al. (2011)
Nile tilapia, *Oreochromis niloticus*	3.56 to 4.38 / 9.89 to 12.17	Pereira et al. (2017)

In an experiment carried out on common carp, the interaction effects of dietary arginine and stocking density on its growth performance, response to stress, and gene expression of immune components of innate immunity were studied. Three different diets with variable arginine levels were formulated as deficient with 4.1% of DA, optimum with 5.3% of dietary protein (OA), and surplus with 8.1% of dietary protein (SA). Common carp were raised on the diets for 30 days at either the 10 g/L or 20 g/L stocking density. Plasma biochemical analytes like plasma cortisol, blood globulin, leukocyte count, and interleukins IL-8, IL-Beta, and TNF-Alpha were determined. The outcome of this study revealed that increased stocking density caused diminished growth performance; similarly, the SA and DA groups significantly lowered the growth performance in comparison to the OA group. The biochemical parameters also were seen as variable in different supplementation groups (Hoseini et al., 2019).

11.4 ANTIOXIDANT EFFECTS OF ARGININE

Both enzymatic and non-enzymatic antioxidant defenses act as protective means for fishes as well as other invertebrates against oxidative stress. The antioxidant compounds obtained through the diet are the most common method to protect fish from oxidative damage (Oliva-Teles, 2012). The various dietary compounds, which include certain AAs, have been reported to show anti-oxidative effects. A similar study revealed the role of dietary arginine in protecting fish against oxidative stress by SOD activity in the gut of *S. aurata* with increased dietary arginine levels (Coutinho et al., 2016). Decreased growth performance as well as antioxidant enzyme activity, while enhanced MDA levels, were seen in the gut of *C. carpio* when fed with a diet deficient in arginine (Wang et al., 2019). However, some studies have not found a significant link between arginine in the diet and antioxidant enzyme activities in some fish species; hence, more studies are needed to get a clear understanding of the antioxidant activities of arginine AA in the diet.

11.5 ARGININE HELPS IN AMMONIA DETOXIFICATION

Water gets contaminated with ammonia in different ways. Various aquatic metabolic and nitrogenous wastes are released in water by aquatic animals. Some fish feeds, which remain unconsumed, decompose and become a source of ammonia. Various anthropogenic activities release ammonia-rich substances in water bodies, and those get accumulated there and cause ammonia toxicity. A diminished water flow rate concerning fish stocking density with poor feeding, as well as water management, enhances ammonia fish toxicity. Ammonia toxicity in water is a risk factor in fish farming. It causes damage to fish tissue, hydromineral imbalance, suppression of the immune system, increased stress, and causes oxidative damage (Peyghan and Takamy, 2002). AA arginine plays a crucial role in the conversion of ammonia to urea in fish during the urea cycle, as will be discussed further in detail. Studies conducted in *P. fulvidraco* showed that supplementation of arginine of 6.6 g per 100 g protein lowered the mortality in *P. fulvidraco* after 72 hours when exposed to ammonia fish water as compared to control with a feeding rate of 5.7 g per 100 g protein (Chen et al., 2016).

11.6 BIOSYNTHESIS OF ARGININE IN FISHES

11.6.1 ARGININE BIOSYNTHESIS IN EARLY DEVELOPMENTAL STAGES

Arginine production occurs through the urea cycle due to the presence of higher expression levels of enzymes needed for the urea cycle during the early life stages of the fish (Andersen et al., 2014). The urea cycle occurs in a series of steps with the employment of a specific enzyme for each step. The first step of the urea cycle involves the formation of glutamine, catalyzed by the enzyme glutamine synthase; it involves the fixing of ammonia and glutamate. Enzyme glutamine synthase is a mitochondrial enzyme. The second step is proceeded by the enzyme carbamoyl phosphate synthase III (CPS III), however, in mammals, this step is catalyzed by the enzyme CPS I that uses glutamine as a nitrogen source instead of ammonia (Shi et al., 2018). As per earlier studies, many teleosts are ureotelic in the early stages of life and produce greater amounts of urea in the early stage, while they are ammoniotelic in adult stages (Fauzi et al., 2019; Terjesen et al., 2001). The activities of CPS III are also reported to exhibit higher activities in the early developmental stages of life in the rainbow trout; these findings provide proof that arginine synthesis is more active in the early developmental stages of fishes (Wright and Land, 1998).

11.6.2 ARGININE BIOSYNTHESIS IN ADULT FISH

Adult teleosts have properly developed gills, which they use for the excretion of ammonia. However, the genes encoding the enzymes necessary for the urea cycle are actively expressed, as detected in adult fish tissues (Wright et al., 2004). Urea cycle enzymes, especially CPS III, have shown much greater activity in the muscle tissue than the liver tissues (LeMoine and Walsh, 2013). The African catfish also showed the localization of these enzymes in the muscle tissues but not in the liver tissues (Tafalla et al., 2001). Similar studies were carried out in trout species, which showed higher CPS III activity in the muscle tissue than in the liver tissues (Buentello and Gatlin, 2000). Comparing the arginine synthesis ability of an adult with the early developmental stage suggested that the arginine biosynthetic ability of adult fishes is low. A new study recently published revealed that the addition of citrulline in the fish diet increased both the muscle and liver arginine levels, thus suggesting that supplementation of citrulline promotes the synthesis of arginine (Clark et al., 2020). The alteration in environmental conditions, such as crowding and higher pH, upregulated the activities of urea cycle enzymes (Laberge et al., 2009). All the results culminate with the findings that arginine synthesis is a physiological process with greater significance despite its lower activity. This is because fishes are not able to synthesize arginine endogenously so efficiently to achieve optimal growth and functioning. Hence, additional arginine supplementation is required in the fish diet for proper growth and development.

11.6.3 ARGININE CATABOLISM IN FISH

A set of four important enzymes are required for the breakdown of arginine into various metabolites having potent physiological roles. These four enzymes include: (a) arginine: glycine aminotransferase (AGAT) which catalyzes the conversion of arginine to creatine; (b) nitric oxide (NO) synthases (3-isoenzymes), which converts arginine to nitric oxide; (c) arginases (2-isoenzymes), which converts arginine into urea and ornithine in fish; and (d) arginine decarboxylase (ADC) which catalyzes the production of agmatine from arginine in fish (Figure 11.1).

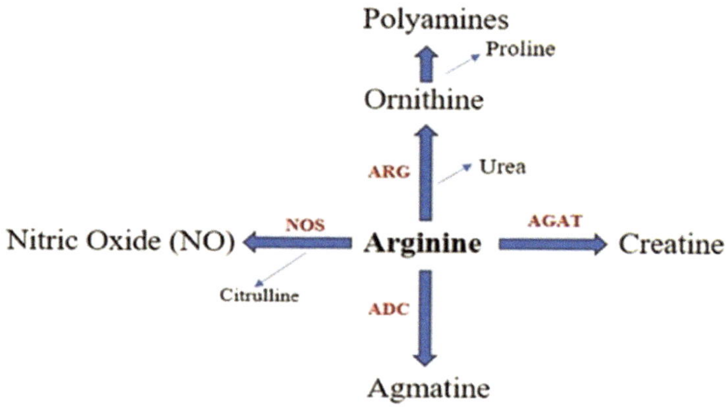

FIGURE 11.1 Catabolism of arginine.

11.6.4 UREA CYCLE IN FISHES

The urea cycle was the first metabolic cycle discovered in the year 1932 by Hans Krebs and Kurt Henseleit. This cycle involves a series of biochemical reactions that result in the production of urea from the toxic compound ammonia. This cycle occurs in two compartments of the cell, that is, mitochondria and cytoplasm, because of the distribution of enzymes in these two compartments.

➢ **The steps involved in the urea cycle are as:**

1. **Synthesis of Carbamoyl Phosphate:** The carbamoyl phosphate synthetase III enzyme catalyzes the condensation of ammonia ions with HCO_3^- to generate carbamoyl phosphate. This is an irreversible reaction and occurs with the mitochondria.

2. **Citrulline Synthesis:** An enzyme ornithine transcarbamoylase converts carbamoyl phosphate to citrulline in the presence of ornithine.

3. **Argininosuccinate Synthesis:** Argininosuccinate synthase enzyme combines citrulline with the aspartate to form argininosuccinate. This reaction is an energy-dependent process and utilizes ATP.

4. **Cleavage of Argininosuccinate:** Argininosuccinate Lyase cleaves the Argininosuccinate into Arginine.

5. **Formation of Urea:** The last enzyme involved in the cycle which cleaves arginine into urea and ornithine. Ornithine generated in the reaction process is reused in the urea cycle.

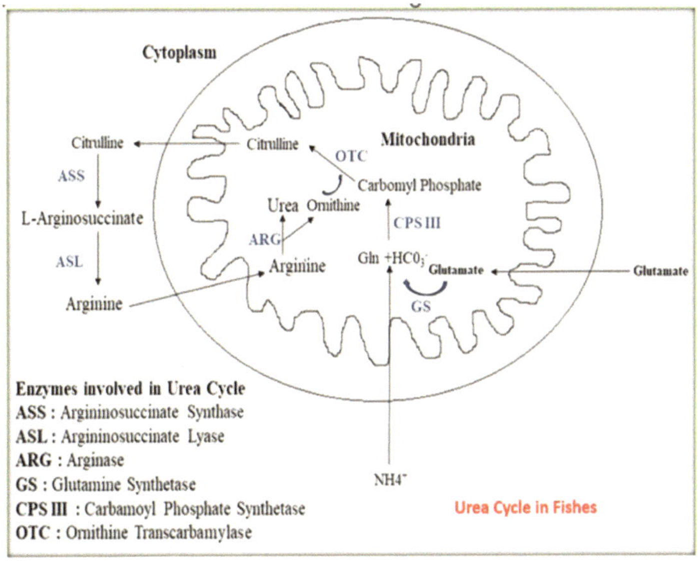

KEYWORDS

- **aquafeed formulation**
- **arginine in fish nutrition**
- **essential amino acids**
- **nitric oxide synthesis**
- **protein metabolism in fish**
- **sustainable aquaculture practices**

REFERENCES

Ahmed, I. (2013). Dietary arginine requirement of fingerling Indian catfish (*Heteropneustes fossilis*, Bloch). *Aquaculture International, 21*(3), 255–271.

Ahmed, I., & Khan, M. A. (2004). Dietary arginine requirement of fingerling Indian major carp, *Cirrhinus mrigala* (Hamilton). *Aquaculture Nutrition, 10*(3), 217–225.

Alam, M. S., Teshima, S., Koshio, S., & Ishikawa, M. (2002). Arginine requirement of juvenile Japanese flounder *Paralichthys olivaceus* estimated by growth and biochemical parameters. *Aquaculture, 205*(3, 4), 127–140.

Andersen, S. M., Taylor, R., Holen, E., Aksnes, A., & Espe, M. (2014). Arginine supplementation and exposure time affects polyamine and glucose metabolism in primary liver cells isolated from Atlantic salmon. *Amino Acids, 46*(5), 1225–1233.

Andersen, S. M., Waagbø, R., & Espe, M. (2016). Functional amino acids in fish health and welfare. *Frontiers in Bioscience, 8*, 143–169.

Aragao, C., Corte-Real, J., Costas, B., Dinis, M. T., & Conceição, L. (2008). Stress response and changes in amino acid requirements in Senegalese sole (*Solea senegalensis* Kaup 1858). *Amino Acids, 34*(1), 143–148.

Ball, R. O., Urschel, K. L., & Pencharz, P. B. (2007). Nutritional consequences of interspecies differences in arginine and lysine metabolism. *The Journal of Nutrition, 137*(6 Suppl 2), 1626S–1641S.

Balment, R. J., Lu, W., Weybourne, E., & Warne, J. M. (2006). Arginine vagotonic: A key hormone in fish physiology and behavior: A review with insights from mammalian models. *General and Comparative Endocrinology, 147*(1), 9–16. https://doi.org/10.1016/j.ygcen.2005.12.022

Balment, R. J., Warne, J. M., Tierney, M., & Hazon, N. (n.d.). Arginine vagotonic and fish osmoregulation. https://doi.org/10.1007/BF00004566

Barziza, D. E., Buentello, J. A., & Gatlin, D. M. III. (2000). Dietary arginine requirement of juvenile red drum (*Sciaenops ocellatus*) based on weight gain and feed efficiency. *The Journal of Nutrition, 130*(7), 1796–1799.

Borlongan, I. G. (1991). Arginine and threonine requirements of milkfish (*Chanos chanos* Forsskal) juveniles. *Aquaculture, 93*(3, 4), 313–322.

Buentello, J. A., & Gatlin, D. M. (2000). The dietary arginine requirement of channel catfish (*Ictalurus punctatus*) is influenced by endogenous synthesis of arginine from glutamic acid. *Aquaculture, 188*(3, 4), 311–321.

Buentello, J. A., & Gatlin, D. M. III. (1999). Nitric oxide production in activated macrophages from channel catfish (*Ictalurus punctatus*): Influence of dietary arginine and culture media. *Aquaculture, 179*(1–4), 513–521.

Buentello, J. A., Reyes-Becerril, M., de Jesús Romero-Geraldo, M., & de Jesús Ascencio-Valle, F. (2007). Effects of dietary arginine on hematological parameters and innate immune function of channel catfish. *Journal of Aquatic Animal Health, 19*(3), 195–203.

Buentello, J., Gatlin, D. M., & Neill, W. H. (n.d.). Effects of water temperature and dissolved oxygen on daily feed consumption, feed utilization and growth of channel catfish (*Ictalurus punctatus*).

Chen, G., Liu, Y., Jiang, J., Jiang, W., Kuang, S., Tang, L., et al. (2015). Effect of dietary arginine on the immune response and gene expression in head kidney and spleen following infection of Jian carp with *Aeromonas hydrophila. Fish & Shellfish Immunology, 44*(1), 195–202.

Chen, G., Liu, Y., Jiang, J., Jiang, W., Kuang, S., Tang, L., Tang, W., Zhang, Y. A., Zhou, X., & Feng, L. (2015). Effect of dietary arginine on the immune response and gene expression in head kidney and spleen following infection of Jian carp with *Aeromonas hydrophila. Fish & Shellfish Immunology, 44*(1), 195–202.

Chen, N., Jin, L., Zhou, H., & Qiu, X. (2012). Effects of dietary arginine levels and carbohydrate-to-lipid ratios on mRNA expression of growth-related hormones in largemouth bass, *Micropterus salmoides. General and Comparative Endocrinology, 179*(1), 121–127.

Chen, N., Jin, L., Zhou, H., & Qiu, X. (2012). Effects of dietary arginine levels and carbohydrate-to-lipid ratios on mRNA expression of growth-related hormones in largemouth bass, *Micropterus salmoides. General and Comparative Endocrinology, 179*(1), 121–127.

Chen, Q., Zhao, H., Huang, Y., Cao, J., Wang, G., Sun, Y., et al. (2016). Effects of dietary arginine levels on growth performance, body composition, serum biochemical indices and resistance ability against ammonia-nitrogen stress in juvenile yellow catfish (*Pelteobagrus fulvidraco*). *Animal Nutrition, 2*(3), 204–210.

Chen, Q., Zhao, H., Huang, Y., Cao, J., Wang, G., Sun, Y., et al. (2016). Effects of dietary arginine levels on growth performance, body composition, serum biochemical indices and resistance ability against ammonia-nitrogen stress in juvenile yellow catfish (*Pelteobagrus fulvidraco*). *Animal Nutrition, 2*(3), 204–210.

Chiu, Y. N., Austic, R. E., & Rumsey, G. L. (1988). Effect of feeding level and dietary electrolytes on the arginine requirement of rainbow trout (*Salmo gairdneri*). *Aquaculture, 69*(1), 79–91.

Clark, T. C., Tinsley, J., Sigholt, T., Macqueen, D. J., & Martin, S. (2020). Supplementation of arginine, ornithine and citrulline in rainbow trout (*Oncorhynchus mykiss*): Effects on growth, amino acid levels in plasma and gene expression responses in liver tissue. *Comparative Biochemistry and Physiology Part A: Molecular & Integrative Physiology, 241*, 110632.

Costas, B., Conceição, L. E., Dias, J., Novoa, B., Figueras, A., & Afonso, A. (2011). Dietary arginine and repeated handling increase disease resistance and modulate innate immune mechanisms of Senegalese sole (*Solea senegalensis* Kaup, 1858). *Fish & Shellfish Immunology, 31*(4), 838–847.

Coutinho, F., Castro, C., Rufino-Palomares, E., Ordóñez-Grande, B., Gallardo, M., & Kaushik, S., et al. (2016). Dietary arginine surplus does not improve intestinal nutrient absorption capacity, amino acid metabolism and oxidative status of gilthead sea bream (*Sparus aurata*) juveniles. *Aquaculture, 464*, 480–488.

Dairiki, J. K., Borghesi, R., Dias, C. T. S., & Cyrino, J. E. P. (2013). Lysine and arginine requirements of *Salminus brasiliensis*. *Pesquisa Agropecuaria Brasileira, 48*(8), 1012–1020.

Fagbenro, O. A., Nwanna, L. C., & Adebayo, O. T. (1999). Dietary arginine requirement of the African catfish, *Clarias gariepinus*. *Journal of Applied Aquaculture, 9*(2), 59–64.

Fauzi, I. A., Haga, Y., Kondo, H., Hirono, I., & Satoh, S. (2019). Effects of arginine supplementation on growth performance and plasma arginine, ornithine and citrulline dynamics of rainbow trout, *Oncorhynchus mykiss*. *Aquaculture Research, 50*(4), 1277–1290.

Floyd, J. D., Jr., Fajans, S. S., Conn, J. W., Knopf, R. F., & Rull, J. A. (1966). Stimulation of insulin secretion by amino acids. *Journal of Clinical Investigation, 45*(10), 1487.

Griffin, M. E., Wilson, K. A., & Brown, P. B. (1994). Dietary arginine requirement of juvenile hybrid striped bass. *The Journal of Nutrition, 124*(6), 888–893.

Han, F., Chi, S., Tan, B., Dong, X., Yang, Q., Liu, H., et al. (2018). Metabolic and immune effects of orange-spotted grouper, *Epinephelus coioides* induced by dietary arginine. *Aquaculture Reports, 10*, 8–16.

Hoseini, S. M., Khan, M. A., Yousefi, M., & Costas, B. (2020). Roles of arginine in fish nutrition and health: Insights for future research. *Reviews in Aquaculture, 12*(4), 2091–2108.

Hoseini, S. M., Vatnikov, Y. A., Kulikov, E. V., Petrov, A. K., Hoseinifar, S. H., & Doan, H. V. (2019). Effects of dietary arginine supplementation on ureagenesis and amino acid metabolism in common carp (*Cyprinus carpio*) exposed to ambient ammonia. *Aquaculture*.

Hosseini, L., Maroufpoor, M., & Kazemi, S. (n.d.). Supplementary description of *Digamasellus punctum* (Berlese) (Mesostigmata, Digamasellidae), and a key to the world species of *Digamasellus* Berlese.

Jobling, M. (2012). National Research Council (NRC): Nutrient requirements of fish and shrimp. *Aquaculture International, 20*(3), 601–602.

Laberge, T., Walsh, P. J., & McDonald, M. D. (2009). Effects of crowding on ornithine-urea cycle enzyme mRNA expression and activity in gulf toadfish (*Opsanus beta*). *Journal of Experimental Biology, 212*(15), 2394–2402.

Lall, S., Kaushik, S., Le Bail, P., Keith, R., Anderson, J., & Plisetskaya, E. (1994). Quantitative arginine requirement of Atlantic salmon (*Salmo salar*) reared in seawater. *Aquaculture, 124*(1, 2), 13–25.

LeMoine, C. M., & Walsh, P. J. (2013). Ontogeny of ornithine-urea cycle gene expression in zebrafish (*Danio rerio*). *American Journal of Physiology-Regulatory, Integrative and Comparative Physiology, 304*(11), R991–R1000.

Liang, H., Ji, K., Ge, X., Ren, M., Liu, B., Xi, B., & Pan, L. (2018). Effects of dietary arginine on antioxidant status and immunity involved in AMPK-NO signaling pathway in juvenile blunt snout bream. *Fish & Shellfish Immunology, 78*, 69–78.

Liang, H., Ren, M., Habte-Tsion, H., Ge, X., Xie, J., Mi, H., et al. (2016). Dietary arginine affects growth performance, plasma amino acid contents and gene expressions of the TOR signaling pathway in juvenile blunt snout bream, *Megalobrama amblycephala*. *Aquaculture, 461*, 1–8.

Lin, H., Tan, X., Zhou, C., Jin, N., Xia, D., Zhong, H., et al. (2015). Effect of dietary arginine levels on the growth performance, feed utilization, non-specific immune response and disease resistance of juvenile golden pompano *Trachinotus ovatus*. *Aquaculture, 437*, 382–389.

Lin, H., Tan, X., Zhou, C., Niu, J., Xia, D., Huang, Z., et al. (2015). Effect of dietary arginine levels on the growth performance, feed utilization, non-specific immune response and disease resistance of juvenile golden pompano *Trachinotus ovatus*. *Aquaculture, 437*, 382–389.

Luzzana, U., Hardy, R. W., & Halver, J. E. (1998). Dietary arginine requirement of fingerling coho salmon (*Oncorhynchus kisutch*). *Aquaculture, 163*(1, 2), 137–150.

Mukhtar, B., Malik, M. F., Shah, S. H., Azzam, A., Slahuddin, & Liaqat, I. (n.d.). Lysine supplementation in fish feed.

Murillo-Gurrea, D., Coloso, R., Borlongan, I. G., & Serrano, A. (2001). Lysine and arginine requirements of juvenile Asian sea bass *Lates calcarifer*. *Journal of Applied Ichthyology, 17*(1), 49–53.

Murthy, H., & Varghese, T. (1995). Arginine and histidine requirements of the Indian major carp, *Labeo rohita* (Hamilton). *Aquaculture Nutrition, 1*(4), 235–239.

National Research Council. (2011). *Nutrient Requirements of Fish and Shrimp* (pp. 67–68). National Academies Press.

Neu, D., Boscolo, W., Zaminhan, M., Almeida, F., Sary, C., & Furuya, W. (2016). Growth performance, biochemical responses, and skeletal muscle development of juvenile Nile tilapia, *Oreochromis niloticus*, fed with increasing levels of arginine. *Journal of the World Aquaculture Society, 47*(2), 248–259.

Nose, T. (1979). Summary report on the requirements of essential amino acids for carp. In K. Tiews & J. Halver (Eds.), *Finfish Nutrition and Fish Feed Technology* (pp. 145–156). Heenemann GmbH.

Oliva-Teles, A. (2012). Nutrition and health of aquaculture fish. *Journal of Fish Diseases, 35*(2), 83–108.

Pereira, R., Rosa, P., & Gatlin, D. M. III. (2017). Glutamine and arginine in diets for Nile tilapia: Effects on growth, innate immune responses, plasma amino acid profiles and whole-body composition. *Aquaculture, 473*, 135–144.

Peres, H., & Oliva-Teles, A. (2009). The optimum dietary essential amino acid profile for gilthead seabream (*Sparus aurata*) juveniles. *Aquaculture, 296*(1, 2), 81–86.

Peyghan, R., & Takamy, G. A. (2002). Histopathological, serum enzyme, cholesterol and urea changes in experimental acute toxicity of ammonia in common carp *Cyprinus carpio* and use of natural zeolite for prevention. *Aquaculture International, 10*(4), 317–325.

Plisetskaya, E. M., Buchelli-Narvaez, L. I., Hardy, R. W., & Dickhoff, W. W. (1991). Effects of injected and dietary arginine on plasma insulin levels and growth of Pacific salmon and rainbow trout. *Comparative Biochemistry and Physiology Part A: Physiology, 98*(1), 165–170.

Pohlenz, C., Buentello, A., Miller, T., Small, B., MacKenzie, D., & Gatlin, D. (2013). Effects of dietary arginine on endocrine growth factors of channel catfish, *Ictalurus punctatus. Comparative Biochemistry and Physiology Part A: Molecular & Integrative Physiology, 166*(2), 215–221.

Rahimnejad, S., & Lee, K. (2014). Dietary arginine requirement of juvenile red sea bream *Pagrus major. Aquaculture, 434*, 418–424.

Rahimnejad, S., & Lee, K. J. (2014). Dietary arginine requirement of juvenile red sea bream *Pagrus major. Aquaculture, 434*, 418–424.

Ren, M., Ai, Q., & Mai, K. (2014). Dietary arginine requirement of juvenile cobia (*Rachycentron canadum*). *Aquaculture Research, 45*(2), 225–233.

Ren, M., Liao, Y., Xie, J., Liu, B., Zhou, Q., Ge, X., et al. (2013). Dietary arginine requirement of juvenile blunt snout bream, *Megalobrama amblycephala. Aquaculture, 414, 415*, 229–234.

Shi, D., Caldovic, L., & Tuchman, M. (2018). Sources and fates of carbamyl phosphate: a labile energy-rich molecule with multiple facets. *Biology (Basel), 7*(2), 34.

Singh, S., & Khan, M. A. (2007). Dietary arginine requirement of fingerling hybrid Clarias (*Clarias gariepinus* × *Clarias macrocephalus*). *Aquaculture Research, 38*(1), 17–25.

Tafalla, C., Figueras, A., & Novoa, B. (2001). Viral hemorrhagic septicemia virus alters turbot (*Scophthalmus maximus*) macrophage nitric oxide production. *Diseases of Aquatic Organisms, 47*(2), 101–107.

Terjesen, B. F., Chadwick, T. D., Verreth, J., Rønnestad, I., & Wright, P. A. (2001). Pathways for urea production during early life of an air-breathing teleost, the African catfish *Clarias gariepinus* Burchell. *Journal of Experimental Biology, 204*(12), 2155–2165.

Tibaldi, E., Tulli, F., & Lanari, D. (1994). Arginine requirement and effect of different dietary arginine and lysine levels for fingerling sea bass (*Dicentrarchus labrax*). *Aquaculture, 127*(3, 4), 207–218.

Tu, Y., Xie, S., Han, D., Yang, Y., Jin, J., & Zhu, X. (2015). Dietary arginine requirement for gibel carp (*Carassius auratus gibelio* var. CAS III) reduces with fish size from 50 g to 150 g associated with modulation of genes involved in TOR signaling pathway. *Aquaculture, 449*, 37–47.

Tu, Y., Xie, S., Han, D., Yang, Y., Jin, J., & Zhu, X. (2015). Dietary arginine requirement for gibel carp (*Carassius auratus gibelio* var. CAS III) reduces with fish size from 50 g to 150 g associated with modulation of genes involved in TOR signaling pathway. *Aquaculture, 449*, 37–47.

Twibell, R. G., & Brown, P. B. (1997). Dietary arginine requirement of juvenile yellow perch. *The Journal of Nutrition, 127*(10), 1838–1841.

Varghese, T., Kumar, V. R., Gopan, A., Valappil, R., Sajina, K. A., Mishal, P., et al. (2020). Dietary arginine modulates nonspecific immune responses in Indian Major Carp, *Cirrhinus mrigala* exposed to hypoxia. *Aquaculture, 529*, 735613.

Wang, L., Li, J., Wang, C., Zhao, Z., Luo, L., Du, X., et al. (2019). Effect of N-carbamoyl glutamate supplementation on the growth performance, antioxidant status and immune response of mirror carp (*Cyprinus carpio*) fed an arginine-deficient diet. *Fish & Shellfish Immunology, 84*, 280–289.

Wang, L., Wu, J., Ca, W., Li, J., Zhao, Z., Luo, L., et al. (2017). Dietary arginine requirement of juvenile hybrid sturgeon (*Acipenser schrenckii* × *Acipenser baerii*). *Aquaculture Research, 48*(11), 5193–5201.

Wright, P. A., & Land, M. D. (1998). Urea production and transport in teleost fishes. *Comparative Biochemistry and Physiology Part A: Molecular & Integrative Physiology, 119*(1), 47–54.

Wright, P. A., Campbell, A., Morgan, R. L., Rosenberger, A. G., & Murray, B. W. (2004). Dogmas and controversies in the handling of nitrogenous wastes: expression of arginase Type I and II genes in rainbow trout: influence of fasting on liver enzyme activity and mRNA levels in juveniles. *Journal of Experimental Biology, 207*(12), 2033–2042.

Xu, Z., Takizawa, F., Elisa, C., Yasuhiro, S., Yang, D., Thomas, J., et al. (2020). Mucosal immunoglobulins of primitive and modern vertebrates operate under the guidance of primordially conserved principles. *Science Immunology*.

Yue, Y., Zou, Z., Zhu, J., Li, D., Xiao, W., Han, J., et al. (2015). Effects of dietary arginine on growth performance, feed utilization, hematological parameters and non-specific immune responses of juvenile Nile tilapia (*Oreochromis niloticus* L.). *Aquaculture Research, 46*(8), 1801–1809.

Zehra, S., & Khan, M. A. (2013). Dietary arginine requirement of fingerling Indian major carp, *Catla catla* (Hamilton). *Journal of the World Aquaculture Society, 44*(3), 363–373.

Zhou, F., Xiong, W., Xiao, J. X., Shao, Q. J., Bergo, O. N., Hua, Y., et al. (2010). Optimum arginine requirement of juvenile black sea bream, *Sparus macrocephalus*. *Aquaculture Research, 41*(11), e418–e430.

Zhou, H., Chen, N., Qiu, X., Zhao, M., & Jin, L. (2012). Arginine requirement and effect of arginine intake on immunity in largemouth bass, *Micropterus salmoides*. *Aquaculture Nutrition, 18*(2), 107–116.

Zhou, Q. C., Zeng, W. P., Wang, H. L., Xie, F. J., & Zheng, C. Q. (2012). Dietary arginine requirement of juvenile yellow grouper *Epinephelus awoara*. *Aquaculture, 350*, 175–182.

Zhou, Q., Jin, M., Elmad, Z., Liang, X., & Mai, K. (2015). Growth, immune response and resistance to *Aeromonas hydrophila* of juvenile yellow catfish, *Pelteobagrus fulvidraco*, fed diets with different arginine levels. *Aquaculture, 437*, 84–91.

Zhou, Q., Zeng, W., Wang, H., Xie, F., Wang, T., & Zheng, C. (2012). Dietary arginine requirement of juvenile yellow grouper *Epinephelus awoara*. *Aquaculture, 350–353*, 175–182.

Index

A

E

Q

R

S

W

Y

Z